2024 IEEE International Conference on Semiconductor Electronics (ICSE 2024)

Kuala Lumpur, Malaysia
19-21 August 2024

IEEE Catalog Number: CFP24421-POD
ISBN: 979-8-3503-7832-0

**Copyright © 2024 by the Institute of Electrical and Electronics Engineers, Inc.
All Rights Reserved**

Copyright and Reprint Permissions: Abstracting is permitted with credit to the source. Libraries are permitted to photocopy beyond the limit of U.S. copyright law for private use of patrons those articles in this volume that carry a code at the bottom of the first page, provided the per-copy fee indicated in the code is paid through Copyright Clearance Center, 222 Rosewood Drive, Danvers, MA 01923.

For other copying, reprint or republication permission, write to IEEE Copyrights Manager, IEEE Service Center, 445 Hoes Lane, Piscataway, NJ 08854. All rights reserved.

*** *This is a print representation of what appears in the IEEE Digital Library. Some format issues inherent in the e-media version may also appear in this print version.*

IEEE Catalog Number: CFP24421-POD
ISBN (Print-On-Demand):   979-8-3503-7832-0
ISBN (Online):   979-8-3503-7831-3

Additional Copies of This Publication Are Available From:

Curran Associates, Inc
57 Morehouse Lane
Red Hook, NY 12571 USA
Phone: (845) 758-0400
Fax: (845) 758-2633
E-mail: curran@proceedings.com
Web: www.proceedings.com

TABLE OF CONTENTS

Chair Message	vi
Organizing Committee	vii
List of Reviewers	viii

Plenary 1: E-mode GaN HEMT with Ferroelectric Gate Material for EV and PV Applications — xi

Professor Dr. Edward (National Yang Ming Chiao Tung University, Taiwan)

Plenary 2: Past, present, and future of integrated circuits technologies and their development — xii

Professor Dr. Hiroshi Iwai (National Yang Ming Chiao Tung University, Taiwan)

Plenary 3: Game Changing Semiconductor Technologies High Power, Miniaturized and Smart LEDs — xiii

Dr. David Lacey (OSRAM Opto Semiconductors (M) Sdn. Bhd.)

Keynote 1: Advances in Electric Field Control and Exploitation in III-N High-Performance Electronic Devices — xiv

Professor Dr. Patrick Fay (University of Notre Dame, Notre Dame, USA)

Keynote 2: High Volume Production of High Efficiency Solar Arrays — xv

Mr. Andree Wibowo (MicroLink Devices, USA)

Keynote 3: Building a Technology Frontrunner Generation — xvi

Professor Dr. Sufian Jusoh (Universiti Kebangsaan Malaysia, Malaysia)

Development of a PVT Verification Methodology for Robust Ultra-Low Power Dynamic Comparators — 1
Julie Roslita Rusli, Suhaidi Shafie, Wan Zuha Wan Hasan, Mohd Amrallah Mustafa, Izanoordina Ahmad, Roslina Mohd Sidek, Hasmayadi Abdul Majid, Haslina Jaafar

Analysis of the Electrical Characteristics for Compact SPICE Modelling of STT-MTJ Device with Physical Parameters Variation — 5
Nurul Ezaila Alias, Melanie Yi Xuan Pai, Michael Loong Peng Tan, Afiq Hamzah, Yasmin Abdul Wahab, Maizan Muhamad

Methodology to Optimize E-Mode GaN HEMT with P-Type Doping under 2DEG Layer — 9
Affendy Muhammad Ridzwan, Mohd Faizol Abdullah, Amir Murtadha Mohamad Yussof, Norazreen Abd Aziz, Hing Wah Lee

Effectiveness of the Heat Spreader in GaN HEMT Studied by a Co-Simulation Approach — 13
Amir Murtadha Mohamad Yussof, Norazreen Abd Aziz, Mohd Faizol Abdullah

Optimizing Methylammonium Tin Iodide-based Perovskite Solar Cells with Different Hole Transport Materials — 17
Budi Mulyanti, Jahril Nur Fauzan, Chandra Wulandari, Lilik Hasanah, Roer Eka Pawinanto

Effect of pH on Electrochemical, Morphological and Optical Properties of Electrodeposited Molybdenum Sulfide Thin Film — 21
Iskandar Dzulkarnain Rummaja, Muhammad Idzdihar Bin Idris, Zul Atfyi Fauzan Mohammed Napiah, Radi Husin Ramlee, Marzaini Rashid, Ahmad Muhajer Abdul Aziz

Early detection of hyperuricemia using hybrid Au-ZnO biosensor and Kretschmann-based SPR at visible optical wavelengths — 25
Muhammad Feidhul Hakim Bin Fatah Yasin, Abdul Halim Abdul Gafor, Noraidatulakma Abdullah, Lee Pey Yee, Syara Kassim, Affa Rozana Abdul Rashid, Kim S Siow, Azrul A Hamzah, Mohd Ambri Mohamed, Siti Nasuha Mustaffa, Nur Akmar Jamil, P. Susthitha Menon

Synthesis of Multi-Spike Gold Nanostar using the Surfactant-Free Method as LSPR Substrate Material — 29
Tita Oktavia Cahya Rahayu, Izzah Hanaanah Ab Aziz, Nur Hidayah Azeman, Tengku Hasnan Tengku Abdul Aziz, Mohd Suzeren Md Jamil, Ahmad Rifqi Md Zain

Reliability Analysis of Multibridge Channel Field Effect Transistor — 33
Nurul Ezaila Alias, Ganlin Huang, Muhammad Naziiruddin Hamzah, Michael Loong Peng Tan, Hanim Hussin, Yasmin Abdul Wahab

An investigation on the interplay between strain and defects on the thermal conductivity of monolayer graphene — 37
Dharma Darren Ram, Muhammad Aniq Shazni Mohammad Haniff, Abdul Manaf Hashim, Mohd Ambri Mohamed

Biomimetic Microstructure Design for Superhydrophobic Structure in Triboelectric Nanogenerator Device — 41
Firdaus Jamal Rashid, Abang Annuar Ehsan, Muhammad Aniq Shazni Mohammad Haniff

Unveiling the photovoltaic properties of ETL-free, lead-free, and graphene-based PSC — 44
Nabilah Ahmad Jalaludin, Fauziyah Salehuddin, Faiz Arith, Anis Suhaila Mohd Zain, Khairil Ezwan Kaharudin, Siti Aisah, Ibrahim Ahmad

Kretschmann-based Surface Plasmon Resonance sensor using Au/Graphene for Glucose Detection — 48
Nur Akmar Jamil, Abdul Halim Abdul Gafor, Noraidatulakma Abdullah, Lee Pey Yee, Syara Kassim, Kim S Siow, Azrul A Hamzah, Mohd Ambri Mohamed, Muhammad Feidhul Hakim Bin Fatah Yasin, P. Susthitha Menon

VCSELs-based Optoelectronics Transceiver for Free Space Photonics Packet Switching Network — 52
Clarence Augustine TH Tee, Hanbin Sun, W P Yeo, Burhanuddin Yeop Majlis, Muhamad Ramdzan Buyong, Ahmad Rifqi Md Zain, Le Song, Zheng Yelong, Sheng Li

Gold Nanorods as Plasmonic-based Optical Biosensor for Creatinine Detection — 56
Basyirah Zulkifli, Ahmad Rifqi Md Zain, Tengku Hasnan Tengku Abdul Aziz, Nur Hidayah Azeman, Mohd Suzeren Md Jamil

Modeling of piezoelectric energy harvester: influence of electrode/piezoelectric width ratio on induced voltage — 60
Ghulam Ali, Feng Xu, Faisal Mohd-Yasin

Investigation of TID effects on electrical characteristics of GaN MIS-HEMT with LPCVD-grown SiN passivation — 64
Chih-Yi Yang, Chin-Han Chung, You-Chen WENG, Jui-Sheng Wu, Tsung-Ying Yang, Edward Chang

The Effects of Al2O3 Interlayer on the Ferroelectric Behavior of Hf0.5Zr0.5O2 Thin Film for E-mode Ferroelectric Charge Trap Gate GaN HEMT — 68
Jui-Sheng Wu, Tsung-Ying Yang, You-Chen WENG, Chih-Yi Yang, Yu-Tung Du, Yi-Hsiang Wei, Edward Chang

Low Noise with High Linearity AlGaN/GaN HEMT Using Γ-Shaped Gate for Ka-Band Applications 72
Howie Tseng, Neng-Da Li, Yueh-Chin Lin, MuYu Chen, Edward-Yi Chang

Threshold Voltage Instability of GaN HEMTs with Thin Barrier AlGaN Technology 75
Tsung-Ying Yang, Jui-Sheng Wu, You-Chen WENG, Chih-Yi Yang, Edward-Yi Chang

1kV Vertical Breakdown Voltage AlGaN/GaN HEMTs on Si with AlN and AlGaN/AlN Superlattice Buffer Engineering 78
You-Chen WENG, Chih-Yi Yang, Tsung-Ying Yang, Chin-Han Chung, Chiang Tsung-Han, Fu-Ching Tung, Shih-Hsiang Lai, Hung-Wei Yu, Edward-Yi Chang

Effect of Silver Nanotriangle Orientation on Localised Surface Plasmon Resonance Sensing Performance 81
Muhammad Asif A. Khushaini, Ahmad Rifqi Md Zain, Nur Hidayah Azeman, Tengku Hasnan Tengku Abdul Aziz, Ahmad Ashrif A. Bakar, Basyirah Zulkifli

Comparative performances of 60nm and 80nm gold nanoparticles based mode-locker for erbium doped fiber laser 85
Noor Zirwatul Ahlam Naharuddin, Mohd. Adzir b. Mahdi, Maisarah Mansor, Nor Hadzfizah Mohd Radi, Rosyati Hamid

Enhancing Localized Surface Plasmon Resonance Response for Albumin Detection by Optimizing the Lateral Size of Hexagonal Gold Nanoparticles 89
Lilik Hasanah, Silva Nurfasha, Chandra Wulandari, Ahmad Aminudin, Yanurita Dwi Hapsari, Mohammad Arifin, Roer Eka Pawinanto, Budi Mulyanti

Bright-dark soliton pairs generation in an erbium-doped fiber laser utilizing gamma alumina saturable absorber 93
Norita Mohd Yusoff, Eng Khoon Ng, Mohd Zul Hilmi Mayzan, Mohd. Adzir b. Mahdi

Physical Characterization of Immersion Method-Based Porous Paper Towards Sensing Application 96
Gan Shin Pyng, Mastura Shafinaz Zainal Abidin

Effects of Growth Parameters on the Morphology of ReS₂ Nanoflakes Prepared by Chemical Vapor Deposition 100
Muhammad Faris Musawwi bin Ruslan, Abdul Rahman Mohmad, Muhammad Hilmi Johari, P. Susthitha Menon, Syahira binti A. Hinayadullah

Impact of Varied Ag-GO Ratios on the Electrochemical Enhancement of Vitamin D Detection 104
Kiki Chan, Nur Azura Mohd Said, Muhammad Aniq Shazni Mohammad Haniff, Mohd Hazani Mat Zaid, Siti Nur Ashakirin Mohd Nashruddin, Farhanulhakim Mohd razip wee, P. Susthitha Menon

Piezoelectric Energy Harvesting from Thermal Vibrations Using Doped Graphene-MXene Heterostructures 108
Kou Lijie, Poh Choon Ooi, Chang Fu Dee, Muhammad Aniq Shazni Mohammad Haniff

AI-based Image Processing Technique for Dielectrophoresis (DEP) In BioMEMs Applications 112
Clarence Augustine TH Tee, EnHao Yu, W P Yeo, Burhanuddin Yeop Majlis, Muhamad Ramdzan Buyong, Ahmad Rifqi Md Zain, Le Song, Zheng Yelong, Sheng Li

Enhancing Dielectrophoresis Analysis via Artificial Intelligence Integration 116
Muhamad Ramdzan Buyong, Muhammad Akmal Suhaimi, Arash Zulkarnain Rozaini, Burhanuddin Yeop Majlis, Farahdiana Wan Yunus, Noratiqah Yaakop, Clarence Augustine TH Tee, Céline Elie Caille, Abdullah Abdulhameed, Aminuddin Ahmad Kayani, Noraziah Zin

ZnO and PZT Thin Film Piezoelectric MEMS Vibrational Energy Harvesters for Cardiac Pacemaker   120
Yap Jia Xui Yap, Avinash Kumaresan, Jumril Yunas, Seri Mastura Mustaza, Huda Abdullah, Iskandar Yahya

Dielectrophoretic Rapid characterization of Antimicrobial resistance bacteria Escherichia coli   124
Arash Zulkarnain Rozaini, Muhammad Akmal Suhaimi, Aminuddin Ahmad Kayani, Abdullah Abdulhameed, Burhanuddin Yeop Majlis, Noraziah Zin, Wan Hanna Melini Wan Mohtar, Muhamad Ramdzan Buyong

Electronic Properties of AB-Stacked Bilayer Graphene Nanoribbons with Zigzag and Armchair Orientations   128
Yuki Wong, Nurul Ezaila Alias, Cheng Siong Lim, Choon Min Cheong, Michael Loong Peng Tan

Reduction of Subthreshold Leakage Current in Resonant Gate Transistor   132
Rhonira Latif, Mohamad Zain Azreen Ramli, Arjuna Marzuki, Masuri Othman

Comparative Electrochemical Performance of Screen Printed Carbon, Gold and Graphene Electrodes   136
Yasmin Abdul Wahab, Mohammad Al Mamun, M. A. Motalib Hossain, Mohd Rafie Johan, Nurul Ezaila Alias, Hanim Hussin, Maizan Muhamad, Hui Yin Nam

Evaluating the Impact of Upright and Inverted Pyramid Microstructures on the Optical Performance of Single Crystalline Silicon Solar Cells   140
Yasmin Abdul Wahab, Md. Yasir Arafat, Sharifah Fatmadiana Wan Muhamad Hatta, Mohammad Aminul Islam, Nurul Ezaila Alias, Mohd Rafie Johan, Hanim Hussin

A Flexible Framework Based on Finite-Element Method for Capacitance Extraction of 3-Dimensional Interconnects   144
Ye Wu, Qiwen Zheng, Zichang Zhang

Bandgap Modification of ZnO Nanorods for Enhanced Photocatalytic Application   148
Aini Ayunni Mohd Raub, Siti Nur Ashakirin Mohd Nashruddin, Mohd Ambri Mohamed, Jumril Yunas

Simulation-Based Approach to Detecting Pulmonary Embolism Using Capacitive Micromachined Ultrasonic Transducers   152
Hussnain Shahid, Dilla Duryha, Rhonira Latif, Poh Choon Ooi, Tehseen Batool

Effect of Seed Layer Cycles on ZnO Nanowires Characteristics for Piezoelectric Nanogenerator Applications   155
Suhana Mohamed Sultan, Izzaty Mohd Zambri, Fong Jun Xian, Pg Rafidah Pg Hj Petra, Khoo Wei How

A Low Power Scan Cell Design using FreePDK3   159
Ahmad Awalluddin Mohd Ghazali, Chia Yee Ooi, Hau Sim Choo, Nordinah Ismail, Siti Rahmah Aid

Exploring the Characterization of Electrodeposited MoS2 as a Hole Transport Layer in Methylammonium Perovskite Solar Cells   163
Ahmad Muhajer Abdul Aziz, Muhammad Idzdihar Bin Idris, Zul Atfyi Fauzan Mohammed Napiah, Radi Husin Ramlee, Muhammad Noorazlan Shah Zainudin, Marzaini Rashid, Mohd Iskandar Dzulkarnain M. Rummaja

Piezoelectric Nanogenerator based on Graphene and MXene Heterostructure   167
Kou Lijie, Rawhan Mohammad Safiul Haque, Poh Choon Ooi, Muhammad Aniq Shazni Mohammad Haniff, Chang Fu Dee

Thermal-Aware Test Scheduling with Floorplanning for Three-Dimensional Stacked Integrated Circuit   171
Patmanathan Ganesan, Chia Yee Ooi, Nordinah Ismail, Siti Rahmah Aid

Deep Learning (DL) based Computer Generated Hologram (CGH) for Beamsteering in Reconfigurable Holographic Switches   175
Clarence Augustine TH Tee, Hanbin Sun, W P Yeo, Burhanuddin Yeop Majlis, Muhamad Ramdzan Buyong, Ahmad Rifqi Md Zain, Le Song, Zheng Yelong, Sheng Li

MESSAGE FROM CHAIR

Prof. Dato' Dr. Burhanuddin Yeop Majlis

Dear esteemed Guests, Distinguished Participants, and Honoured Sponsors,

On behalf of the ICSE 2024 organizing committee, I am honored and delighted to welcome you 2024 IEEE International Conference on Semiconductor Electronics (ICSE2024).

This is the 16th ICSE organized by the **Electron Devices Chapter of IEEE Malaysia Section**, co-organized by the **Institute of Microengineering and Nanoelectronics (IMEN), Universiti Kebangsaan Malaysia** and technically co-sponsored by the **IEEE Electron Devices Society Malaysia Chapter.** Over the last 32 years, the ICSE conference series has become the prominent international forum on semiconductor electronics embracing all aspects of semiconductor technology from circuit device, modelling and simulation, photonics and sensor technology, MEMS technology, process and fabrication, packaging technology and manufacturing, failure analysis and reliability, material, and devices and nanoelectronics.

On behalf of the organizing committee, we thank you for your active participation in ICSE 2024. Your strong continuous support in selecting ICSE 2024 as the platform to publish your latest research in semiconductor electronics is greatly appreciated. During the 2-day conference, 46 oral presentations will be delivered across a broad spectrum of technical sessions. These include three plenary speakers which are Prof. Dr. Edward Yi Chang (National Yang Ming Chiao Tung University, Taiwan), Prof. Dr. Hiroshi Iwai (National Yang Ming Chiao Tung University, Taiwan) and Dr. David Lacey (OSRAM, Malaysia), and three keynote speakers which are Prof. Dr. Patrick Fay (University of Notre Dame, USA), Mr. Andree Wibowo (Microlink Devices, USA) and Prof. Dr. Sufian Jusuh (Universiti Kebangsaan Malaysia, Malaysia).

I would like to express my gratitude to members of the organizing committee, secretarial staff, and everyone who has worked hard to make this conference into reality. Finally, I hope that ICSE 2024 will be successful and enjoyable to all participants.

Thank you and Terima kasih.

Prof. Dato' Dr. Burhanuddin Yeop Majlis
Chairman
2024 IEEE International Conference on Semiconductor Electronics (ICSE2024)

ORGANIZING COMMITTEE

Chair: Prof. Dato' Dr. Burhanuddin Yeop Majlis

Vice-Chair 1: Ts. Dr. Nurul Ezaila Alias

Vice-Chair 2: Ir. Dr. Azrif Manut

Secretary: Ts. Dr. Suhana Mohamed Sultan

Treasurer: Dr. Haslina Jaafar

Technical & Publication Chair 1: Dr. Yasmin Abdul Wahab

Technical & Publication Chair 2: Ir. Dr. Maizan Muhamad

Technical & Publication Chair 3: Ir. Dr. Hanim Hussin

Program Chair: Prof. Dr. Ahmad Rifqi Md Zain

Program Co-Chair: Dr. Sharifah Fatmadiana Wan Muhammad Hatta

Promotional Chair: Assoc. Prof. Dr. Aliza Aini Md Ralib

Logistic Chair: Prof. Dr. Norhayati Soin

Logistic Co-Chair: Ir. Dr. Maizatul Zolkapli

International Editor: Prof. Dr. Eur Ing. Ir. Clarence Augustine Th Tee (Zhejiang Normal University)

SECRETARIAT COMMITTEE

Leader: Dr. Iskandar Yahya
Member:

> Dr. Mohd Zulhakimi Ab. Razak
> Dr. Rhonira Latif
> Arash Zulkarnain Bin Ahmad Rozaini
> Roharsyafinaz Binti Roslan
> Siti Nasuha Binti Mustaffa
> Muhammad Faris Musawwi Bin Ruslan
> Nur Ariena Hanis Binti Mohd Nor
> Muhammad Izzuddin Bin Abd Samad

COMMITTEE MEMBERS

Ir. Bernard Lim
Ir. Dr. Hazian Mamat
Prof. Ir. Dr. Ahmad Sabirin Zoolfakar
Assoc. Prof. Dr. P. Susthitha Menon

LIST OF REVIEWERS

Name	Affiliation	Country
Abang Annuar Ehsan	Universiti Kebangsaan Malaysia	Malaysia
Affa Rozana Abdul Rashid	USIM	Malaysia
Ahmad Alabqari Ma' Radzi	Universiti Tun Hussein Onn Malaysia	Malaysia
Ahmad Rifqi Md Zain	UKM	Malaysia
Ahmad Sabirin Zoolfakar	Universiti Teknologi MARA	Malaysia
Alireza Ghasempour	University of Applied Science and Technology	USA
Aliza Aini Md Ralib	International Islamic University Malaysia	Malaysia
Amiza Rasmi	TM Research & Development	Malaysia
Azli Yahya	Universiti Teknologi Malaysia	Malaysia
Azrif Manut	Universiti Teknologi MARA Shah Alam	Malaysia
Azura Hamzah	Universiti Teknologi Malaysia	Malaysia
Badariah Bais	Universiti Kebangsaan Malaysia	Malaysia
Badrul Hisham Ahmad	Universiti Teknikal Malaysia Melaka	Malaysia
Burhanuddin Yeop Majlis	Universiti Kebangsaan Malaysia	Malaysia
Chang Fu Dee	Universiti Kebangsaan Malaysia (UKM)	Malaysia
Chau Yuen	Nanyang Technological University	Singapore
China Venkateswarlu Sonagiri	Institute of Aeronautical Engineering	India
Chutisant Kerdvibulvech	National Institute of Development Administration	Thailand
Dan Ciulin	E-I-A	Switzerland
Dan L. Milici	University of Suceava	Romania
Datta Chavan	Bharati Vidyapeeth Deemed University College of Engineering, Pune	India
David Forsyth	UTM	United Kingdom (Great Britain)
Dilla Duryha	IMEN, UKM	Malaysia
Domenico Ciuonzo	University of Naples Federico II	Italy
Duu Sheng Ong	Multimedia University	Malaysia
EDS Malaysia Malaysia	Universiti Kebangsaan Malaysia	Malaysia
Ekaterina Pshehotskaya	Moscow State University	Russia
Faizah Abu Bakar	Universiti Malaysia Perlis	Malaysia
Haidawati Nasir	Universiti Kuala Lumpur	Malaysia
Hanim Hussin	Universiti Teknologi MARA	Malaysia
Harikrishnan Ramiah	Universiti Malaya	Malaysia
Haslina Jaafar	Universiti Putra Malaysia	Malaysia
Heydar Toossian Shandiz	Ferdowsi University of Mashhad	Iran
Hing Keung Lau	Vocational Training Council	Hong Kong
Ibrahim Ahmad	Universiti Tenaga Nasional	Malaysia
I-Cheng Chang	National Dong Hwa University	Taiwan
Ir. Hazian Bin Mamat	Mimos Berhad	Malaysia
Iskandar Yahya	Universiti Kebangsaan Malaysia	Malaysia
Ismail Saad	Universiti Malaysia Sabah	Malaysia
Iwan Adhicandra	Bakrie University	Indonesia
John Ojur Dennis	Universiti Teknologi PETRONAS	Malaysia
Josip Music	University of Split	Croatia
Jumril Yunas	Universiti Kebangsaan Malaysia	Malaysia

Name	Affiliation	Country
Li Wah Thong	Multimedia University	Malaysia
Li-Cheng Wu	Taiwan Power Research Institute	Taiwan
Maizan Muhamad	Universiti Teknologi MARA	Malaysia
Maizatul Zolkapli	Universiti Teknologi MARA	Malaysia
Marinah Othman	Universiti Sains Islam Malaysia	Malaysia
Md Ali Rani	Universiti Putra Malaysia	Malaysia
Md Nabil	UNITEN	Malaysia
Md Rafiqul Islam	International Islamic University Malaysia	Malaysia
Mohamed Atef	United Arab Emirates University, AlAin	United Arab Emirates
Mohamed B Abdelhalim	Arab Academy for Science, Technology & Maritime Transport	Egypt
Mohammad Faiz Liew Abdullah	Universiti Tun Hussein Onn Malaysia (UTHM)	Malaysia
Mohd Amrallah Mustafa	Universiti Putra Malaysia	Malaysia
Mohd Khairuddin Md Arshad	Universiti Malaysia Perlis	Malaysia
Mohd Nazim Mohtar	Universiti Putra Malaysia	Malaysia
Mohd Tafir Mustaffa	Universiti Sains Malaysia	Malaysia
Montadar Abas Taher	University of Diyala	Iraq
Muhamad Ramdzan Buyong	UKM	Malaysia
Muhammad Ibn Ibrahimy	International Islamic University Malaysia	Malaysia
Muhammad Mokhzaini Azizan	Universiti Sains Islam Malaysia	Malaysia
Muzamir Isa	Universiti Malaysia Perlis	Malaysia
Nafarizal Nayan	Universiti Tun Hussein Onn Malaysia	Malaysia
Noor Ain Kamsani	Universiti Putra Malaysia	Malaysia
Nor Farahidah Za'bah	International Islamic University Malaysia	Malaysia
Nor Hafizah Ngajikin	Universiti Tun Hussein Onn Malaysia	Malaysia
Norhana Arsad	Universiti Kebangsaan Malaysia	Malaysia
Norhayati Soin	University of Malaya	Malaysia
Norhisam Misron	Universiti Putra Malaysia	Malaysia
Nowshad Amin	Universiti Kebangsaan Malaysia	Malaysia
Nurul Ezaila Alias	Universiti Teknologi Malaysia	Malaysia
P. Susthitha Menon	Universiti Kebangsaan Malaysia	Malaysia
Pin-Yu Chen	IBM Research	USA
Poming Lee	NCTU	Taiwan
Pooya Ghani	MAPNA Electric & Control, Engineering & Manufacturing Co. (MECO)	Iran
Puteri Sarah Mohamad Saad	Universiti Teknologi MARA	Malaysia
Rachit Patel	National Institute of Technology Patna	India
Ratheesh Kumar Meleppat	University of California Davis	USA
Razali Ismail	Universiti Teknologi Malaysia	Malaysia
Robiah Ahmad	Universiti Teknologi Malaysia	Malaysia
Rosario Morello	University Mediterranea of Reggio Calabria	Italy
Rosaura Palma-Orozco	Instituto Politécnico Nacional	Mexico
Rosminazuin Ab Rahim	International Islamic University Malaysia	Malaysia
Rubita Sudirman	Universiti Teknologi Malaysia	Malaysia

Name	Affiliation	Country
S. M. A. Motakabber	International Islamic University Malaysia	Malaysia
Saadah Abdul Rahman	University of Malay	Malaysia
Sabrin Samsudin	Universiti Teknologi Mara	Malaysia
Samir Ladaci	Ecole Nationale Polytechnique	Algeria
Satya Nagabhushana Rao Kamisetti	JNTUK	India
Sergey B. Biryuchinskiy	Vigitek, Inc.	USA
Sew Sun Tiang	Universiti Sains Malaysia	Malaysia
Shaharin Fadzli Bin Abd Rahman	Universiti Teknologi Malaysia	Malaysia
Shahrir Rizal Kasjoo	Universiti Malaysia Perlis	Malaysia
Sharifah Fatmadiana Wan Muhamad Hatta	University of Malaya	Malaysia
Sharifah Md Yasin	Universiti Putra Malaysia	Malaysia
Sheroz Khan	Onaizah College of Engineering	Saudi Arabia
Siti Azlida Ibrahim	Multimedia University	Malaysia
Siti Nooraya Mohd Tawil	Universiti Pertahanan Nasional Malaysia	Malaysia
Siti Noorjannah Ibrahim	International Islamic University Malaysia	Malaysia
Smain Femmam	University UHA	France
Sotiris Karachontzitis	Independent Authority for Public Revenue	Greece
Sreedharan Pillai Sreelal	Indian Space Research Organization	India
Suhana Mohamed Sultan	Faculty of Electrical Engineering, Universiti Teknologi Malaysia	Malaysia
Sulaiman Wadi Harun	Uni Malaya	Malaysia
Syahrul Ashikin Azmi	Universiti Malaysia Perlis	Malaysia
Syarifah Abd. Rahim	Universiti Malaysia Pahang	Malaysia
Teddy Surya Gunawan	International Islamic University Malaysia	Malaysia
Ulas Kilic	Ege University	Turkey
Umapathy Eaganathan	School of Technology	Malaysia
Wan Zuha Wan Hasan	Universiti Putra Malaysia	Malaysia
Wei Wei	Xi'an University of Technology	China
Wira Hidayat bin Mohd Saad	Universiti Teknikal Malaysia Melaka	Malaysia
Xiaoce Feng	Wayne State University	USA
Yaareb M.Basheer Ismael Al-Khashab	Ministry of Water Resources/Badush Dam	Iraq
Yasmin Abdul Wahab	Universiti Malaya	Malaysia
Ying-Khai Teh	San Diego State University	USA
Zaharah Johari	Universiti Teknologi Malaysia	Malaysia
Zubaida Yusoff	Multimedia University	Malaysia

PLENARY SPEAKER 1

E-mode GaN HEMT with Ferroelectric Gate Material for EV and PV Applications

Professor Dr. Prof. Dr Edward

National Yang Ming Chiao Tung University, Taiwan

Abstract: AlGaN/GaN high-electron-mobility transistors (HEMTs) have high saturation velocity, high mobility, and high current density with high breakdown electric field, which make them promising for next-generation high power and high frequency device applications. However, AlGaN/GaN HEMTs always demonstrate normally-on operation, which is not desirable for Electrical vehicle applications due to safety concerns. To achieve normally-off operation for GaN HEMT device, several methods have been proposed to achieve high positive threshold voltage. For example, p-GaN, recessed-gate, and fluorine treated AlGaN technologies have been demonstrated. In this work, a new charge storage structure with hybrid ferroelectric charge-trapping gate stack is used to realize the GaN E-mode MIS-HEMT for power switching applications. This gate stack structure combines ferroelectric film with charge-trapping layer and current blocking film, resulting in a polarization charge field against the applied gate voltage, and thus achieves a large positive shift of the threshold voltage. As a result, the device has a high Vth, meanwhile maintain low on Resistance, high $I_{(DS,MAX)}$, and high breakdown voltage are maintained, perfect for EV applications.

Prof. Dr. Edward Y. Chang received his B.S. degree from Materials Science and Engineering, National Tsing Hua University, Hsinchu, Taiwan in 1977, and his Ph.D. degree from Materials Science and Engineering, University of Minnesota Minneapolis, MN in 1985.Dr. Chang was with Unisys Corporation GaAs Component Group, Eagan, MN, 1985 to 1988 and Comsat Labs Microelectronic Group from 1988 to 1992. He worked on the GaAs MMIC programs on both groups. He joined National Chiao Tung University (NCTU), Hsinchu, Taiwan, in 1992. In 1994, he helped set up the first GaAs MMIC production line in Taiwan, and become president of Hexawave Inc., Hsinchu, Taiwan, in 1995. He returned to the teaching position at NCTU in 1999. Dr. Chang is currently the Dean of Research and Development and professor of the Department of Materials Science and Engineering and Dept of Elecrical Engineering at NCTU. He is also the director of Diamond Lab and the director of NCTU-TSMC research centre. His research interests include new device and process technologies for Compound Semiconductors for wireless communication and high-power electronic applications. Dr. Chang is a senior member and a DL of the IEEE EDS. Currently, his research activities include InP, GaAs based compound materials and devices (HEMT, HBT) for wireless communication and sub-milimeterwave imaging applications. GaN based materials (MBE, MOCVD) and High frequency & High power electronic (HEMT) applications. III-V/ Si integration (Ge, SiGe, GaAs, InP) for logic applications. Dr. Chang research Award and Distinguished Contribution for Technical Transfer to Industry, both from National Science Council, Taiwan. National Award for Academic Contribution to Industry and Industry and Economy Contribution Award, both from Ministry of Economic Affairs, Taiwan. Distinguished Electrical Engineering Professor Award from Chinese Institute of Electrical Engineering. He is also Distinguished Lecturer of IEEE, Fellow of Taiwanese Materials Society, and Fellow of IEEE.

PLENARY SPEAKER 2

Past, present, and future of integrated circuits technologies and their development

Professor Dr. Hiroshi Iwai

National Yang Ming Chiao Tung University, Taiwan

Abstract: Currently, the widespread adoption of chat-type AI has made it possible for anyone to easily create texts, images, and music. As AI continues to evolve, it is expected that in the near future, AI will increasingly take over many intellectual tasks currently performed by humans. This represents a crucial turning point for humanity, hinting at the advent of a superintelligent society. The arrival of this superintelligent society is driven by the advancement of ICT technology, particularly the remarkable performance improvement of nano-CMOS integrated circuits capable of processing vast amounts of information instantly. However, due inability to handle higher frequencies like microwaves as well as large power consumption because of its large size, research was initiated to realize semiconductor solid-state amplifiers. As a result of the dedicated efforts of numerous researchers spanning over 20 years, the transistor was invented in 1947, marking the beginning of microelectronics. Subsequently, in 1969, MOS LSI made its debut in the market. Over the past 50 years, miniaturization has continued to progress, and significant performance improvements have been achieved now through nano CMOS integrated circuits. What will be the progress of integrated circuit technology after reaching the limits of miniaturization and 3D integration? Even after reaching these limitations, it will take a considerable amount of time for the technology to mature and achieve optimization. This talk explains the history, current state, and future of ICT technology and integrated circuit technology.

Hiroshi Iwai is a semiconductor device engineer, who contributed to the development of integrated circuits technologies and products. He received BE and Doctor Degree in Engineering from the University of Tokyo in 1972 and 1993, respectively. He worked at Toshiba, Tokyo Institute of Technology, and National Yang Ming Chiao Tung University for 50 years since 1973. Especially, he has contributed to the miniaturization of the LSIs since the 8 μm PMOS generation. He developed Toshiba's first NMOS LSI technology in 1975, used 1k bit SRAMs and 12-bit microprocessors used for the first automobile engine control in the world. For the development of a 64k bit DRAM, he introduced stepper lithography and dry process into production for the first time in the world in 1980. He broke the 0.1 μm limit of CMOS miniaturization by developing 40 nm MOSFET technologies in 1993, which was more than 2 generations earlier than competitive companies. He published new scaling scheme or roadmap of the CMOS scaling including double gate MOSFETs and high-k gate insulator in 1993. He developed world-first 0.15 μm RFCMOS technologies in the middle of 1990s and published a roadmap of RFCMOS technology in 1999, which contributed to the realization of Bluetooth. Dr. Iwai has contributed to IEEE and EDS activities and administrations over 30 years. He served as the IEEE EDS president and the IEEE Division I director. He was appointed to the first IEEE EDS eminent lecturer. Currently, he is a vice dean and a distinguished chair professor of National Yang Ming Chiao Tung University and a professor emeritus of Tokyo Institute of Technology. He is a life fellow of IEEE, a fellow and awarded life member of ECS, and a fellow of JSAP, IEICE, and IEEJ.

PLENARY SPEAKER 3

Game Changing Semiconductor Technologies High Power, Miniaturized and Smart LEDs

Dr. David Lacey

OSRAM Opto Semiconductors (M) Sdn. Bhd.

Abstract: With a strong background in academics and research, Dr. David Lacey has two-and-a-half decades of experience in the semiconductor industry, specifically in producing and enhancing organic and inorganic lighting solutions. With 24 publications and about a dozen patents under his belt, Dr. Lacey's foray into innovation stretches across multiple fields; from creating new devices to enhancing the durability and robustness of safety encapsulations to coming up with new methods of increasing production volume in efficient ways. In the area of new device creation, Dr. Lacey holds patents in creating electroluminescent devices, organic light-emitting components comprising an electroluminescent layer and an optoelectronic component including a flexible carrier strip and an optoelectronic semiconductor chip. A large area of his focus over the years has also been centred on the bettering of the protective encapsulation for electronic components, making them less susceptible to damage from moisture and external influences. His patents in this area include the invention of a lead frame for a radiation-emitting component containing a protective layer for the reflective coating and enhancing protective encapsulation for electronic components, particularly electro-optical or optoelectronic components such as an organic light emitting diode (OLED), and producing such products efficiently by reshaping the encapsulation element using heat. They also cover the creation of an electroluminescent device having a protection layer in the cap bonding region that protects the layers below from damage during removal of polymer materials. Aside from that, Dr. Lacey has created different methods of superimposing a plane encapsulation element and a drying agent on top of each other as well as a method of encapsulating an OLED by producing the organic optoelectronic component on a substrate wherein the organic optoelectronic component has an active region and regions with contact pads or scribe / rupture regions. He has also contributed in the improving of encapsulation for electroluminescent devices in relation to homogeneous or uniform deposition of active organic materials.

David Lacey, Director ADS R&D, Osram Opto Semiconductors - ams OSRAM group. Dr. Lacey obtained his DPhil in Chemistry/Materials Science from the University of Sussex, UK and has held R&D, Engineering and Business responsibilities in semiconductor manufacturing companies across Europe, USA and Asia. He has more than 30 years experience in semiconductor & display technology development & manufacturing, in both start-up and multi-national environments. He has been based in Malaysia since 2001 and is currently Chairman of FREPENCA - the Penang Trade Zone Companies Association & President-elect of SFAM - the Semiconductor Fabrication Association of Malaysia. He was one of the exco committee members involved in the earlier phase of conceptualisation and inception of CREST since March 2012. Dr. Lacey was bestowed with an honorary Doctor of Science by Universiti Sains Malaysia at the Arau Palace in 2021, an accolade reflecting his profound impact on advancing LED technologies and expanding opportunities in Malaysia.

KEYNOTE SPEAKER 1

Advances in Electric Field Control and Exploitation in III-N High-Performance Electronic Devices

Professor Dr. Patrick Fay

University of Notre Dame, Notre Dame, USA

Abstract: Wide bandgap III-N semiconductors are promising for electronic devices operating at high power levels and in harsh environments due to the combination of their excellent carrier transport properties and the ability to operate at high internal electric fields. However, the performance of many current-generation devices is below the fundamental performance limits expected from the material properties. This can be addressed through novel device design concepts. In this talk, recent work on polarization-graded structures for performance enhancement in mm-wave HEMTs, cost-effective edge termination strategies for vertical power devices, and devices exploiting impact ionization and avalanche in GaN will be reviewed. For example, the use of polarization-grading has been shown to decrease the peak electric field in the channel, increase the breakdown voltage, and improve the power scaling of III-N based HEMTs, without the use of field plates that limit high-frequency performance; experimentally-validated power-added efficiency of 50% at 94 GHz has been achieved. In vertical devices, device high-field operation is often limited by edge effects; we report a strategy for edge termination that provides a large process window that is tolerant of both fabrication processing and epitaxial layer thickness and doping variations, and enables robust avalanche operation to be achieved in practice. In addition to increased breakdown voltage, the ability to harness impact ionization and avalanche for device functionality is also critical for avalanche photodiodes and negative-resistance oscillators such as IMPATT diodes. We report the recent demonstration of experimentally-measured negative resistance at microwave frequencies from GaN-based IMPATT diodes, illustrating direct exploitation of the high-field operation of GaN pn junctions for advanced functionality.

Patrick Fay's research interests include the design, fabrication, and characterization of microwave and millimeter-wave electronic devices and circuits, as well as power electronic devices. His recent work focuses on the use of polarization-engineered III-N heterostructures for highperformance, high linearity applications, ferroelectric-augmented III-N devices for non-traditional applications such as RF/mm-wave switching and routing for advanced communication systems, and exploitation of wide- and ultra-wide band gap semiconductors for RF and high-power applications. His research also includes the use of micromachining techniques for the fabrication of RF through sub-millimeter-wave packaging. He established the High Speed Circuits and Devices Laboratory at Notre Dame, which includes device and circuit characterization capabilities at frequencies up to 1 THz. He also oversaw the design, construction, and commissioning of the 9000 sq. ft. class 100 cleanroom housed in Stinson-Remick Hall at Notre Dame, and has served as the director of this facility since 2003. Prof. Fay is a fellow of the IEEE, is an IEEE Electron Devices Society Distinguished Lecturer, and has published 11 book chapters and more than 400 articles in scientific journals and conference proceedings.

KEYNOTE SPEAKER 2

High Volume Production of High Efficiency Solar Arrays

Mr. Andree Wibowo

MicroLink Devices, USA

Abstract: MicroLink Devices began the investigation of Epitaxial Lift Off (ELO) of solar cells in 2007. The initial work was funded by NSF where a stack of solar cells separated by multiple release layers was proposed. The results from this investigation indicated that it was not possible to perform multiple lift-offs simultaneously due to the great difficulties in handling ultrathin semiconductor layers. The work was then progressed towards a single large wafer 4-inch GaAs lift off with a high efficiency inverted metamorphic solar cell device (IMM). There was a large solar cell initiative (Solar America Initiative) which funded numerous solar cell technologies and MicroLink was selected for IMM solar cells for CPV applications. The Solar America initiative enabled MicroLink to further advance the IMM technology where many of the key problems on the high efficiency IMM solar cells were solved. Large interest was generated from our large area IMM solar cells which drove numerous programs from many funding agencies. The additional funding enabled MicroLink to produce an array of solar cells, where they were integrated with bypass diodes and interconnects and packaged. One of the very first application of our solar arrays was to increase the endurance of small electric UAV where solar arrays were attached to the top surface of the wing. This enabled the UAV to double its endurance in comparison to a standard UAV that is operating on battery alone. This effort led to the integration of MicroLink's solar array onto a stratospheric UAV (Airbus Zephyr) where specific power (W/kg) is the key metric to enable true stratospheric operation throughout its entire mission. Currently, the solar arrays for Zephyr has reached TRL 9 and is in production. MicroLink is in the process of transitioning the standard Zephyr solar arrays towards Space satellites and is currently undergoing qualification.

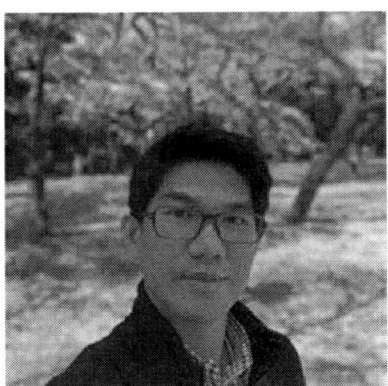

Mr. Andree Wibowo, Vice President of Corporate Strategy joined MicroLink Devices in 2001 after receiving his B.S in Chemical Engineering from University of Wisconsin at Madison. He has over 20 years experience in the compound semiconductor industry specifically in MOCVD systems and volume manufacturing of GaAs materials including HBT and solar cells. Mr. Wibowo was responsible for transitioning HBT and BiFET devices from R&D to volume manufacturing, successfully earning ISO 9001:2000 certification for high-volume production at MicroLink Devices. In this capacity, he was responsible for establishing solar cell epitaxy production expansion of lightweight, high efficiency solar cells for space and terrestrial application. Mr. Wibowo obtained an MBA from the University of Chicago in 2012.

KEYNOTE SPEAKER 3

Building a Technology Frontrunner Generation

Professor Dr. Sufian Jusoh

Universiti Kebangsaan Malaysia, Malaysia

Abstract: The Association of Southeast Asian Nations (ASEAN) is embarking on a new regional integration plan, i.e. the ASEAN Vision 2045. According to the ASEAN Secretariat, there are about 213 million youths in ASEAN in 2021, which is about 30 percent of the population. The youths in 2025 will be the leaders of ASEAN and instrumental for the ASEAN Community Building in the next 20 years. Evidence from countries like India, Korea and Singapore show that they have benefited from the technology transfer of hardware and software elements of technology, through a collaborative process that allows scientific findings, knowledge and intellectual property to flow from creators such as universities and research institutions to public and private users. Technology transfers coupled with domestic-driven innovations by local talents have turned Korea and Singapore into a leading technology creators and frontrunners, whereas Malaysia and Vietnam remained as technology adopters. Moving forward to the ASEAN 2045, ASEAN Member States and the youths must be ready to turn ASEAN as a technology frontrunner, instead of a technology adopter. This is to support ASEAN's vision to become a major player in the trade, investment and innovation in the advanced and digital technologies.

Professor Dr. Sufian Jusoh is the Director and a Professor of International Trade and Investment of the Institute of Malaysian and International Studies (IKMAS UKM), Universiti Kebangsaan Malaysia. He is a globally re-known expert and scholar in the realm of International Trade and Investment, ASEAN and Regional Integration, and International Development. Professor Dr. Sufian Jusoh is an FDI Advisor, ArtNet, under the United Nations Economic Commission for Asia and the Pacific (UNESCAP), an International Investment Policy Consultant at the World Bank Group, an External Fellow at the World Trade Institute, University of Bern, Switzerland, Malaysia's representative to ERIA RIN and a Co-Founder of the ASEAN Economic Integration Forum. Professor Sufian used to be a Malaysian delegate to the ASEAN High Level Task Force for ASEAN Community Vision Post 2025. In APEC, Professor Dr. Sufian Jusoh played a key role in the success of Malaysia's chairmanship of APEC 2020, leading to the APEC Putrajaya Vision 2040, Malaysian representative to the preparation of the APEC Food Security Roadmap 2030, and a delegate to the APEC Investment Expert Group. Sufian played a key role in the reform of the investment policies in Bangladesh, Malaysia, Laos, Myanmar, Timor-Leste, the Federated States of Micronesia and Pakistan. Sufian is a Barrister-at-Law (England and Wales) of Lincoln's Inn, London, and holds an LL.B (Hons.) from Cardiff Law School; LL.M (Merit), University College London, and a Doctor of Law (summa cum Laude) from University of Bern, Switzerland. Professor Sufian Jusoh has also been admitted as a Member of the Churchill College, University of Cambridge.

ICSE 2024

19 -21 August 2024 | Kuala Lumpur, Malaysia

2024 IEEE International Conference on
Semiconductor Electronics

PROCEEDINGS

19 -21 August 2024 | Grand Millenium, Kuala Lumpur

The 16th ICSE

Sponsored by:

PUBLICATION CONTACT

Nurul Ezaila Alias
Faculty of Electrical Engineering
Universiti Teknologi Malaysia
81310 UTM Johor Bahru
Johor, MALAYSIA
Email: ezaila@utm.my /
edsmalaysia@gmail.com

Development of a PVT Verification Methodology for Robust Ultra-Low Power Dynamic Comparators

Julie Roslita Rusli
Electronics Technology Section
British Malaysian Institute
Universiti Kuala Lumpur
Selangor, Malaysia
julie@unikl.edu.my

Suhaidi Shafie
Department of Electrical and
Electronics Engineering
Universiti Putra Malaysia
Selangor, Malaysia
suhaidi@upm.edu.my

Roslina Sidek
Department of Electrical and
Electronics Engineering
Universiti Putra Malaysia
Selangor, Malaysia
roslinams@upm.edu.my

Mohd Amrallah Mustafa
Department of Electrical and
Electronics Engineering
Universiti Putra Malaysia
Selangor, Malaysia
amrallah@upm.edu.my

Izanoordina Ahmad
Electronics Technology Section
British Malaysian Institute
Universiti Kuala Lumpur
Selangor, Malaysia
izanoordina@unikl.edu.my

Wan Zuha Wan Hassan
Department of Electrical and
Electronics Engineering
Universiti Putra Malaysia
Selangor, Malaysia
wanzuha@upm.edu.my

Haslina Jaafar
Department of Electrical and
Electronics Engineering
Universiti Putra Malaysia
Selangor, Malaysia
jhaslina@upm.edu.my

Hasmayadi Abdul Majid
Analog IC Design
SMD Semiconductor Sdn. Bhd.
Sarawak Malaysia
*hasmayadia@smdsemiconductor
.com*

Abstract— Achieving high manufacturing yield for high-resolution successive approximation register (SAR) analog-to-digital converters (ADCs) faces significant challenges due to variations in oxide thickness and dopant concentration during fabrication. These variations can lead to metastability errors, impacting performance and reliability. This paper presents a novel verification method designed to evaluate the performance of a proposed double-tail dynamic comparator circuit in practical scenarios. A dedicated test scheme is introduced to detect and validate worst-case input voltage transitions across 45 process, voltage, and temperature (PVT) corner simulations, considering variations in VDD, operating temperature, and fabrication processes. The proposed PVT test scheme has been implemented and validated on both the JRR1 comparator and the conventional Double-tail comparator, successfully identifying metastability errors in specific test corner simulations. The results demonstrate the effectiveness of the test scheme in assessing delay and power performance, highlighting its potential to ensure high manufacturing yield and reliability in SAR ADC designs.

Keywords—double-tail comparator, SAR ADC, robust comparator, ultra-low power dynamic comparator.

I. INTRODUCTION

Dynamic comparators are essential components within SAR ADCs, playing a critical role in converting analog signals to digital representations. The dynamic comparator block is particularly significant due to its substantial influence on the ADC's overall resolution. Ensuring robustness in low-power dynamic comparators is increasingly challenging (1,2). The primary function of the comparator in SAR ADCs is to differentiate the LSB from the differential output of the DAC within a specified timeframe. Errors in each comparator cycle can disrupt the conversion process, emphasizing the need for precise comparator operation. While previous research has focused on optimizing dynamic comparators for accuracy under nominal conditions, the effects of metastability errors during PVT variations remain underexplored(3). Detailed verification results for 45 PVT corner variations are lacking in existing literature, indicating a need for further research in

addressing PVT variations in low-voltage comparator designs. Proposed calibration circuits to mitigate these issues typically increase power consumption and require additional space (4). To ensure the robustness of comparator designs, process corner simulations are essential during the design phase. These simulations account for extreme parameter variations inherent in integrated circuit designs fabricated on semiconductor wafers. PVT variations significantly impact comparator decisions in low-voltage comparators, potentially causing metastability errors in LSB quantization. Additionally, variations in oxide thickness and dopant concentration during fabrication affect the threshold voltage (V_{TH}) of CMOS transistors, further challenging comparator accuracy (5,6).

Process corner simulations are crucial at the design stage to ensure robustness, representing the extreme parameter variations of integrated circuit designs fabricated on semiconductor wafers (7,8,9). These variations include process transistor properties, supply voltages, and die temperatures (10,11). The 45-corner PVT variation simulations verify the robustness of the designed comparator circuit with a 10% variation of V_{DD} and temperature in Typical-Typical (TT), Fast-Fast (FF), Fast-Slow (FS), Slow-Fast (SF), and Slow-Slow (SS) fabrication processes (2). This research focuses on developing a PVT verification method for a robust ultra-low-power dynamic comparator on a single chip with SAR ADC using 180 nm CMOS technology with an 800mV voltage supply in the analog block.

II. RESEARCH METHODOLOGY

This section presents the development of the PVT Test Scheme to detect and verify worst-case input voltage transitions during all 45 PVT corners simulations. In the procedure, the input test pattern procedure is determined based on the arrangement of voltage input different (ΔVin) considering all possible input sequences in SAR ADC operation, alongside the 45 process corner setup. The flow of the PVT Test Scheme development process is illustrated in Fig. 1.

979-8-3503-7832-0/24 $31.00 © 2024 IEEE

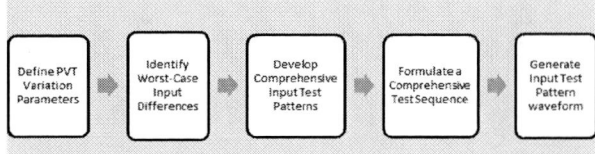

Fig. 1. PVT Test Scheme development process.

A. Novel PVT Test Scheme Development

A dedicated PVT test scheme has been accurately developed to ensure systematic validation of a new comparator design. The scheme incorporates a novel input test pattern database, carefully specifying PVT variations with a 10% margin, which serves as performance and robustness evaluation benchmarks. Key parameters such as voltage supply (V_{DD}) and temperature corners are defined with similar variations to cover a broad range of operating conditions. The development process starts by defining parameters for process variations (FS, SS, SF, TT, FF), voltage supply (720mV, 800mV, 880mV), and temperature range (0˚C, 30˚C, 100˚C). Subsequently, worst-case input voltage differences are identified, including a small ΔVin of 0.8mV and a large ΔVin of 800mV.

To ensure comprehensive testing, input test patterns are accurately created for various scenarios: Big Positive (BP), Small Negative (SN), Small Positive (SP), and Big Negative (BN). These patterns are then used to formulate a test sequence encompassing all possible transitions, thus covering a wide spectrum of operational scenarios. Finally, the input test patterns are translated into waveforms for simulation, allowing for evaluation under worst-case conditions over a 13μs period. This systematic approach guarantees a thorough assessment of the comparator's performance across process, voltage, and temperature variations.

The input test pattern is carefully designed to sensitize the circuit to operate under specified conditions, with each sequence tailored to represent essential circuit operations. Table 1 provides a detailed list of input test patterns used to verify the transitions of worst-case input sequences, ensuring comprehensive coverage. Notably, the input test pattern includes BN, BP, SN, and SP, representing Big Negative, Big Positive, Small Negative, and Small Positive, respectively. By analyzing this sequence, the comparator's performance under worst-case conditions is effectively evaluated, providing valuable insights into its behavior across various operating conditions.

In accordance with the comparator design specification, the clock frequency was set to 2 MHz, and the period of each test sequence was set to 1 μs. To verify the performance of the proposed comparator, the INP voltage and INM voltage of input transistors are set in the worst-case condition of ΔVin. The small ΔVin is 0.8mV, and the large ΔVin is 800mV, with the value of ΔVin defined based on ½ LSB of a 10-bit digital output. For a small ΔVin of 0.8mV, INP is set to 400.4mV and INM to 399.6mV. For a large ΔVin, INP is set to the maximum voltage for logic '1' at 800mV, and INM is set to the maximum voltage for logic '0' at 0V. The input signal toggles through various states, encompassing all possible scenarios during SAR ADC operation.

TABLE I. INPUT TEST PATTERN

Test Sequence	Period (μs)	Input		Voltage Input Different (ΔVin)	Remarks
		INM (mV)	INP (mV)		
1	0-1	800.0	000.0	- 800 mV	Big Negative (BN)
2	1-2	000.0	800.0	+ 800 mV	Big Positive (BP)
3	2-3	399.6	400.4	+ 0.8 mV	Small Positive (SP)
4	3-4	000.0	800.0	+ 800 mV	Big Positive (BP)
5	4-5	400.4	399.6	- 0.8 mV	Small Negative (SN)
6	5-6	399.6	400.4	+ 0.8 mV	Small Positive (SP)
7	6-7	800.0	000.0	- 800 mV	Big Negative (BN)
8	7-8	399.6	400.4	+ 0.8 mV	Small Positive (SP)
9	8-9	400.4	399.6	- 0.8 mV	Small Negative (SN)
10	9-10	800.0	000.0	- 800 mV	Big Negative (BN)
11	10-11	400.4	399.6	- 0.8 mV	Small Negative (SN)
12	11-12	000.0	800.0	+ 800 mV	Big Positive (BP)
13	12-13	800.0	000.0	- 800 mV	Big Negative (BN)

This detailed approach allows for a comprehensive evaluation of the comparator's performance, shedding light on its behavior under varying process, voltage, and temperature conditions. Finally, the input test pattern is translated into waveforms as shown in Fig. 2, providing visual representation within 13μs, aiding in further analysis and validation. To validate the effectiveness of the proposed PVT test scheme, simulations were conducted on the JRR1 comparator Fig. 3 and the conventional Double-tail dynamic comparator using the input test pattern waveform shown in Fig. 2. The comparators were tested with 13 sequences of the differential input waveform (INM and INP) to evaluate their performance under worst-case input conditions.

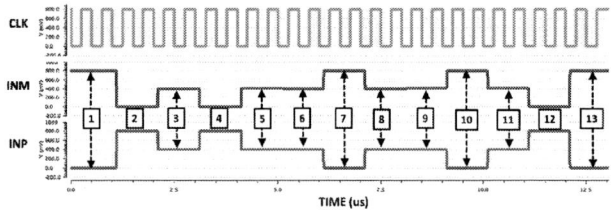

Fig. 2. Waveform of Input Test Pattern

III. PVT SIMULATION RESULT

This section presents the execution of 45-corner simulations during the schematic phases of the proposed comparator design. It encompasses an analysis of Delay and Power Consumption parameters for both double-tail dynamic comparators JRR1 and Double-Tail across these 45 corners in PVT simulation.

A. PVT Schematic Simulation

The performance of the proposed comparator circuit under PVT variation was verified through schematic design simulations. The results from the PVT corner simulations for the JRR1 comparator circuit are depicted in Fig. 4, utilizing the input test pattern waveform shown in Fig. 2. Simulation results indicate that the differential outputs, OUTP_JRR1 and OUTM_JRR1, successfully compared the 13 sequences of ΔVin. The JRR1 comparator demonstrated accurate functionality across the 45-process corners test setup. Conversely, the Double-tail comparator circuit

979-8-3503-7832-0/24 $31.00 © 2024 IEEE

schematic failed under certain test conditions, as illustrated in Fig. 5.

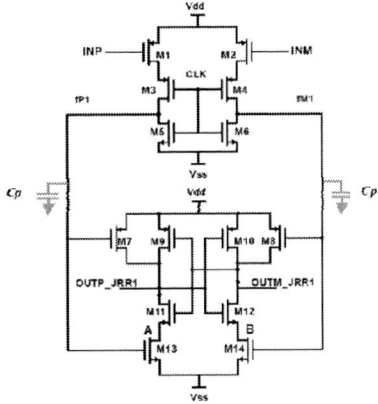

Fig. 3. Architecture of proposed JRR1 comparator circuit

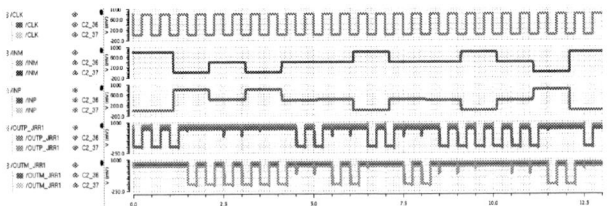

Fig. 4. PVT simulation for schematic JRR1 comparator

As shown in Fig. 5, 9 types of test process corners failed in test sequence numbers 3, 5, 6, 8, 9, and 11. For test sequences 3 and 5, the transition of ΔVin is from Big Positive (BP) +800mV to Small Positive (SP) +0.8mV and Small Negative (SN) -0.8mV. For test sequences 8 and 11, the transition of ΔVin is from Big Negative (BN) -800mV to SP and SN. For test sequences 6 and 9, the transition of ΔVin is from SN to SP and vice versa.

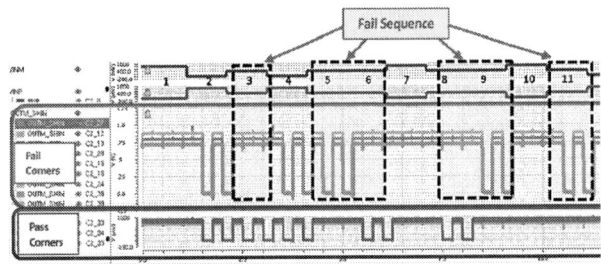

Fig. 5. Schematic Double-Tail comparator fails PVT corners simulation

Table II shows the process corner test No. 1 until No. 4 exhibit failures at the FS process corner when the NMOS transistor operates at high mobility and the PMOS transistor at low mobility. In this weak condition, the PMOS transistor requires longer operation time. In test sequences 3, 5, 6, 8, 9, and 11, the transition of ΔVin causes the PMOS transistor to fail to switch within 0.5µs due to metastability. For test configurations 6 and 7, NMOS and PMOS are set at low mobility, with V_{DD} supplies set 10% lower than the nominal V_{DD} of 720mV and a low temperature of 0°C. The transistors take longer to operate in this weak condition, resulting in low gain at the input latch. The existence of offset voltage in the Double-Tail comparator, detectable in

test sequences 3, 5, 6, 8, 9, and 11, affects performance in FS and SS process corners where V_{DD} is 800mV and 720mV with a temperature of 0°C for test configurations 6 and 7. Test configuration 8 at V_{DD} = 720mV and a temperature of 100°C, together with test configuration 9 at V_{DD} = 800mV and a temperature of 0°C, show failures at the typical process corner for both configurations.

TABLE II.   DOUBLE-TAIL COMPARATOR FAIL PVT

No	Process Corner Test	NMOS	PMOS
1	C2_12_FS; 800 mV; 0 °C	High mobility	Low mobility
2	C2_13_FS; 800 mV; 30 °C	High mobility	Low mobility
3	C2_15_FS; 880 mV; 0 °C	High mobility	Low mobility
4	C2_16_FS; 880 mV; 30 °C	High mobility	Low mobility
5	C2_24_SF; 880 mV; 0 °C	Low mobility	High mobility
6	C2_27_SS; 720 mV; 0 °C	Low mobility	Low mobility
7	C2_28_SS; 720 mV; 30 °C	Low mobility	Low mobility
8	C2_38_TT; 720 mV; 100 °C	Nominal	Nominal
9	C2_39_TT; 800 mV; 0 °C	Nominal	Nominal

B. Energy Consumption Performance in PVT Schematic Simulation

Power consumption is the main parameter measured in the proposed design. This section discusses the effect of V_{DD} on power consumption performance, considering variations in process, voltage, and temperature. Figure 6 illustrates the variation in power consumption performance in the PVT simulation for JRR1 comparators. In the 45-process corner simulations, power consumption increases from 37.57nW to 76.12nW.

Table III shows the energy comparison in PVT simulation of the proposed comparator with a conventional Double-Tail comparator. The maximum average power consumption for JRR1 is 76.12nW at (V_{DD} = 880mV, Temp. = 100°C, Process Corner = FF), which is 70% lower than that of the Double-Tail comparator. The minimum average power consumption is 3.6nW at (V_{DD} = 720mV, Temp. = 0°C, Process Corner = SS), 57% lower than that of the Double-Tail comparator.

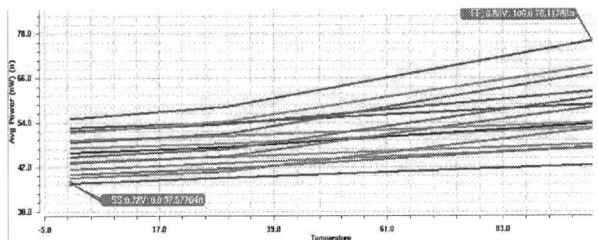

Fig. 6. Power Consumption of JRR1 comparator

TABLE III.   ENERGY COMPARISON IN PVT SIMULATION

Comparator Configuration	Double-Tail	JRR1
Max. Average Power (880 mV; 100 °C; FF)	249.23 nW	76.12 nW
Min. Average Power (720 mV; 0 °C; SS)	86.42 nW	37.58 nW

C. Delay Performance in PVT Schematic Simulation

From the simulated results in Fig. 7, a longer regeneration delay occurs at (V_{DD} = 720mV; Temp. = 0°C;

Process Corner = SS) with 1.22ns due to limited voltage headroom, slow electron mobility at 0°C, and slow mobility in NMOS and PMOS transistors in the SS process corner. However, the regeneration delay is reduced to 188.64ps at (V_{DD} =880mV; Temp. = 100°C; Process Corner = FF).

As shown in Fig.8, with ΔVin = 0.8mV, the regeneration delay becomes critical as the input transistors M1 and M2 operate in subthreshold conditions with INP = 400.4mV and INM = 399.6mV, limiting current flow and increasing the regeneration process duration. Therefore, at (V_{DD} = 720 mV; Temp. = 0°C; Process Corner = SS), the regeneration delay worsens to 27ns.

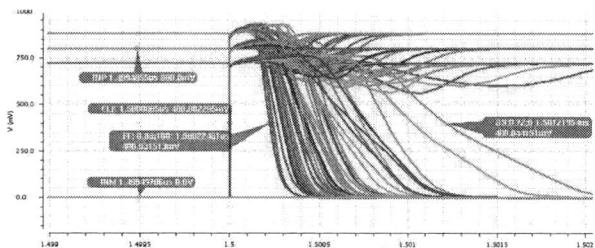

Fig. 7. Regeneration delay of JRR1 comparator with ΔVin=800 mV in PVT schematic simulation.

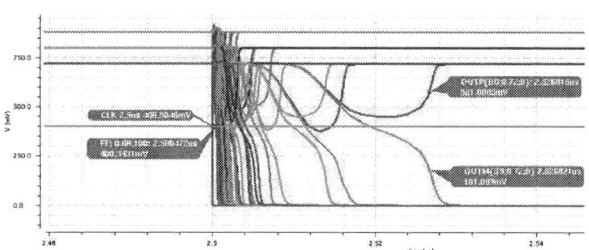

Fig. 8. Regeneration delay of JRR1 comparator with ΔVin=0.8 mV in PVT schematic simulation.

TABLE IV.   DELAY COMPARISON IN PVT SIMULATION

Comparator Configuration	Double-tail	JRR1
Max Regeneration Delay ΔVin = 800 mV (720 mV; 0 °C; SS)	1.39 ns	1.22 ns
Min Regeneration Delay ΔVin = 800 mV; (880 mV;100 °C; FF)	296 ps	220 ps
Max Regeneration Delay ΔVin = 0.8 mV (720 mV; 0 °C; SS)	Fail corner	27 ns
Min Regeneration Delay ΔVin = 0.8 mV; (880 mV;100 °C; FF)	Fail corner	472 ps

The comparison in Table IV shows that the proposed JRR1 comparator significantly outperforms the conventional Double-tail comparator in terms of regeneration delay across various process, voltage, and temperature conditions. At an input voltage difference (ΔVin) of 800 mV, the JRR1 comparator has lower maximum and minimum delays compared to the Double-tail comparator. Notably, at a smaller ΔVin of 0.8 mV, the Double-tail comparator fails to regenerate, while the JRR1 still functions effectively with a maximum delay of 27 ns and a minimum delay of 472 ps. This indicates the superior robustness and efficiency of the JRR1 comparator, making it more reliable for precise applications under varied conditions.

CONCLUSION

This paper has introduced a novel PVT verification method designed to evaluate the performance of dynamic comparators in practical scenarios. The proposed PVT test scheme has been successfully implemented and validated on both JRR1 and Double-tail dynamic comparators. By using a specific input test pattern, the scheme effectively detects metastability errors in the Double-tail comparator across 45 PVT corner simulations. Additionally, the simulated results confirm the capability of the test scheme to assess delay and power performance comprehensively. The successful implementation and validation of this method highlight its robustness and reliability, making it a valuable tool for thorough PVT evaluation in dynamic comparator design.

ACKNOWLEDGMENT

The authors acknowledge the Universiti Kuala Lumpur British Malaysian Institute for sponsoring this research through the STRG Grant (STR19085), and University Putra Malaysia for supervising the research funded by the FRGS Grant (5540076) from the Ministry of Education Malaysia.

REFERENCES

[1] Folla, J. K., Crespo, M. L., Wembe, E. T., Bhuiyan, M. A. S., Cicuttin, A., Essimbi, B. Z., & Reaz, M. B. I. (2021). A low-offset low-power and high-speed dynamic latch comparator with a preamplifier-enhanced stage. IET Circuits, Devices and Systems, 15(1), 65–77.

[2] Abdelmagid, B. A., & Mohieldin, A. N. (2022). Digitally assisted dynamic comparator with reduced offset across process, voltage, and temperature variations. AEU - International Journal of Electronics and Communications, 146(January), 154116.

[3] Roslita Rusli, J., Shafie, S., Mohd Sidek, R., Abdul Majid, H., Wan Hassan, W., & Mustafa, M. (2020). Optimized low voltage low power dynamic comparator robust to process, voltage and temperature variation. Indonesian Journal of Electrical Engineering and Computer Science, 17(2), 783-792.

[4] Bindra, H. S., Lokin, C. E., Schinkel, D., Annema, A. J., & Nauta, B. (2018). A 1.2-V Dynamic Bias Latch-Type Comparator in 65-nm CMOS With 0.4-mV Input Noise. IEEE Journal of Solid-State Circuits, 53(7), 1902–1912.

[5] Chevella, S., O'Hare, D., & O'Connell, I. (2020). A Low-Power 1-V Supply Dynamic Comparator. IEEE Solid-State Circuits Letters, 3, 154–157.

[6] de La Fuente-Cortes, G., Espinosa Flores-Verdad, G., Gonzalez-Diaz, V. R., & Diaz-Mendez, A. (2017). A new CMOS comparator robust to process and temperature variations for SAR ADC converters. Analog Integrated Circuits and Signal Processing, 90(2), 301–308.

[7] Benschwartz, R., & Sakthivel, P. (2016). A Process Variation Tolerant OTA Design for Low Power ASIC Design. Circuits and Systems, 07(06), 956–970.

[8] Nagy, L., Arbet, D., Kovac, M., Potocny, M., Sovcik, M., & Stopjakova, V. (2020). Dynamic Properties of Ultra Low-Voltage Rail-To-Rail Comparator Designed in 130 nm CMOS Technology. Proceedings - 2020 23rd International Symposium on Design and Diagnostics of Electronic Circuits and Systems, DDECS 2020, 15–18.

[9] Stopjakova, V., Rakus, M., Kovac, M., Arbet, D., Nagy, L., Sovcik, M., & Potocny, M. (2018). Ultra-low voltage analog IC design: Challenges, methods and examples. Radioengineering,27(1),171–185.

[10] Shoniker, M., Oleynikov, O., Cockburn, B. F., Han, J., Rana, M., & Pedrycz, W. (2018). Automatic Selection of Process Corner Simulations for Faster Design Verification. IEEE Transactions on Computer-Aided Design of Integrated Circuits and Systems, 37(6), 1312–1316.

[11] J. R. Rusli, S. Shafie, W. Z. W. Hassan, H. A. Majid, I. Ahmad and M. A. Mustafa, "A Post-Silicon Validation Method for Low-Power 180 nm Dynamic Comparator in Differential 10-bit SAR ADC," 2023 IEEE 9th International Conference on Smart Instrumentation, Measurement and Applications (ICSIMA), Kuala Lumpur, Malaysia, 2023, pp. 199-204.

Analysis of the Electrical Characteristics for Compact SPICE Modelling of STT-MTJ Device with Physical Parameters Variation

M. Y. Xuan Pai
Faculty of Electrical
Engineering,
Universiti Teknologi Malaysia,
81310 Johor Bahru, Malaysia
melaniepaiyi@graduate.utm.my

N. Ezaila Alias*
Faculty of Electrical
Engineering,
Universiti Teknologi Malaysia,
81310 Johor Bahru, Malaysia
ezaila@fke.utm.my

M. L. Peng Tan
Faculty of Electrical
Engineering,
Universiti Teknologi Malaysia,
81310 Johor Bahru, Malaysia
michael@utm.my

Afiq Hamzah
Faculty of Electrical
Engineering,
Universiti Teknologi Malaysia,
81310 Johor Bahru, Malaysia
mafiq@fke.utm.my

Yasmin Abdul Wahab
Nanotechnology & Catalysis
Research Centre,
Universiti Malaya,
50603 Kuala Lumpur, Malaysia
yasminaw@um.edu.my

Maizan Muhamad
School of Electrical
Engineering,
College of Engineering,
Universiti Teknologi MARA,
40450 Shah Alam, Malaysia
maizan@uitm.edu.my

Abstract—**Spin-Transfer Torque Magnetoresistive Random Access Memory (STT-MRAM) operates on the principle of magnetic anisotropy energy to retain information and magnetoresistance to retrieve information. STT-MRAM consists of an MTJ (Magnetic Tunnel Junction) and a transistor device. The MTJ comprises two layers of ferromagnetic metal separated by an insulator. A major concern in evaluating STT-MRAM technology is developing a scalable MTJ compact model capable of incorporating real variable effects across numerous technical nodes. Therefore, this work involves simulating the STT-MTJ device compact SPICE modeling at the device level using a SPICE subcircuit and a mathematical model to analyze the electrical characteristics. The I-V characteristics of parallel and anti-parallel orientations of the STT-MTJ are simulated. The parallel resistance (RP) and anti-parallel resistance (RAP) of the STT-MTJ device are varied from their original values to observe the I-V characteristic graph for each case. The I-V characteristics for different resistance values and variations in width and length are analyzed. It is observed that the Tunneling Magnetoresistance (TMR) increases by 230.77% when the resistance for parallel current (IP) is reduced, whereas the TMR increases by 105.56% when the resistance for anti-parallel current (IAP) is reduced. Moreover, as the width and length of the Free Layer (FL) increase (by 222.22%), the write current for the MTJ also increases by 492.55%. The parameters used in the STT-MTJ can be adjusted for different MTJ materials to achieve higher performance efficiency.**

Keywords— Spin-transfer torque-magnetic tunnel junction (STT-MTJ), compact modelling, device-level.

I. INTRODUCTION

Due to the rising of leakage currents caused by the complementary metal-oxide semiconductor (CMOS) fabrication node shrinking below 90nm, the memory and logic circuits have high static power [1]. The non-volatile memory (NVM) implementation in logic circuits may offer a number of significant benefits, including run-time reconfigurability, built-in function programmability (in LUT, PLD, or FPGA), and instantaneous on/off. Additionally, instantaneous on/off switching of circuitry on demand could be made possible by the incorporation of non-volatile flip-flop registers that are evenly spaced throughout logic chips, this can provide a practical way to counteract the increasing dissipation caused by leakage currents in current CMOS logic circuits [1]. Some examples of NVM are ferroelectric random-access memory (FRAM), phase-change random-access memory (PRAM), and resistive random-access memory (RRAM). For the upcoming generation of NVM memory devices, STT-MRAM is seen as a potential memory technology [2] due to their potential for high write/read speed, good integration with CMOS processes, and technology maturity [3-4]. Hybrids MTJ/CMOS circuits for memory and logic applications such as Magnetic RAM (MRAM) and Magnetic FPGA [5-6] have been suggested. The MRAM makes use of MTJ where the STT current changes its resistance state and data can be read through the sensing of TMR difference [7]. The STT-MRAM technology can be evaluated by the process of developing a user self-defined MTJ compact model. There are studies have been carried out and the MTJ behaviors were took as a model with SPICE subcircuits such as curve fitting circuit and bi-stable circuit in terms of the experimental input specifications (parallel and anti-parallel resistance, critical switching current, and thermal stability).

The STT-MTJ device is represented by a SPICE micromodel that shows how the applied bias affects the device current in each of the different resistance states [8]. A perpendicular Magnetic Tunnel Junction (pMTJ) mainly consists of three layers which are; a thin oxide layer, MgO and two ferromagnetic (FM) layers, CoFeB. One of the most vital areas of study is the MTJ switching techniques, since they have a direct impact on the performance of hybrid

---

This work was supported and funded by the Universiti Teknologi Malaysia (UTM) Fundamental Research Grant Project No.Q.J130000.3823.22H52. The authors acknowledge the Research Management Centre (RMC), School of Graduate Studies (SPS), and Faculty of Electrical Engineering (FKE) of Universiti Teknologi Malaysia (UTM) for providing excellent support and a stimulating research environment.

979-8-3503-7832-0/24 $31.00 © 2024 IEEE

MTJ/CMOS circuits in terms of power, speed, and area. A revolutionary STT switching technique has been presented and developed quickly by many firms during the past four years [9-10]. By allowing a low current (120µA) to reverse the state of the MTJ with the decrease in MTJ size, it significantly enhances its scalability and promises to accelerate the extensive commercial applications of hybrid MTJ/CMOS circuits. These hybrid circuits exhibit outstanding performance in terms of power, speed, and chip area by combining the benefits of CMOS and MTJ technology [11].

In this work, the objective is to simulate the compact SPICE modelling of STT-MTJ device at device level using a SPICE subcircuit and mathematical model to analysis the electrical characteristics. An oxide layer (MgO) is positioned between two ferromagnetic (FM) layers (CoFeB) in a perpendicular MTJ (pMTJ). A pMTJ is considered in this work due to its lower switching current compared to in-plane MTJ (iMTJ). The parallel (R_P) or anti-parallel (R_{AP}) resistance state that represents the binary states of STT-MRAM is determined by taking the Free Layer (FL) which will give either a parallel (P) or an anti-parallel (AP) orientation.

II. COMPACT SPICE MODELLING OF STT-MTJ DEVICE

A SPICE micromodel of the STT-MTJ device illustrates how applied bias influences device current in each of the several resistance states [8]. The STT-MTJ physical device modelled is made up of two ferromagnetic (FM) layers (CoFeB) and an oxide layer (MgO). The suggested SPCE model [8] is based on three already existing mathematical models that are pertinent to the STT-MTJ; the critical current switching Slonczewski model, TMR effect voltage dependency model, and the Brinkman physical model of MTJ conductance. Each model's behaviour is initially examined using MATLAB before being built as a small SPICE model. The crucial finding from the simulation is that the STT-MTJ resistance is constant where it is not relying on the bias when device is in P state and resulting in the current grows linearly with the applied bias. On the other hand, when the device is in AP state, the current and bias have a non-linear connection. The suggested STT-MTJ subcircuit enables immediate MTJ model integration into other intricate designs such as non-volatile SRAM and non-volatile flip-flop [8].

A. Proposed STT-MTJ Subcircuit Design

A STT-MTJ subcircuit is constructed and verified using LTspice, shown in Fig. 1. V_1 is powered with pulse wave function of 5 V. On the other hand, V_2 is supplied with piecewise linear (PWL) function of 10 V. In order to manage the transition between the P and AP stages, a switch called MYSW is connected in parallel with R_a. The comparator subcircuit of the voltage-controlled switch MYSW controls the switch's operation [4]. The switch will be off when $V_b \geq$ 80 mV, R_P and R_a will be connected in series (R_{AP}), showing the status of the AP resistance. On the other hand, when $V_b <$ 80 mV, the switch will be on and the total terminal resistance will drop to R_P.

The switching condition is determined by TMR effect bias voltage dependency model and Slonczewski model, whereas the Brinkman physical model of the MTJ conductance controls the values of R_P and R_a. The MYSW switch's input signal is sent through the port V_b, where it is compared with the reference threshold voltage.

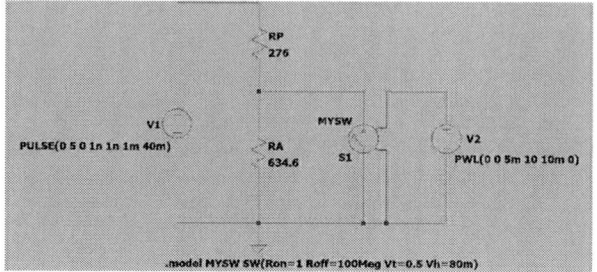

Fig. 1. The STT-MTJ subcircuit is developed using LTspice.

B. Mathematical Modelling and Simulation of the STT-MTJ

The resistors, R_a and R_P (Fig. 1) are connected in series to represent the equation (1) and (2) respectively. The Brinkman Physical Model of MTJ conductance modelling provides an explanation for the MTJ device's resistance, which is caused by an oxide barrier layer that interfaces with FM layers. The tunnel barrier's height and the interface effect are the primary factors influencing resistance in Brinkman model. The suggested model resistance at P state is calculated using a simplified version of Brinkman model as follows:

$$R_P(V) = \frac{R_P(0)}{(1 + \frac{t_{ox}^2 \times q^2 \times m}{4\hbar^2 \times \Phi \times V_b^2})} \tag{1}$$

where
t_{ox}: oxide thickness
q: electron charge
m: electron mass
$\hbar$: Constant Planck
Φ: potential difference for MgO
V_b: voltage across the two terminals of the device
$R_P(0)$: bulk resistance of the apparatus, as indicated by the subsequent formula:

$$R_P(0) = \frac{t_{ox}}{(223.76 \times \Phi^{\frac{1}{2}} \times sr)\exp(1.025 \times t_{ox} \times \Phi^{\frac{1}{2}})} \tag{2}$$

where sr: surface of the MTJ (width × length).

The modelling of Brinkman Physical Model of MTJ uses the following equations to determine the resistance at P state and AP state.

$$R_a(V) = R_p(V) \times TMR(V) \tag{3}$$

$$R_{AP} = R_a(V) + R_P(V) = R_P(V)[1 + TMR(V)] \tag{4}$$

where   TMR(V): tunnel magnetoresistance.

The simulation result of $R_P(V)$ and $R_a(V)$ with a bias ranged from -0.2V to +0.2V is presented in Fig. 2. The following settings were applied to the simulation: $t_{ox} = 0.75$nm

979-8-3503-7832-0/24 $31.00 © 2024 IEEE

and $\Phi = 0.4$ for the MgO tunnel barrier. From Fig. 2, it is clear that the $R_P(V)$ (blue line) has a constant value of 276 Ω, which is equal to $R_P(0)$ and independent of applied bias. Additionally, $R_a(V)$ exhibits a quadratic behaviour with a peak resistance value of 634.6 Ω at 0 V, as seen in Fig. 2.

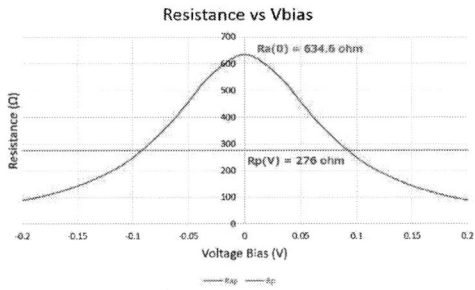

Fig. 2. Brinkman physical model of MTJ conductance.

TMR lowers as a function of applied bias (increase of the barrier height) by increasing the junction voltage in MTJ TMR. The formula for TMR is as follows: V_h is the threshold voltage that corresponds to the voltage at TMR(V) = 0.5 × TMR(0), and TMR(0) is the TMR(V) ratio without any bias applied.

$$TMR(V) = \frac{TMR(0)}{1 + (\frac{V_b}{V_h})^2} \qquad (5)$$

The TMR(0) equals 230% and V_h equals 80 mV respectively. From Fig. 3, a quadratic TMR(V) - V_b curve is calculated with a peak value of TMR(0) = 230% with V_b = 0V. Due to the square factor of $(\frac{V_b}{V_h})^2$, TMR(V) is the quadratic nature. In Fig. 3, it can be observed that when V_h reaches $V_h \approx$ 80mV, the TMR(V) has decreased to TMR(V) = 0.5 × TMR(0) that is approximately 115%, this demonstrates the accuracy of the simulation and aligns with the TMR definition.

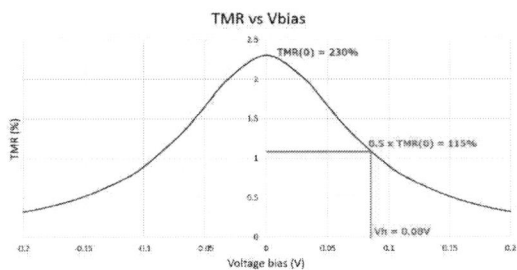

Fig. 3. TMR effect bias dependence model.

III. RESULT AND DISCUSSION

In order to verify the outcome of proposed SPICE model, the AP and P states are checked individually by flipping the switch MYSW on and off, respectively. After performing DC voltage sweep between -200 mV to 200 mV, the current passing through R_P is computed. The simulation result in HPSICE is shown in Fig. 4. It can be observed that the current increase linearly for both I_P and I_{AP} with respect to voltage bias. I_P represents the current when STT-MTJ operates in P state, whereas I_{AP} represents it operates in AP state.

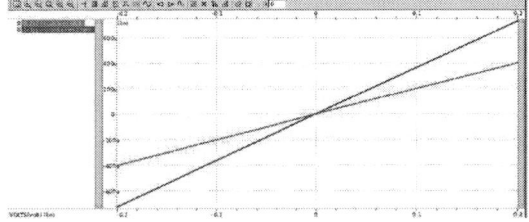

Fig. 4. DC analysis of I_P (blue line) and I_{AP} (red line) vs voltage bias.

Additionally, a transient simulation of 40 ms is also performed for both I_P and I_{AP} using HSPICE. Transient analysis is performed to understand the system response for both I_P and I_{AP}. It also provides insights into the dynamic behaviour of the SPICE model. The result is presented in Fig. 5. Blue line represents the I_{AP} while the red line represents the I_P. The transient simulation indicates that when current is passed through the device, the current is spin-polarized and exerting a torque on the magnetization of the free layer (FL). The torque can overcome the energy barrier associated with the magnetization orientation if the applied current is greater than the critical switching current.

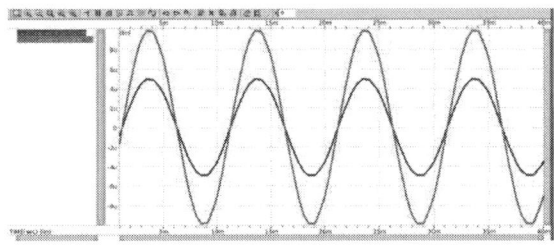

Fig. 5. Transient analysis for I_P (red line) and I_{AP} (blue line) for 40ms.

The STT-MTJ's parallel and anti-parallel resistance, R_P and R_{AP} are varied to obeserve the electrical characteristic of the SPICE model. The simulation result in HSPICE for both I_P and I_{AP} with parallel resistance variation are shown in Fig. 6 and Fig. 7 respectively. It can be observed in both figures, as the R_P and R_{AP} increases, the I_P and I_{AP} becomes lower than the original value, as according to Ohm's law, V = IR. When R_P in both cases are set to 100Ω and 200Ω respectively regardless the value of R_{AP}, the value of I_P and I_{AP} are almost similar. Thus, it can be concluded that R_P value has the most significant effect on the switching current.

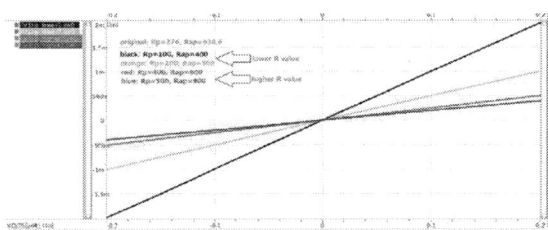

Fig. 6. I-V characteristic graph for I_P with parallel resistance variation.

979-8-3503-7832-0/24 $31.00 © 2024 IEEE

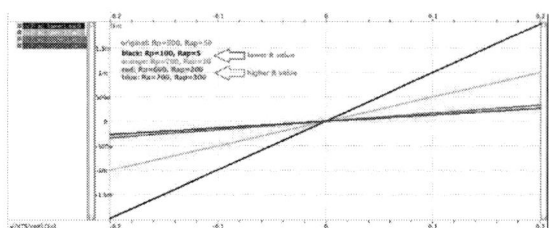

Fig. 7. I-V characteristic for I_{AP} with parallel resistance variation.

IV. CONCLUSION

In conclusion, STT-MRAM emerges as a promising memory technology for the next generation of Non-Volatile Memory (NVM) devices, owing to its potential for high write/read speeds, seamless integration with CMOS processes, and technological maturity. The successful development and simulation of the STT-MTJ circuit model using HSPICE underscore its efficacy. The electrical characteristics derived from the SPICE model align closely with reference values from research papers, affirming its accuracy. By varying resistance, the switching current of STT-MTJ can be observed, revealing significant improvements in Tunneling Magnetoresistance (TMR) — a 230.77% increase for parallel current (IP) reduction and a 105.56% increase for anti-parallel current (IAP) reduction. Moreover, as the width and length of the Free Layer (FL) increase by 222.22%, the write current for the MTJ also escalates by 492.55%. The adaptability of parameters within the STT-MTJ, tailored to different MTJ materials, offers avenues for enhanced performance efficiency. STT-MTJ presents a compelling proposition for designers, facilitating the simulation of STT-MTJ/CMOS controller memory circuits. This approach not only achieves high chip density but also ensures energy efficiency, surpassing traditional MOSFETs in these regards.

ACKNOWLEDGMENT

This work was supported and funded by the Universiti Teknologi Malaysia (UTM) Fundamental Research Grant Project No.Q.J130000.3823.22H52. The authors acknowledge the Research Management Centre (RMC), School of Graduate Studies (SPS), and Faculty of Electrical Engineering (FKE) of Universiti Teknologi Malaysia (UTM) for providing excellent support and a stimulating research environment.

REFERENCES

[1] N. S. Kim, T. Austin, D. Baauw, T. Mudge, K. Flautner, J. S. Hu, M. J. Irwin, M. Kandemir, and V. Narayanan, "Leakage current: Moore's law meets the static power," Computer, vol. 36, no. 12, pp. 68–75, Dec. 2003.

[2] W. Kang, L. Zhang, J.-O. Klein, Y. Zhang, D. Ravelosona, and W. Zhao, "Reconfigurable Codesign of STT-MRAM Under Process Variations in Deeply Scaled Technology," IEEE Trans. Electron Devices, vol. 62, no. 6, pp. 1769– 1777, Jun. 2015.

[3] C. Chappert, A. Fert, and F. Nguyen Van Dau, "The emergence of spin electronics in data storage," Nature Mater, vol. 6, no. 11, pp. 813–823, Jun. 2007.

[4] International Technology Roadmap for Semiconductors (ITRS), Process Integration, Devices and Structures. (2007)

[5] W. Zhao, E. Belhaire, C. Chappert, François Jacquet and Pascale Mazoyer "New Non-Volatile logic based on Spin-MTJ" Physica Status Solidi a-applications and materials science, 250, Vol.6, Mai 2008, pp.1373-1377.

[6] S.Ikeda, J.Hayakawa, Y.M.Lee, F.Matsukura, Y.Ohno, T.Hanyu and H.Ohno "Magnetic Tunnel Junctions for Spintronic Memories and Beyond" IEEE Transactions on Electron Devices, Vol.54. 2007, pp.991- 1002.

[7] J. Kim, A. Chen, B. Behin-Aein, S. Kumar, J. -P. Wang and C. H. Kim, "A technology-agnostic MTJ SPICE model with user-defined dimensions for STT-MRAM scalability studies," 2015 IEEE Custom Integrated Circuits Conference (CICC), San Jose, CA, USA, 2015, pp. 1-4, doi: 10.1109/CICC.2015.7338407.

[8] Chong Jian Yee, F. O. Hatem, T. N. Kumar and H. A. F. Almurib, "Compact SPICE modeling of STT-MTJ device," 2015 IEEE Student Conference on Research and Development (SCOReD), Kuala Lumpur, Malaysia, 2015, pp. 625-628, doi: 10.1109/SCORED.2015.7449413.

[9] M.Hosomi et al, "A Novel nonvolatile memory with Spin Torque Transfer Magnetization Switching: Spin-Ram" IEEE International Electron Devices Meeting Technical Digest, Dec. 5, 2005, Page(s): 459 – 462.

[10] T.Kawahara, R.Takemura, K.Miura, J.Hayakawa, S.Ikeda, Y.Lee, R.Sasaki, Y.Goto, K.Ito, T.Meguro, F.Matsukura, H.Takahashi, H.Matsuoka, H.Ohno "2Mb Spin-Transfer Torque RAM (SPRAM) with Bit-by-Bit Bidirectional Current Write and Parallelizing-Direction Current Read" International Solid State Circuits Conference, USA, pp.480-482.(2007).

[11] H.Ohno, "Spin manipulation in semiconductors and metals-Roadmap for Spintroincs", Intermag, May, 2008.

Methodology to Optimize E-Mode GaN HEMT with P-Type Doping under 2DEG Layer

M.N.A. Muhammad Ridzwan
Faculty of Engineering & Built Environment, Universiti Kebangsaan Malaysia, Selangor, Malaysia
a181425@siswa.ukm.edu.my

M.F. Abdullah
*Center for Semiconductor & Thin Film Research, MIMOS Berhad.
Kuala Lumpur, Malaysia*
faizol.abdullah@mimos.my

A.M. Mohamad Yussof
Faculty of Engineering & Built Environment, Universiti Kebangsaan Malaysia, Selangor, Malaysia
a181491@siswa.ukm.edu.my

N. Abd Aziz
Faculty of Engineering & Built Environment, Universiti Kebangsaan Malaysia, Selangor, Malaysia
norazreen@ukm.edu.my

H.W. Lee
*Center for Semiconductor & Thin Film Research, MIMOS Berhad,
Kuala Lumpur, Malaysia*
hingwah.lee@mimos.my

Abstract—This article presents the methodology to optimize the enhancement-mode (E-mode) GaN high electron mobility transistor (HEMT). The focus is on the recessed Schottky gate with a p-type GaN under the two-dimensional electron gas (2DEG) layer. Design of experiment 3 factors (t_{AlGaN}, L_{gate}, P_{GaN}) at 4 levels for technology computer-aided design (TCAD) simulation produced 2 responses (V_T, R_{ON}). 128 sets of observations are fed into the analysis of variance (ANOVA) and artificial neural network (ANN) models, where they narrowed down the list of potential optimum E-mode GaN HEMTs. The best device with t_{AlGaN} = 4 nm, L_{gate} = 2 μm, and P_{GaN} = 1.2 ×10^{19} cm^{-3} is predicted V_T = 1.263 V and R_{ON} = 3.317 Ω. The verification by TCAD gives V_T = 1.224 V and R_{ON} = 3.235 Ω which is very close to the ANOVA-ANN prediction. The p-type GaN under 2DEG created a local GaN p-n junction that depletes electrons in 2DEG at the thermal equilibrium.

Keywords—*E-mode HEMT, p-type GaN, DOE, ANOVA, TCAD, ANN*

I. INTRODUCTION

The existence of two-dimensional electron gas (2DEG) at the wide bandgap AlGaN/GaN heterostructure is fundamental for building GaN high electron mobility transistors (HEMTs) that are in demand for power converters [1] and RF power amplifiers [2]. Electrons in the 2DEG layer are at a high concentration of ~10^{13} cm^{-2} and mobility of ~1500 cm^{2}V^{-1}s^{-1} even at room temperature [3]. Therefore, in default, GaN HEMT is a normally-ON transistor. The use of the Schottky gate, e.g. Ni/Au on AlGaN/GaN can deplete electrons in 2DEG at the applied negative gate bias V_{gate} to turn OFF the HEMT. This normally-ON HEMT is also known as depletion-mode (D-mode) HEMT.

This study resumes the effort to provide a normally-OFF HEMT, where electrons in 2DEG in default were depleted at V_{gate} = 0 V. Positive threshold voltage V_T requires positive V_{gate} to turn ON the HEMT, so this device also can be known as enhancement-mode (E-mode) HEMT. For power converters, E-mode is always preferred over D-mode for fail-safe operation and minimum static power consumption. Hamady et al. using technology computer-aided design (TCAD) investigated how p-type GaN buried under 2DEG can convert D-mode HEMT into E-mode HEMT [4]. Note this concept is different from a gate injection transistor, where p-type GaN is above the AlGaN/GaN layers [5].

Buried p-type GaN can be obtained either by i) introducing Mg dopant during growth, etching to leave only p-type GaN region, then regrowth [6], or ii) Mg implantation into the selected area under the intended gate [7]. Understanding the behavior of E-mode HEMT with p-type GaN under 2DEG is crucial to rely on TCAD (with calibration to some actual samples) as it will save much time and money on the technology development.

In this article, we propose a methodology to optimize E-mode GaN HEMT with p-type doping under Schottky gate and 2DEG layer using i) design of experiment (DOE), ii) TCAD simulation iii) analysis of variance (ANOVA), and iv) artificial neural network (ANN). This technology is based on a recessed Schottky gate, T-gate, AlN spacer, and p-type GaN under the 2DEG layer as explained above to convert D-mode HEMT into E-mode HEMT.

II. EXPERIMENTAL

Fig. 1a guides the workflow to obtain an optimized E-mode HEMT. Essentially, we have to i) define the original structure of D-mode GaN HEMT, ii) perform the DOE for the selected gate features, and iii) conduct full-scale TCAD simulation using Silvaco Victory Device to collect the responses. The responses from TCAD simulation are then going to iv) ANOVA using OriginPro to identify the main effect, v) fed into an ANN model built using MATLAB Neural Net Fitting, vi) predict output based on a refined search on the main effect, and finally vii) run the TCAD simulation on the optimized variables for verification.

Fig. 1b shows the schematic of GaN HEMT in this study. It is a T-shaped Schottky gate-type of HEMT on Al$_{0.3}$Ga$_{0.7}$N/AlN/GaN. We vary three factors, which are i) thickness of remaining AlGaN under gate recess t_{AlGaN}, ii) length of the gate in recess L_{gate}, and iii) doping concentration of p-type GaN under the gate P_{GaN}. The original structure is a normally-ON HEMT made of Schottky gate contact on AlGaN/AlN/GaN/sapphire substrate. This original device has t_{AlGaN} = 24 nm, L_{gate} = 2 μm, and no region/concentration of P_{GaN}. It is necessary to calibrate the model built by Victory Device with the experimental device to apply physical models precisely, e.g. SRH, Auger, DopingDep, high field saturation, and polarization models [8], but we omit the calibration in this work to only emphasize on the proposed methodology. Table 1 is the DOE for the 3 factors at 4 levels and the number of required runs is $4^3 = 64$. The simulation is repeated to obtain a total of 128 responses.

III. RESULTS AND DISCUSSION

We tracked two outputs of HEMT from the Victory Device; V_T and channel ON resistance R_{ON} based on the HEMT transfer characteristic. Analyzing the significance of t_{AlGaN}, L_{gate}, and P_{GaN} factors on the responses of V_T and R_{ON} requires two separate three-way ANOVAs. Fig. 2a shows the box plot and plots of the main effects of t_{AlGaN}, L_{gate}, and P_{GaN}

on V_T. The middle 50% of V_T from all studied factors in the box plot is from −3.144 V to −3.321 V, while the min and max values are −7.611 V and 9.455 V, respectively. We found at the 0.05 level, the population means of both t_{AlGaN} and P_{GaN} are significantly different, while the population means of L_{gate} are not significantly different. Fig. 2b shows the box plot and plots of the main effects of t_{AlGaN}, L_{gate}, and P_{GaN} on V_T. The middle 50% of R_{ON} from all studied factors in the box plot is from 3.304 Ω to 3.537 V, while the min and max values are 3.117 Ω and 3.830 V, respectively. At the 0.05 level, we found that the population means of all t_{AlGaN}, L_{gate}, and P_{GaN} are significantly different.

It is easy to reject the null hypothesis based on F-values and P-values in Figs. 2a and 2b. Variations in t_{AlGaN}, L_{gate}, and P_{GaN} are indeed statistically significant. At this point, we can safely lock the value of t_{AlGaN} = 4 nm as it always leads to positive V_T and lowest R_{ON}. L_{gate} and P_{GaN} are optimized using ANN, where 128 observations of 3 factors and 2 responses are fed into an ANN model. Fig. 3a shows the ANN model with 50 hidden layers trained using Bayesian regularization backpropagation. We allocated 90% of observations for training and 10% for testing. These results to a satisfyingly good fit on training (R = 0.99045) and test (R = 0.96867) as indicated by regression in Fig. 3b. Additionally, the error histogram for 20 bins is provided in Fig. 3c. Based on ANOVA, we created 20 sets of HEMT to predict their responses with refined values of P_{GaN} = 1.2 − 2.0 ×10^19 cm^−3. With the right combination of t_{AlGaN} = 4 nm, L_{gate} = 0.5 − 2 µm, and P_{GaN} = 1.2 − 2 ×10^19 cm^−3, the optimum GaN HEMT with positive V_T ~1 V and good R_{ON} should be found within this range. Too low V_T is not good as it is prone to spurious turn-ON during high-speed operation, while too high V_T is also not good as it increases switching losses. Table 2 provides the values of V_T and R_{ON} from the 20 sets of HEMT after the prediction by the trained ANN model. We predicted the optimum device with t_{AlGaN} = 4 nm, L_{gate} = 2 µm, and P_{GaN} = 1.2 ×10^19 cm^−3 to give V_T = 1.263 V (E-mode HEMT) and R_{ON} = 3.317 Ω.

The predicted optimum device with t_{AlGaN} = 4 nm, L_{gate} = 2 µm, and P_{GaN} = 1.2 ×10^19 cm^−3 is verified by Victory Device simulation. Figs. 4a and 4b compare the original D-mode HEMT and the optimized E-mode HEMT at the thermal equilibrium. At the interface of the AlN spacer and GaN channel of D-mode HEMT, conduction band E_C is slightly below the Fermi level approximately −0.136 eV.

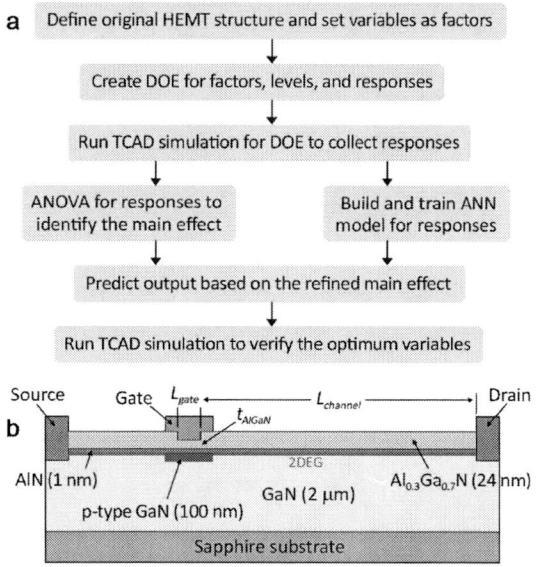

Fig. 1. a) Workflow for the proposed methodology. b) Schematic of GaN HEMT to indicate the device components and their dimension.

Table 1. Factors and levels for setting up the TCAD simulation in Victory Device. The created DOE of 4^3 requires a minimum of 64 simulations to be run.

Level	t_{AlGaN} (nm)	L_{gate} (µm)	P_{GaN} (cm^{-3})
1	4	0.5	1 ×10^18
2	8	1.0	5 ×10^18
3	12	1.5	1 ×10^19
4	16	2.0	5 ×10^19

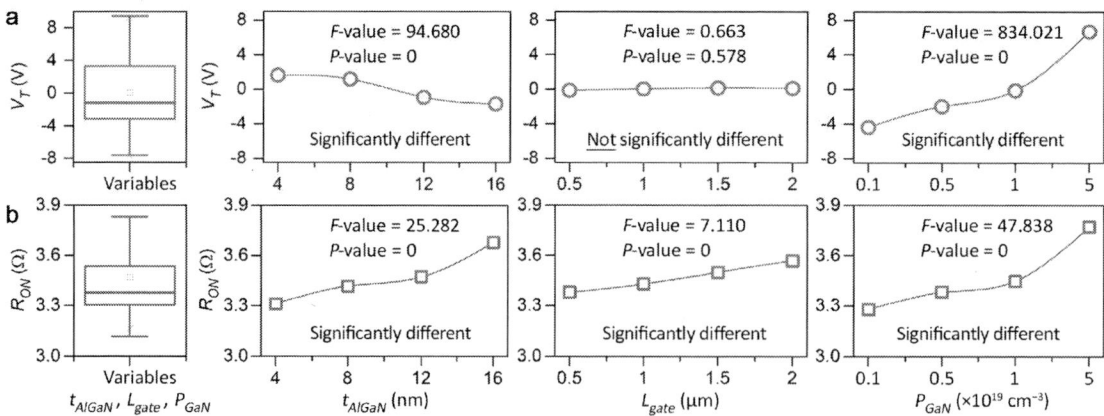

Fig. 2. Box plots and plots of main effects from the three-way ANOVA for the outputs of: a) threshold voltage V_T, and b) channel on resistance R_{ON}. The F-value and P-value for each factor of t_{AlGaN}, L_{gate}, and P_{GaN} are provided.

979-8-3503-7832-0/24 $31.00 © 2024 IEEE

This results in a known accumulation of electrons $\sim 4 \times 10^{20}$ cm^{-3} as a 2DEG layer and forms a normally-ON HEMT. We observed the p-type doping of GaN under the 2DEG layer does not shift the E_C at the interface to a positive upward direction above the Fermi level, which makes it different from gate injection transistor technology that employs p-type GaN above the AlGaN/GaN heterostructure [9]. This p-type doping under the gate forms a local GaN p-n junction with the 2DEG layer, thus depleting the electrons to the point

device becomes a normally-OFF HEMT. Moving further from the interface, the holes are the majority carrier and it is not our concern anymore as GaN HEMT operation focuses on the 2DEG layer at the interface. The recessed Schottky gate did influence how the local GaN p-n junction determined the positive V_T of the E-mode HEMT.

In Figs. 5a – 5h, the performance of the original D-mode HEMT is compared against the optimized E-mode HEMT. Based on transfer characteristic in Fig. 5a, D-mode HEMT has $V_T = -6.629$ V and $R_{ON} = 6.666$ Ω at $V_{drain} = 1$ V. We applied $V_{gate} = -7, -5, -3,$ and -1 V for every sweep of V_{drain} to generate output characteristic in Fig. 5b. We recorded the max output power at saturation of $V_{drain} = 20$ V and $V_{gate} = -1$ V (equivalent to $V_T + 5.629$ V) is $P_{out} = 33.820$ W. Figs. 5c and 5d show the contour for current density during the OFF- and ON-state, respectively for the original D-mode HEMT. The control of OFF and ON is well understood for a typical Schottky-gated GaN HEMT, where the negative bias $V_b < V_T$ is required to deplete the electrons in the 2DEG layer under the gate. While for the predicted optimum device, based on transfer characteristic in Fig. 5e, the verified optimum E-mode HEMT has $V_T = 1.224$ V and $R_{ON} = 3.235$ Ω at $V_{drain} = 1$ V. We applied $V_{gate} = 1, 3, 5,$ and 7 V for every sweep of V_{drain} to generate output characteristic in Fig. 5f. We recorded the max output power at saturation of $V_{drain} = 20$ V and $V_{gate} = 7$ V (equivalent to $V_T + 5.776$ V) is $P_{out} = 49.549$ W. Figs. 5g and 5h show the contour for current density during the OFF- and ON-state, respectively for the optimized E-mode HEMT. The contours of current density also suggest the optimized E-mode HEMT can give better gate control and

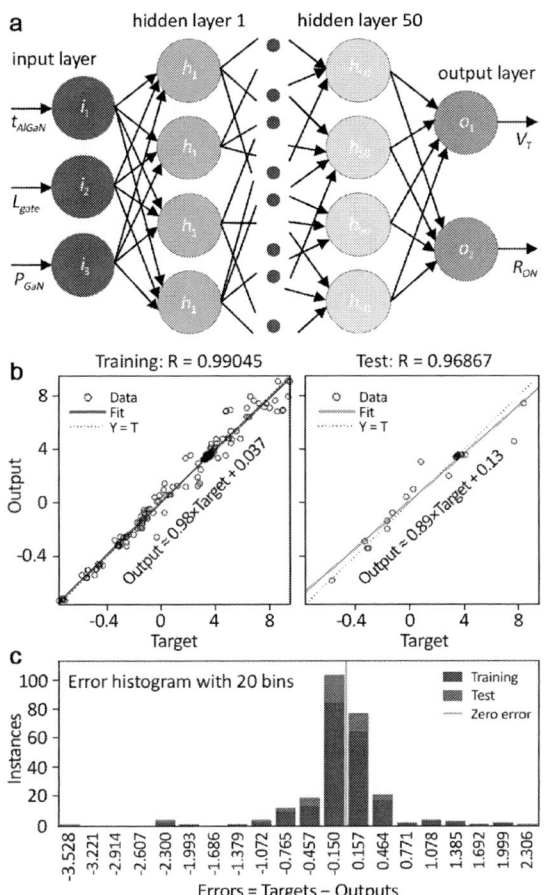

Fig. 3. a) Schematic of ANN model of 3 inputs, 2 outputs, and 50 hidden layers, using Bayesian regularization backpropagation. b) ANN regression plots for the training of 90% observation and test of 10% observation. c) Error histogram.

Table 2. The prediction on V_T and R_{ON} is based on the refinement of P_{GaN} as the main effect. Note that we kept the constant of $t_{AlGaN} = 4$ nm and coarsely vary the L_{gate}. Not all predictions are shown.

No	User input		ANN prediction	
	L_{gate} (μm)	P_{GaN} (cm^{-3})	V_T (V)	R_{ON} (Ω)
1	0.5	1.2×10^{19}	2.950	3.387
2	1.0	1.2×10^{19}	3.586	3.356
3	1.5	1.2×10^{19}	3.822	3.358
4	2.0	1.2×10^{19}	1.263	3.317
5	0.5	1.4×10^{19}	3.210	3.404
.	.	.	.	.
.	.	.	.	.
10	1.0	1.6×10^{19}	4.357	3.392
11	1.5	1.6×10^{19}	4.212	3.384
.	.	.	.	.
.	.	.	.	.
19	1.5	2.0×10^{19}	3.863	3.397
20	2.0	2.0×10^{19}	0.956	3.349

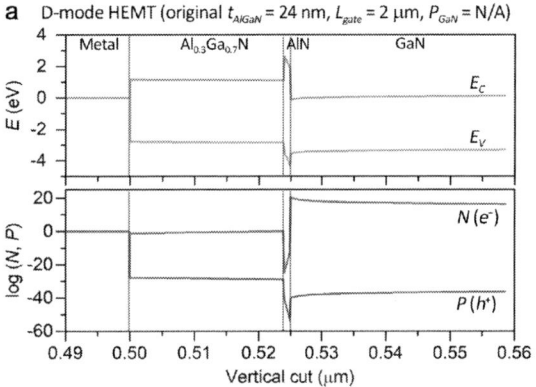

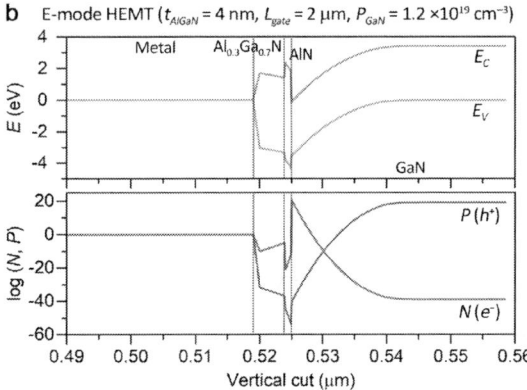

Fig. 4. Vertical cutline at gate inspecting energy bands and carriers concentration for: a) Original D-mode HEMT, and b) optimized E-mode HEMT. Both devices in are in thermal equilibrium.

979-8-3503-7832-0/24 $31.00 © 2024 IEEE

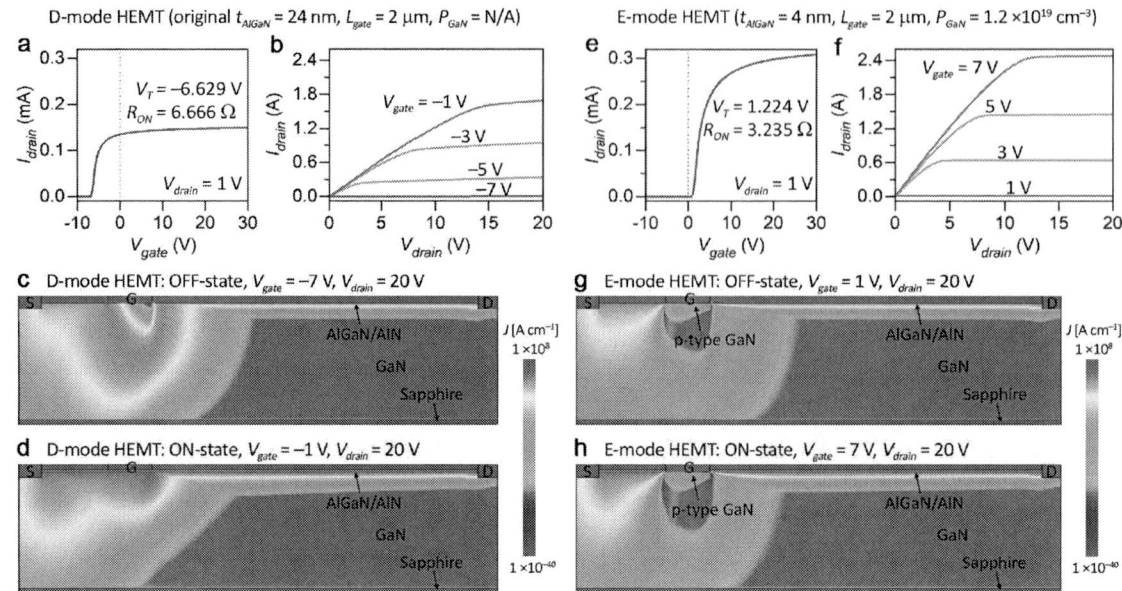

Fig. 5. Comparison between original D-mode HEMT with the optimized E-mode HEMT. On the left-hand side: a) Transfer, and b) output characteristics of D-mode HEMT. c) Contour of conduction current density in Victory Visual at OFF-state, and d) ON-state for D-mode HEMT. On the right-hand side: e–h) Similar plots and contours for the optimized E-mode HEMT.

higher breakdown at OFF-state compared to the original D-mode HEMT, but more TCAD simulation works are needed to explore this hypothesis fully. The use of p-type GaN within the GaN channel layer is a technique that we can find in current aperture vertical electron transistors [10]. However, in this work, the function of p-type GaN is only intended to control the gate instead of creating the aperture for vertical GaN HEMT. This methodology can be expanded for larger DOE than 3 factors at 4 levels to develop new HEMT technology, but the bottleneck within the implementation will be at the stage of running a full-scale TCAD simulation. Mesh quality and physic models in Victory Device cannot be compromised to reduce computational resources for the large DOE, as the obtained HEMT characteristics will be erroneous.

IV. CONCLUSION

The proposed methodology in this work is basically gathering the common methods of DOE, ANOVA, TCAD simulation, and ANN to obtain an optimum E-mode GaN HEMT. Data reliability can be improved by calibrating the several (selected) TCAD models with the actual measured device. Nevertheless, the methodology will not be affected. The defined 'optimum device' may vary depending on the application and its driver, so users can easily predict their 'optimum device' using the methodology explained in this article. In summary, we found the optimized E-mode HEMT is better than the original D-mode HEMT for power converter applications since it is normally OFF, has lower R_{ON}, and has higher P_{out} at a comparable ON voltage.

ACKNOWLEDGMENT

Student authors received financial support from 'Final Year Project Fund' allocated by UKM. This work is supported by MIMOS and CEDEC USM through the RMKe-12 Development Expenditure (DE) fund for Project P.10 under the Ekonomi Madani initiative.

REFERENCES

[1] A. Matallana, E. Ibarra, I. Lopez, J. Andreu, J.I. Garate, X. Jorda, and J. Rebollo, "Power module electronics in HEV/EV applications: New trends in wide-bandgap semiconductor technologies and design aspects," Renew. Sust. Energ. Rev., vol. 113, pp. 109264, 2019.

[2] K.H. Hamza and D. Nirmal, "A review of GaN HEMT broadband power amplifiers," AEU-Int. J. Electron. C., vol. 116, pp. 153040, 2020.

[3] M.F. Abdullah, M.R.M. Hussin, M.A. Ismail, and S.K.W. Sabli, "Chip-level thermal management in GaN HEMT: Critical review on recent patents and inventions," Microelectron. Eng., vol. 273, pp. 111958, 2023.

[4] S. Hamady, F. Morancho, B. Beydoun, P. Austin, and M. Gavelle, "P-doped region below the AlGaN/GaN interface for normally-off HEMT," 2014 16th European Conference on Power Electronics and Applications, pp. 1–8, 2014.

[5] Z. Wang, J. Nan, Z. Tian, P. Liu, Y. Wu, and J. Zhang, "Review on main gate characteristics of P-type GaN gate high-electron-mobility transistors," Micromachines, vol. 15, 80, 2024.

[6] H. Lv, Y. Cao, M. Ma, Z. Wang, X. Zhang, C. Chen, L. Wu, L. Lv, X. Zheng, Y. Wang, W. Tian, and X. Ma, "Effect of p-type GaN buried layer on the temperature of AlGaN/GaN HEMTs," Micromachines, vol. 14, pp. 1457, 2023.

[7] P. Doring, M. Sinnwell, S. Muller, H. Czap, R. Driad, P. Bruckner, K. Kohler, L. Kirste, M. Mikulla, and R. Quay, "A study on the performance of AlGaN/GaN HEMTs regrown on Mg-implanted GaN layers with low channel thickness," IEEE Trans. Electron Devices, vol. 70, pp. 947–952, 2023.

[8] Y. Wang, M.T. Bao, F. Cao, J.X. Tang, and X. Luo, "Technology computer aided design study of GaN MISFET with double P-buried layers," IEEE Access, vol. 7, pp. 87574–87581, 2019.

[9] Y. Uemoto, M. Hikita, H. Ueno, H. Matsuo, H. Ishida, M. Yanagihara, T. Ueda, T. Tanaka, and D. Ueda, "Gate injection transistor (GIT)—A normally-off AlGaN/GaN power transistor using conductivity modulation," IEEE Trans. Electron Devices, vol. 54, pp. 3393–3399, 2007.

[10] P. Doring, R. Driad, R. Reiner, P. Waltereit, S. Leone, M. Mikulla, and O. Ambacher, "Technology of GaN-based large area CAVETs with co-integrated HEMTs," IEEE Trans. Electron Devices, vol. 68, pp. 5547–5552, 2021.

Effectiveness of the Heat Spreader in GaN HEMT Studied by a Co-Simulation Approach

A.M. Mohamad Yussof
Faculty of Engineering & Built Environment, Universiti Kebangsaan Malaysia, Selangor, Malaysia
a181491@siswa.ukm.edu.my

N. Abd Aziz
Faculty of Engineering & Built Environment, Universiti Kebangsaan Malaysia, Selangor, Malaysia
norazreen@ukm.edu.my

M.F. Abdullah
Center for Semiconductor & Thin Film Research, MIMOS Berhad. Kuala Lumpur, Malaysia
faizol.abdullah@mimos.my

Abstract— This article examines thermal management using a co-simulation approach with Silvaco and Ansys software to analyze heat distribution and thermal profiles in GaN high electron mobility transistor (HEMT) devices within transistor packages. The model is constructed based on a simplified GaN-on-Si HEMT structure. The efficacy of various materials specifically sapphire, SiC, and polycrystalline diamond (PCD) with different thermal conductivities as heat spreaders is assessed at different device locations. For the original Si-GaN-Si$_3$N$_4$ chip, the peak junction temperature, T_{j_max} is 175.0 °C at an output power density of 5 W/mm². The relative increase in T_{j_max} compared to the non-junction area on the GaN surface, ΔT_{j_max} is 5.5%, indicating the thermal crosstalk effect. When a 10 μm-thick PCD (thermal conductivity of 1200 W/m·K) heat spreader is added to the Si-GaN-Si$_3$N$_4$ chip, T_{j_max} drops to 171.5 °C, and ΔT_{j_max} reduces to 3.4% (38% reduction). Replacing the Si substrate with PCD further reduces T_{j_max} to 167.7 °C and ΔT_{j_max} to 1.5% (73% reduction). The addition of PCD layers mitigates the "spike-like" thermal profile due to enhanced heat distribution from the hotspot. Even though the reduction in T_{j_max} is minimal, PCD primarily aids in heat spreading, redistributing heat from hotspots across the chip. This improves heat distribution, mitigates thermal gradients, and reduces localized overheating, which can prevent thermal crosstalk and potential damage to the device

Keywords—GaN HEMT, heat spreader, polycrystalline diamond, Silvaco, Ansys

I. INTRODUCTION

In recent years, the GaN high electron mobility transistor (HEMT) has emerged as a promising option for high-voltage power converters exceeding 600 V. GaN, belonging to the family of III-V semiconductor and a wide-bandgap material, possesses a wide bandgap of approximately 3.4 eV at room temperature, enabling it to withstand high electric fields of around 3.3 MV/cm. This capability exceeds that of Si by approximately tenfold before reaching its breakdown voltage. Unlike metal-oxide-semiconductor field-effect transistors fabricated on Si and SiC, the structure of GaN HEMT involves creating a two-dimensional electron gas (2DEG) channel through spontaneous piezoelectric polarization at the AlGaN/GaN heterostructure. The trapped electrons in this 2DEG layer, with a concentration of about 10^{13} cm^{-2}, exhibit high mobility even at room temperature, making GaN HEMT a normally ON transistor. During the ON mode, the 2DEG experiences self-heating, forming hotspot regions under the gate metals towards the drain, which directly influences the device's reliability and output power density.

Effective thermal management is crucial for optimizing GaN HEMT performance and reliability. Thermal management strategies for GaN HEMT typically fall into two categories: chip-level and package-level. Chip-level strategies involve utilizing a substrate or thin film heat spreader in direct contact with hotspot regions. Effectively designed heat spreaders can significantly reduce localized hotspots, ensuring uniform temperature distribution across the chip. Previous studies have demonstrated the effectiveness of various materials and configurations for heat spreaders.

Cu is favored for its excellent thermal conductivity, as shown by Mohanty et al. [1], who demonstrated a copper-filled micro-trench structure that reduces 17% heat dissipation. AlN is notable for its high thermal conductivity and electrical insulation, effectively reducing the thermal resistance in hybrid III-V laser on Si [2]. Diamond heat spreaders, with exceptional thermal properties, significantly improve heat dissipation in high-frequency devices, according to Mahrokh et al. [3]. Al, while less thermally conductive, offers a promising heat spreader candidate when coated with Cu-Gr composite for maintaining uniform temperature in battery packs [4]. These advancements highlight heat spreaders' crucial role in managing thermal loads and enhancing semiconductor device reliability.

Various software tools can simulate thermal management in GaN HEMT devices. Technology computer-aided design (TCAD) software such as Silvaco ATLAS and finite element analysis (FEA) software like Ansys and COMSOL Multiphysics are commonly used. A recommended approach is to first identify hotspot positions using Silvaco Victory Device, then later rebuild its model using Ansys Icepak for a comprehensive thermal co-simulation. The insight provided by co-simulation using Silvaco and Ansys is useful for improving future inventions related to thermal management in GaN HEMT.

II. METHODOLOGY

This section outlines the simulation procedures conducted to analyze the thermal behavior of a GaN HEMT chip during operation. The methodology encompasses several key areas: simulations using Silvaco software suite to analyze temperature distribution and identify hotspots, development of a simplified 2D and 3D model of the GaN HEMT chip for thermal using Ansys Electronics Desktop (EDT) - Icepak, exploration of alternative substrate materials to improve heat dissipation, and evaluation of the optimal location and thermal conductivity variations of the selected materials.

979-8-3503-7832-0/24 $31.00 © 2024 IEEE

Fig. 1 displays temperature changes observed in a GaN HEMT chip during operation, depicted in a 2D format. The temperature distribution within the chip was revealed, pinpointing areas of significant heat generation and dissipation. The hottest spot reached approximately 420 K, while cooler regions were identified at the source and drain of the devices, registering around 300 K [5].

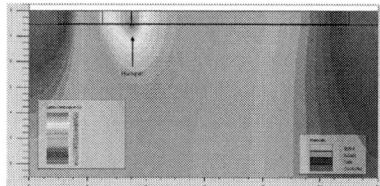

Fig. 1. Lattice temperature contour from the Silvaco Victory Device.

Analysis indicated that the hottest spot mainly occurred at the gate of the device, affecting its operational performance and long-term reliability. Although the AlGaN layer was present in the device structure, it was not prominently visible in the visualization due to its small size compared to other layers [6]. Fig. 2 shows a simulation model that provides a simplified representation of the GaN HEMT chip model, focusing on key layers and components essential for understanding heat dissipation dynamics. It aims to offer insights into heat dissipation mechanisms within the device under study.

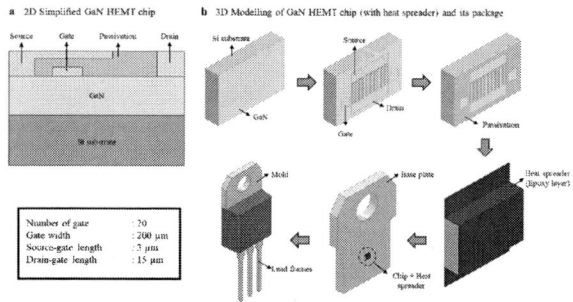

Fig. 2. a) Simplified structure of the GaN HEMT structure for thermal co-simulation in Ansys EDT. b) Steps for building the 3D model of the 20 gates GaN HEMT starting from the initial GaN-on-Si substrate into the packaged chip inside the TO-220.

Initially, the structure of a typical GaN-on-Si HEMT was complex, involving multiple layers and materials. To simplify the simulation process, the structure was streamlined. Key layers such as the nucleation, buffer, channel, and barrier layers were combined into a single 6 μm-thick GaN layer. Similarly, metal layers like source, drain, and gate electrodes were simplified to gold. The dimensions of these components were based on previous reports. Additionally, the Si substrate thickness was set at 150 μm. The 2D representation of the GaN HEMT was transformed into a 3D model with 20 gates, each 200 μm wide. The process involved assembling the chip and packaging it into a TO-220 package using Ansys EDT Icepak software. A polycrystalline diamond (PCD) heat spreader was optionally applied to the chip's front and sides. The chip was then encapsulated in epoxy and the ambient air is set to 20 °C.

An 'air cabinet' is utilized to encompass the model and boundary conditions are set at the limits of $\pm x$, $\pm y$, and $\pm z$, designated as 'opening' boundaries. Thermal boundary resistances between GaN-Si (substrate), GaN-Si$_3$N$_4$ (passivation), and GaN-PCD (passivation) are set to 4, 3, and 18 m^{2}K/GW, respectively, for the Si-GaN-Si$_3$N$_4$ and Si-GaN-PCD models. To simulate hotspots, 20 patches of 1 μm-length 2D rectangles are placed under each gate, slightly shifted by 0.2 μm towards the drain. These patches are specified as 'source' boundary conditions, with each receiving a heat flux, Q of 10, 20, 30, 40, and 50 mW, corresponding to total heat inputs, Q_{total} of 0.2, 0.4, 0.6, 0.8, and 1.0 W, respectively. Considering the chip is constructed with 20 gates and a gate width of 200 μm, these variations can also be translated into output power densities, P_d of 1, 2, 3, 4, and 5 W/mm, respectively. It's noteworthy that the reported maximum value of P_d (DC power) for GaN-on-Si is approximately ~4.5 W/mm, while GaN-on-SiC can reach as high as ~15 W/mm, contingent upon the efficiency of heat dissipation from the hotspot regions [7].

The study explored substituting the Si substrate with alternative materials like SiC, sapphire, and PCD to improve heat dissipation. For SiC and sapphire, materials were selected from the Ansys EDT library. However, PCD requires additional material properties, including thermal conductivity, mass density, and specific heat. Simulations were conducted at a consistent power density, P_d of 5 W/mm, and the outcomes were compared. In this phase, the study varied the location of the optimal material identified earlier. Three locations were considered: Chip-level passivation heat spreader (CPHS), package-level encapsulation heat spreader (PEHS), and chip-level substrate heat spreader (CSHS) as shown in Fig. 3. Simulations were performed at a consistent P_d of 5 W/mm to determine the most effective location based on the results. Using data from previous stages, the study further varied the thermal conductivity of the selected material while keeping other parameters fixed. Thermal conductivity variations were examined across power density increments ranging from 1 W/mm to 5 W/mm. This aimed to evaluate the material's impact on device performance across different power density levels, categorizing performance into poor, moderate, and high categories.

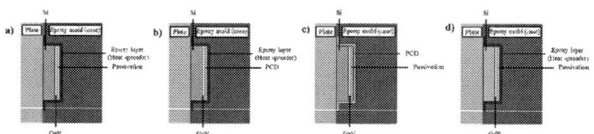

Fig. 3. PCD locations in GaN HEMT a) Conventional GaN HEMT device (reference). b) Chip-level passivation heat spreader (CPHS). c) Package-level encapsulation heat spreader (PEHS). d) Chip-level substrate heat spreader (CSHS).

III. RESULTS AND DISCUSSION

The original Si-GaN-Si$_3$N$_4$ configuration, depicted in Fig. 4a, served as the benchmark in this study, representing conventional semiconductor industry practices. This reference design intentionally omitted the implementation of a PCD heat spreader to facilitate comparison with various approaches. Within Ansys EDT, the reference configuration underwent meshing and analysis processes, with all components mounted on the convective boundary to ambient temperature for heat dissipation. Strategic use of vertical and horizontal planes facilitated measurements from both top and side views to calculate parameters such as junction temperature, T_j at specific gates, and plate temperature, T_p at the back frame.

979-8-3503-7832-0/24 $31.00 © 2024 IEEE

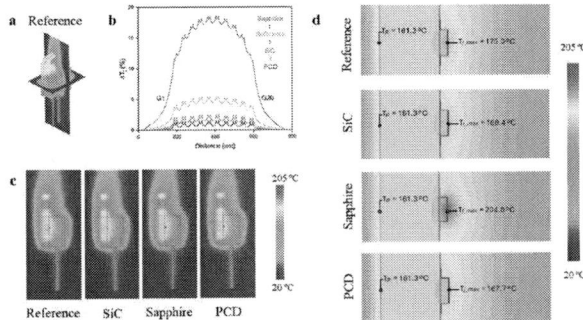

Fig. 4. a) Temperature distribution on the package surface. b) The percentage change of surface temperature in the form of ΔT_j for the four different materials at a similar P_d at 5 W/mm. c) Side view temperature distribution of the package for the four different materials at a similar P_d at 5 W/mm. d) Magnified side view cross-center of the package of four different materials with a similar P_d at 5 W/mm.

Three alternative materials i.e. SiC, sapphire, and PCD were proposed to replace Si in the reference design. These materials were positioned instead of Si bulk and simulated through the software, with results compared against the reference design to optimize thermal dissipation. In Fig. 4c, side views of the simulated designs showcase heat dissipation through the devices, all subjected to a power density, P_d of approximately 5 W/mm. Meanwhile, Fig. 4d shows a magnified side view revealing sapphire as the hottest spot at the junction with a temperature of approximately 204.6 °C, while PCD exhibited the coldest spot at around 167.7 °C, lower than Si, SiC, and sapphire. Plate temperatures, T_p for all materials remained comparable, hovering around 161.3 °C. Subsequently, a comparison graph in Fig. 4b depicted the percentage difference of surface temperature, ΔT_j for each material. PCD showed the lowest ΔT_j (1.5%), compared to Si (5.5%), SiC (2.5%), and sapphire (18.5%) suggesting its effectiveness in thermal optimization. This is in line with findings by Hu et al. [8] that the implementation of PCD gives an improvement in the GaN HEMT device compared to other materials as it can improve heat dissipation while reducing the temperature of the device due to its stability, high thermal conductivity and low thermal resistance.

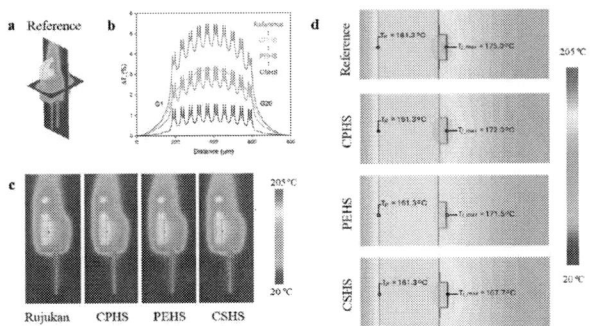

Fig. 5. a) Temperature distribution on the package surface. b) The percentage change of surface temperature in the form of ΔT_j for the four different locations at a similar P_d at 5 W/mm. c) Temperature distribution at the side view of the package for the four different locations at a similar P_d at 5 W/mm. d) Magnified the side view cross-center of the package of four different locations with a similar P_d at 5 W/mm.

Once the optimal material for heat dissipation has been identified, the study further investigates its placement within

the device, using the reference design as a benchmark, as depicted in Fig. 5a. Three distinct approaches were proposed to determine the most efficient positioning of the selected material, PCD. These approaches included the CPHS, PEHS, and CSHS.

In the CPHS approach, PCD replaced the Si_3N_4 layer on the passivation layer while in the PEHS approach involved adding an epoxy layer onto the chip to serve as a heat spreader, subsequently replaced by PCD to assess any improvements in heat dissipation. Finally, the CSHS approach entailed substituting the Si bulk with PCD while maintaining the same volume of material to gauge the impact of variations in PCD location.

Fig. 5c provides a side view of the packages with different PCD placement variations compared to the reference while a closer examination is in Fig. 5d. It is revealed that the CPHS approach exhibited the highest temperature at approximately 172.0 °C, whereas the CSHS approach demonstrated the lowest temperature at about 167.7 °C, a difference of 7.3 °C compared to the reference design. Additionally, Fig. 5b presents the percentage change of the package surface, indicating that the CSHS approach had the lowest ΔT_j value, suggesting optimized heat dissipation from the chip to the outer back frame of the package. Consequently, the CSHS approach emerges as the most effective placement variation, exhibiting the lowest maximum junction temperature and demonstrating superior heat dissipation performance compared to both the reference design and the alternative approaches (CPHS and PEHS). These results are in agreement with those from Kagawa et al. [9] which implemented the same approach as it uses PCD as the heat spreader material while located at the substrate, which can improve the thermal management of the GaN HEMT device.

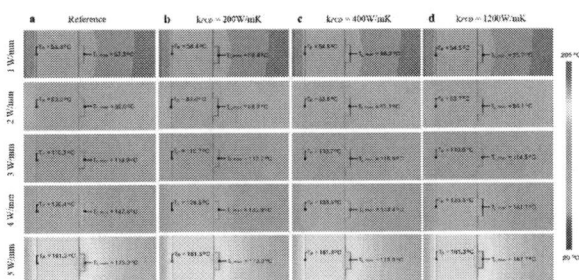

Fig. 6. PCD heat spreader with different thermal conductivity values compared to the reference design at different and varied P_d. Temperature distribution inside the package. Images are magnified from the top view cross-center of the GaN HEMT chip, varied based on the values of k_{PCD} and compared to the reference: a) Reference, b) 200, c) 400, and d) 1200 W/m·K.

Fig. 6 compares the thermal conductivity variations of PCD material at the CSHS location across different power densities ranging from 1 W/mm to 5 W/mm. The results reveal a consistent trend: as power density increases, junction temperature, T_j rises, while it decreases with higher thermal conductivity of PCD material. Particularly, the highest thermal conductivity tested, 1200 W/m·K, exhibited the lowest temperatures compared to other variations and the reference design. This indicates that a thermal conductivity of 1200 W/m·K could effectively address thermal management challenges, offering optimal performance and improved heat dissipation efficiency at the designated location within the device.

These findings suggest that a higher thermal conductivity value, specifically 1200 W/m·K, may serve as a viable solution to enhance device performance and reliability. Implementing this optimal thermal conductivity value could significantly improve heat dissipation efficiency, particularly at critical locations within the device. Therefore, understanding the impact of thermal conductivity variations on device performance is crucial for guiding the selection and optimization of thermal management strategies in GaN HEMT devices. The high thermal conductivity of PCD shows the ability to dissipate heat more efficiently as the increment of thermal conductivity value gave rise to the heat dissipation performance of the device [7]. The implementation of PCD at a substrate with high thermal conductivity also has a positive impact on the device as there were increments of the performances of the device such as enhancement of drain current and transconductance of the GaN HEMT device [10].

Fig. 7a serves as a reference plot showcasing the thermal conductivity of the benchmark design, setting the stage for comparing variations in the thermal conductivity of the PCD material at the CSHS location. Fig. 7b – Fig. 7d depict thermal conductivities of 200 W/m·K, 400 W/m·K, and 1200 W/m·K, respectively. Across all these levels, the percentage change in surface temperature, ΔT_j consistently increases with the escalation of power density, P_d from 1 W/mm to 5 W/mm. ΔT_j, reflecting chip heat dissipation efficiency, is derived from the comparison between junction temperature, T_j and plate temperature, T_p, offering insights into the device's overall thermal resistance.

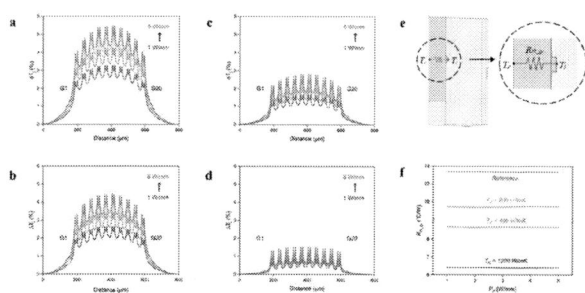

Fig. 7. Thermal conductivity comparison at different power densities. The percentage change of surface temperature in the form of ΔT_j for four different thermal conductivity with the varied P_d and compared to the reference. a) Reference, b) 200, c) 400, and d) 1200 W/m·K. e) Schematic of the simple R_{th}. f) Plot of R_{th_jp} versus P_d.

Fig. 7e highlights the thermal resistance of the device, R_{th_jp} between the plate temperature, T_p at the back frame of the package and the junction temperature, T_j. Conversely, Fig. 7f compares thermal resistance, R_{th_jp} across variations of thermal conductivity from 1 W/mm to 5 W/mm, with the reference design exhibiting the highest thermal resistance. This indicates lower effectiveness in dissipating heat from the chip to the package's back frame. Notably, the thermal conductivity of 1200 W/m·K showcases the lowest thermal resistance, suggesting superior heat dissipation efficiency. Therefore, among the tested thermal conductivity variations, 1200 W/m·K emerges as the optimal choice, offering promising solutions for enhancing heat dissipation within the device and addressing heating issues effectively. Therefore, TABLE I shows the comparison between the thermal conductivity applied at power density, P_d at 5 W/mm.

TABLE I   COMPARISON OF THERMAL CONDUCTIVITY AT 5 W/MM

Thermal Conductivity (W/m·K)	Junction Temperature (°C)	Reduction Temperature (%)	Thermal Resistance (°C/W)
200	173.0	20.0	11.8
400	170.0	40.1	8.9
1200	167.7	72.7	6.4

IV.   CONCLUSION

In conclusion, this study highlights the effectiveness of PCD in dissipating heat in GaN HEMT devices. The CSHS method, employing PCD at the substrate level, showed the most efficient heat dissipation. Higher thermal conductivity values of PCD, notably 1200 W/m·K, offered promising improvements in heat dissipation efficiency. These findings emphasize the importance of selecting appropriate materials and placement strategies for heat spreaders in GaN HEMT devices to enhance performance and reliability in high-voltage power applications.

ACKNOWLEDGMENT

The student author received financial support from 'Final Year Project Fund' allocated by UKM. This work is supported by MIMOS and CEDEC USM through the RMKe-12 Development Expenditure (DE) fund for Project P.10 under the Ekonomi Madani initiative.

REFERENCES

[1] S.K. Mohanty, Y.Y. Chen, P.H. Yeh, and R.H. Horng, "Thermal management of GaN-on-Si high electron mobility transistor by copper filled micro-trench structure," Sci. Rep., vol. 9, pp. 19691, 2019.

[2] S. Lei, I. Matthews, J. Camus, S. Bensalem, M.A. Djouadi, A. Shen, D.H. Juan, and R. Enright, "AlN thin-films as heat spreaders in III–V photonics devices Part 2: Simulations," in 2016 15th IEEE Intersociety Conference on Thermal and Thermomechanical Phenomena in Electronic Systems (ITherm), 2016, pp. 1024–1029.

[3] M. Mahrokh, H. Yu, and Y. Guo, "Thermal modeling of GaN HEMT devices with diamond heat-spreader," Int. J. Therm. Eng. Des. Syst., vol. 8, pp. 986–991, 2020.

[4] S.S. Sainudeen, A. Joseph, M. Joseph, and V. Sajith, "Heat transfer phenomena of copper-graphene nanocomposite coated aluminium heat spreaders: An interferometric study," Appl. Therm. Eng., vol. 212, pp. 118545, 2022.

[5] H.P. Chou, S. Cheng, C.H. Cheng, and C.W. Chuang, "Development and characterization of the thermal behavior of packaged cascode GaN HEMTs," Mater. Sci. Semicond. Process., vol. 41, pp. 304, 2016.

[6] L. Sang, "Diamond as the heat spreader for the thermal dissipation of GaN-based electronic devices," Funct. Diamond, vol. 1, pp. 174–188, 2022.

[7] M.F. Abdullah, M.R.M. Hussin, M.A. Ismail, and S.K.W. Sabli, "Chip-level thermal management in GaN HEMT: Critical review on recent patents and inventions," Microelectron. Eng., vol. 273, pp. 111958, 2023.

[8] X. Hu, L. Ge, Z. Liu, M. Li, Y. Wang, S. Han, Y. Peng, M. Xu, X. Hu, and G. Tang, "Diamond-SiC composite substrates: A novel strategy as efficient heat sinks for GaN-based devices," Carbon, vol. 218, pp. 118755, 2024.

[9] R. Kagawa, Z. Cheng, K. Kawamura, Y. Ohno, C. Moriyama, Y. Sakaida, S. Ouchi, H. Uratani, K. Inoue, and Y. Nagai, "High thermal stability and low thermal resistance of large area GaN/3C-SiC/diamond junctions for practical device processes," Small, vol. 20, pp. 2305574, 2024.

[10] F.Z. Tijent, M. Faqir, P.L. Voss, J.P. Salvestrini, and A. Ougazzaden, "Reduction of self-heating effects in GaN HEMT via h-BN passivation and lift-off transfer to diamond substrate: A simulation study," Mater. Sci. Eng. B, vol. 301, pp. 117185, 2024.

Optimizing Methylammonium Tin Iodide-based Perovskite Solar Cells with Different Hole Transport Materials

Budi Mulyanti
Study Program of Electrical Engineering
Universitas Pendidikan Indonesia
Bandung, Indonesia
bmulyanti@upi.edu

Jahril Nur Fauzan
Study Program of Electrical Engineering
Education
Universitas Pendidikan Indonesia
Bandung, Indonesia
jahrilnurfauzan@upi.edu

Chandra Wulandari
Study Program of Physics
Institut Teknologi Bandung
Bandung, Indonesia
33322303@mahasiswa.itb.ac.id

Lilik Hasanah
Study Program of Physics
Universitas Pendidikan Indonesia
Bandung, Indonesia
lilikhasanah@upi.edu

Roer Eka Pawinanto
Study Program of Industrial Automation and
Robotics Engineering Education
Universitas Pendidikan Indonesia
Bandung, Indonesia
roer_eka@upi.edu

Abstract— Perovskite materials have garnered significant attention in the field of solar cell applications due to their highly promising performance. However, the utilization of lead (Pb) as the primary material in perovskite solar cells has been associated to health and environmental hazards. Consequently an alternative material for lead is necessary in order to develop environmentally friendly and high-efficiency solar cells. The reported efficiency value still necessitates further enhancement, indicating that the development of lead-free solar cells requires improvement. Methylammonium tin iodide (MaSnI₃), a lead-free perovskite compound, can be utilized as a substitute for lead halide-based solar cells. This substitution is possible since MaSnI₃ is stable, non-toxic, and has a high efficiency. The simulation was conducted by the Solar Cell Capacitance Simulator (SCAPS-1D), employing the AM 1.5G spectrum. The primary objective of this research is to enhance the performance of the hole transport layer materials in order to achieve a high power conversion efficiency (PCE) using MaSnI₃ material. Organic and inorganic HTL such as CuSBS, PEDOT:PSS+WO₃, Spiro-OMeTAD, P3HT, NiO, CuI, SrCu₂O₂ and PTAA are required in our simulations. These materials were selected to explore their potential in optimizing device performance due to their distinct properties, such as conductivity, stability, and compatibility with MaSnI3. Regarding HTM performance for device construction, our modelling results indicate that the 100 nm CuSBS₂ layer works better. We have found that by using the configuration of ITO/TIO₂/MASnI₃/CuSBS₂/Au, a PCE of 27.67% with a Voc of 0.97 V can be achieved.

Keywords— Perovskite, solar cell, hole transport material, power conversion efficiency, SCAPS

I. INTRODUCTION

The very promising perovskite solar cell (PSC) technology may compete with conventional silicon solar cells thanks to using a single junction design and higher than 25% power conversion efficiency (PCE) [1]. It has been developed in less than ten years and is rapidly approaching commercialization [2]. However, a few significant obstacles stand in the way of PSC's industrialization journey, including the toxicity of the materials used and the gadget's stability in the air when illuminated [3]. The optimal mixture for perovskite solar cells to date consists of in the ABX3 perovskite structure, the A cation site can be occupied by methylammonium ($CH_3NH_3^+$) (MA), formamidinium ($NH_2CHNH_2^+$) (FA), and cesium (Cs^+); the B cation site is occupied by lead (Pb); and the X anion site is occupied by iodine (I) and bromine (Br) [4]. Nevertheless, the main environmental concern is the presence of heavy and toxic elements such as lead, which are connected to the whole perovskite solar energy lifecycle cells [5]. Consequently, in spite of its high efficiency, the scientific and commercial communities have less faith in its future, which presents an opportunity for lead-free perovskite materials. Researchers are exploring lead-free perovskite compounds for potential use in solar cells. Due to their large band gaps, various lead-free perovskite absorber materials can serve as alternatives to harmful lead-containing perovskites [6]. These consist of direct bandgap methylammonium tin iodide (MASnI₃: 1.3 eV), formamidinium tin iodide (1.41 eV), and cesium tin iodide (CsSnI₃: 1.22 eV) [7]. However, with these devices malfunctioning outdoors, there is a significant problem that needs to be addressed. A lead-free MASnI₃-based perovskite solar cell with a 6% power conversion efficiency (PCE) was reported by Neol et al. To prevent exposure to air, the device was created in a sealed nitrogen glove box utilizing a spin-coating technique and a cell configuration of FTO/c-TiO₂/mp-TiO₂/MASnI₃/Spiro-OMeTAD/Au [8]. When the light-harvesting material comes into contact with the environment, Sn^{2+} is oxidized to Sn^{4+}, which is a more stable form. Consequently, this leads to the destruction of the light-absorbing material's charge neutrality, this generates methyl ammonium iodide (MAI) and SnO2. Experimental results showed that, in terms of characteristics, due to its effect on spin-orbit coupling, the energy splitting was more pronounced in FASnI3. However, it was less significant in MASnI₃ because of the influence of hydrogen bonds. Studies

979-8-3503-7832-0/24 $31.00 © 2024 IEEE

published in 2014 showed efficiencies of 5.73% for MASn($I_{1-x}Br_x$)$_3$ and 6.4% for MASnI$_3$, using the mesoscopic device architecture of FTO/c-TiO$_2$/mp-TiO$_2$/absorber/spiro-OMeTAD/Au [9]. Prior work by the Yang Yang group produced MASnI$_3$, which was employed in the production of MAPbI$_3$ films [10]. Yokoyama et al. created perovskite solar cells with the use of a two-step deposition procedure and an MAI vapour reaction [11]. A power conversion efficiency (PCE) of 9% for pure FaSnI3 in a 2D/3D arrangement was reported by Shao et al. By employing methods for trap site passivation and band energy alignment to control the A-site cation, Kohei et al. increased this to 13% and achieved a tolerance factor of roughly 1 [12]. However, compared to lead-based perovskite, the PCE is much lower and it has stability problems, therefore additional investigation is required to produce toxic-free, extremely stable, and effective PSCs.

Even though testing on tin-based perovskite solar cells has advanced quickly, more advancements can be achieved by fine-tuning different parameters and device designs. These advancements can then steer subsequent experimental developments [13]. Using various parameters, this simulation work has investigated, examined, and studied perovskite based on methylammonium tin iodide (MASnI$_3$). This research employed a 1D solar cell capacitance simulator (SCAPS, version 3.3.07) to simulate the device's behavior under AM15G light conditions. To optimize performance more effectively, adjustments were made to variables related to the doping concentrations and thicknesses of different layers, including the electron transport layer (ETL), perovskite absorber layer, and hole transport layer (HTL), and their effects were evaluated [20]. Additionally, fault densities were taken into account. After evaluating the impacts of various HTLs, an optimized configuration was identified, attaining the maximum efficiency of 27.67% the highest recorded outcome for this configuration by employing any simulation technique. This was achieved through a combination of analytical techniques and theoretical modeling.

II. DEVICE STRUCTURE AND SIMULATION METHODOLOGY

A. Device Structure

In this simulation experiment, MASnI3 was used as the light-absorbing material and was thoroughly investigated. Unlike lead-based perovskites, this material's notable advantage, its non-toxicity, has contributed to its recent surge in popularity [13]. MASnI$_3$ could serve as an effective and alternative substitute for MAPbX$_3$. One of the most critical aspects of the simulation is the chosen device configuration. As shown in Fig. 1a, the device simulation in this study operates in the n-i-p configuration of ITO/TiO$_2$/MASnI$_3$/HTL/Au.

MASnI$_3$, placed between the ETL and HTL, has been selected as the primary light-absorbing layer. The device's front contact consists of indium-doped tin oxide (FTO), while the back contact is made of gold (Au). The fluorescent effect occurs through the FTO termination. The matching energy band diagram for several HTLs is displayed in Figure 1b. The goal of the study is to simulate a device with the configuration shown in Figure 1a. The sun spectrum of AM 1.5 G is used to simulate it. Characteristics used in simulations include layer thickness, doping concentration, electron and hole mobility,

effective density of states, conduction band (CB), valence band (VB), dielectric permittivity, electron affinity, energy gap, and thermal velocities.

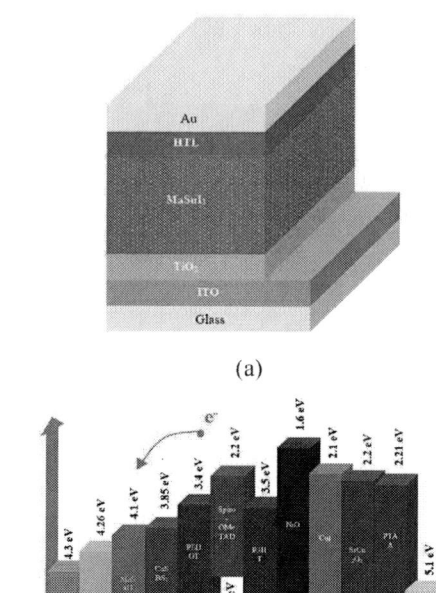

Fig 1. (a) Schematic structure of the simulated PSCs. (b) Energy band diagram of the simulated structure with different HTL

B. Simulation Study

Ideal electrical and optical characteristics for high power conversion efficiency were investigated through simulations using the SCAPS-1D tool. The University of Ghent in Belgium's Department of Electronics and Information Systems (ELIS) developed it, the SCAPS-1D program employs the continuity equation and Poisson's equation for both charge carriers [16]. It is possible to assemble up to seven layers of various solar cell types, and simulations may be run in both light and dark settings. The following is a possible expression for the equations used by the software to calculate output results: Poisson's semiconductor equation can be written as:

$$\nabla^2 \Psi = \frac{q}{\varepsilon}(n - p + N_A - N_D) \qquad (1)$$

where N_A stands for the acceptor concentration, N_D for the donor concentration, and Ψ for the electrostatic potential. The semiconductor continuity equation can be written as:

$$\nabla^2 J_n - q\frac{\partial n}{\partial t} = +qR \qquad (2)$$

$$\nabla^2 J_p + q\frac{\partial n}{\partial t} = -qR \qquad (3)$$

J_p stands for the current density for holes, J_n for electrons, and R regarding the carrier recombination rate. Two main processes affect current flow in semiconductors: the diffusion current generated by the concentration gradient and the drift of minority charge carriers caused by the electric field. The continuity equations describe the relationships between these drift and diffusion currents [25].

$$J_n = qn\mu_n E + qD_n \nabla n \qquad (4)$$

$$J_p = qn\mu_p E + qD_p \nabla p \qquad (5)$$

D_P denotes the hole diffusion coefficient, while D_n represents the electron diffusion coefficient.

Three primary parts make up any photovoltaic device: the carriers are collected by the carrier collector, transferred to external circuitry through the metal contacts, and converted into charge carriers, such as electrons and holes, by the light-absorbing layer [17]. We tried to mimic the configuration in this study using MASnI$_3$ absorber material so that we could clearly comprehend how the input parameters affected the device's performance parameters. The simulation software incorporated references to a range of other research articles from which the attributes of the different layers were extracted.

III. RESULT AND DISCUSSION

SCAPS simulations were conducted using tabulated parameters from various theoretical and experimental studies. Every component of a perovskite solar cell has a significant impact on how well it performs. In the ITO/TiO$_2$/MASnI$_3$/HTL/Au device configuration, different hole transport layers were utilized while keeping all other parameters constant. The optimum material for HTL is chosen after the variation indicates how well the device model performs with varying HTL. The J-V characteristics of the simulated PSCs with PEDOT:PSS+WO$_3$, Spiro-OMeTAD, P3HT, CuSnS$_2$, NiO, CuI, SrCu$_2$O$_2$, and PTAA are shown in Fig. 2a. It has been observed that the device containing CuSbS$_2$ performs better than the others. The performance metrics obtained from simulations using various HTLs are presented in Table 1. NiO's efficiency is comparable to CuSbS$_2$. Figure 2b displays the matching QE spectra of the PSCs with various HTLs, and they agree with the J-V features. Compared to other PSCs, the one using CuSbS$_2$ as HTL shows a greater short circuit current in the 350–650 nm range. The superior performance of perovskite solar cells featuring CuSbS$_2$, SrCu$_2$O$_2$, and the main reasons NiO is a superior HTL compared to other HTLs are its excellent mobility and conductivity as an inorganic p-type semiconductor based on copper, as well as its favorable alignment of the active perovskite absorber layer with the valence band maximum [18].

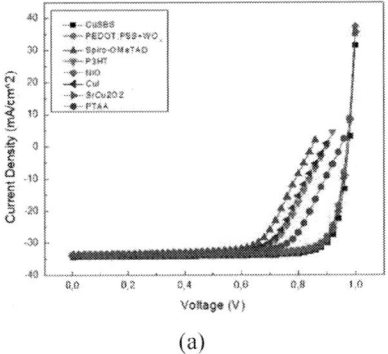

(a)

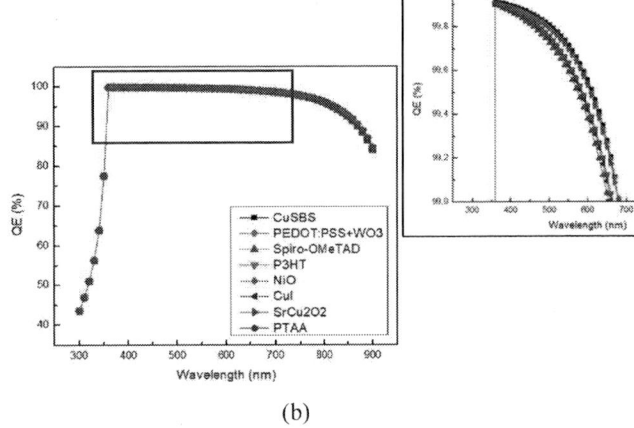

(b)

Fig. 2. (a) J-V chracteristic, and (b) correpond QE spectra of PSCs with different HTL

TABLE I.   PERFORMANCE PARAMETERS OF DEVICE WITH VARIOUS HTLS.

HTL	V_{OC} (V)	J_{sc} (mA.cm^{-2})	FF (%)	PCE (%)
CuSBS	0.97	34.13	83.01	27.67
PEDOT:PSS+WO$_3$	0,97	33,93	81,02	26.68
Spiro-OMeTAD	0.84	33.64	69.24	19.80
P3HT	0.89	33.70	70.49	21.34
NiO	0.97	34.04	82.23	27.21
CuI	0.89	33.68	69.71	21.01
SrCu$_2$O$_2$	0.97	33.99	81.58	26.94
PTAA	0.94	33.75	71.89	23.03

NiO or CuSbS$_2$ as HTL HTL layer than other HTLs primarily due to its high mobility and conductivity as an inorganic p-type semiconductor based on copper, and its excellent alignment of the valence band maximum with the active perovskite absorber layer [19].

The V_{OC} of solar cell depends on the following equation:

$$V_{OC} = \frac{nkT}{q}\ln\left(\frac{J_{SC}}{J_O} + 1\right) \qquad (6)$$

Where T represents the temperature, q is the electronic charge, J_O denotes the reverse saturation current density, n is the ideality factor, and K is the Boltzmann constant. It is clear that J_{SC} and the reverse saturation current density influence the V_{OC} value. It is apparent that as doping levels increase, the short-circuit current density decreases. Consequently, a quicker rate of decrease in the reverse saturation current is required than J_{SC} with higher doping levels for the V_{OC} to increase despite the decrease in J_{SC}. Increased doping concentration can lead to higher conductivity and a higher built-in potential[20].

979-8-3503-7832-0/24 $31.00 © 2024 IEEE

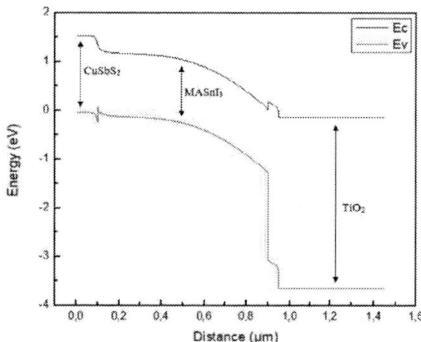

Fig. 3. The energy band structure diagram of PSCs

Lastly, an estimate of the device's overall performance is made using the improved characteristics of every PSC part. The energy band structure of a device with layers of CuSbS$_2$, MASnI$_3$, and TiO$_2$ is shown in Figure 3. The valence band offset (ΔEv) is roughly 0.5 eV and the conduction band offset (ΔEc) is approximately 1.0 eV at the CuSbS$_2$/MASnI$_3$ interface. ΔEc is approximately 1.1 eV and ΔEv is approximately 2.1 eV at the MASnI$_3$/TiO$_2$ contact. These offsets affect charge carrier dynamics, making them crucial. For instance, the ΔEc at the CuSbS$_2$/MASnI$_3$ interface stops electrons from traveling from CuSbS$_2$ to MASnI$_3$ in order to reduce recombination. Similarly, to further reduce recombination, the ΔEv at the MASnI$_3$/TiO$_2$ interface stops hole passage from MASnI$_3$ to TiO$_2$. The band offsets are essential for facilitating effective charge carrier transport and separation, which enhances the device's performance and power conversion efficiency (PCE) [21]. The optimized parameters include I_{SC} = 34.13 mA/cm2, V_{OC} = 0.97 V, FF = 83.01%, and PCE = 27.67%. Simulation results indicate that all components and parameters play significant roles in enhancing the functionality of the device.

IV. CONCLUSION

In conclusion, a detailed analysis was performed using SCAPS-1D numerical simulations to enhance the performance of lead-free MASnI$_3$ perovskite solar cells. To get the maximum efficiency, the impacts of several HTLs on the device's performance were investigated. It was discovered that devices with TiO$_2$ as the ETL and CuSBS as the HTL performed better than the others. Up until now, this simulator has produced the maximum PCE of 27.67% using MASnI$_3$ materials. This modeled device structure can assist research communities in guiding their future studies towards the experimental creation of perovskite solar cells free of lead.

ACKNOWLEDGMENT

This work is supported by grant from Directorate of Higher Education, Ministry of Education and Culture, Republic of Indonesia.

REFERENCES

[1] H. Arbouz, "Optimization of lead-free CsSnI3-based perovskite solar cell structure," vol. 33, no. 1, 2023, doi: doi:10.1515/arh-2022-0138.

[2] S. S. Nair, P. Thakur, F. Wan, A. V Trukhanov, L. V Panina, and A. Thakur, "Performance evaluation and the optimization of an inverted photo-voltaic cell with lead-free double perovskite material and inorganic transport layer materials," Solar Energy, vol. 262, p. 111823, 2023, doi: https://doi.org/10.1016/j.solener.2023.111823.

[3] P. Saha, S. Singh, and S. Bhattacharya, "Performance optimization of MASnI3 perovskite solar cells: Insights into device architecture," Micro and Nanostructures, vol. 191, p. 207827, 2024, doi: https://doi.org/10.1016/j.micrna.2024.207827.

[4] S. Vaishnavi and G. Seetharaman, "Device modelling and optimization of various n-i-p configurations for MASnI3 perovskite solar cells to achieve high efficiency using Solar Cell Capacitance Simulator-1D," Mater Today Commun, vol. 37, p. 107007, 2023, doi: https://doi.org/10.1016/j.mtcomm.2023.107007.

[5] P. Liu et al., "High-performance perovskite solar cells based on passivating interfacial and intergranular defects," Solar Energy Materials and Solar Cells, vol. 212, Aug. 2020, doi: 10.1016/j.solmat.2020.110555.

[6] Y. Raoui, H. Ez-Zahraouy, N. Tahiri, O. El Bounagui, S. Ahmad, and S. Kazim, "Performance analysis of MAPbI3 based perovskite solar cells employing diverse charge selective contacts: Simulation study."

[7] H. MallaHasan and Ö. Onay, "Investigation of the effect of different factors on the performance of several perovskite solar cells: a simulation study by SCAPS," European Journal of Engineering Science and Technology, vol. 5, no. 1, pp. 20–38, Apr. 2022, doi: 10.33422/ejest.v5i1.927.

[8] S. Abdelaziz, A. Zekry, A. Shaker, and M. Abouelatta, "Investigating the performance of formamidinium tin-based perovskite solar cell by SCAPS device simulation," Opt Mater (Amst), vol. 101, Mar. 2020, doi: 10.1016/j.optmat.2020.109738.

[9] L. Peng and W. Xie, "Theoretical and experimental investigations on the bulk photovoltaic effect in lead-free perovskites MASnI3 and FASnI3," RSC Adv, vol. 10, no. 25, pp. 14679–14688, 2020, doi: 10.1039/D0RA02584D.

[10] L. Liang, P. Gao, Adv. Sci. 5 (2018)

[11] X. Liu, K. Yan, D. Tan, X. Liang, H. Zhang, W. Huang, ACS Energy Lett. 3 (2018) 2701–2707.

[12] S. Shao, J. Liu, G. Portale, H.H. Fang, G.R. Blake, G.H. ten Brink, L.J.A. Koster, M. A. Loi, Adv. Energy Mater. 8 (2018)

[13] S. Yasin, T. Al Zoubi, and M. Moustafa, "Design and simulation of high efficiency lead-free heterostructure perovskite solar cell using SCAPS-1D," Optik (Stuttg), vol. 229, Mar. 2021, doi: 10.1016/j.ijleo.2021.166258.

[14] F. Jannat, S. Ahmed, and M. A. Alim, "Performance analysis of cesium formamidinium lead mixed halide based perovskite solar cell with MoOx as hole transport material via SCAPS-1D," Optik (Stuttg), vol. 228, Feb. 2021, doi: 10.1016/j.ijleo.2020.166202.

[15] S. Ahmed, F. Jannat, M. A. K. Khan, and M. A. Alim, "Numerical development of eco-friendly Cs2TiBr6 based perovskite solar cell with all-inorganic charge transport materials via SCAPS-1D," Optik (Stuttg), vol. 225, Jan. 2021, doi: 10.1016/j.ijleo.2020.165765.

[16] A. C. P. Reyes et al., "Study of a lead-free perovskite solar cell using CZTS as HTL to achieve a 20% PCE by SCAPS-1D simulation," Micromachines (Basel), vol. 12, no. 12, Dec. 2021, doi: 10.3390/mi12121508.

[17] M. S. Islam et al., "Defect study and modelling of SnX3-based perovskite solar cells with SCAPS-1D," Nanomaterials, vol. 11, no. 5, May 2021, doi: 10.3390/nano11051218.

[18] A. Tara, V. Bharti, S. Sharma, and R. Gupta, "Device simulation of FASnI3 based perovskite solar cell with Zn(O0.3, S0.7) as electron transport layer using SCAPS-1D," Opt Mater (Amst), vol. 119, Sep. 2021, doi: 10.1016/j.optmat.2021.111362.

[19] H. Alipour and A. Ghadimi, "Optimization of lead-free perovskite solar cells in normal-structure with WO3 and water-free PEDOT: PSS composite for hole transport layer by SCAPS-1D simulation," Opt Mater (Amst), vol. 120, Oct. 2021, doi: 10.1016/j.optmat.2021.111432.

[20] S. Karthick, J. Bouclé, and S. Velumani, "Effect of bismuth iodide (BiI3) interfacial layer with different HTL's in FAPI based perovskite solar cell – SCAPS – 1D study," Solar Energy, vol. 218, pp. 157–168, Apr. 2021, doi: 10.1016/j.solener.2021.02.041.

[21] M. Vishnuwaran, K. Ramachandran, P. Roy, and A. khare, " SCAPS simulated FASnI 3 and MASnI 3 based PSC solar cells: A comparison of device performance ," IOP Conf Ser Mater Sci Eng, vol. 1219, no. 1, p. 012048, Jan. 2022, doi: 10.1088/1757-899x/1219/1/012048.

Effect of pH on Electrochemical, Morphological and Optical Properties of Electrodeposited Molybdenum Sulfide Thin Film

Iskandar Dzulkarnain Rummaja
Faculty of Electronic and Computer Technology and Engineering,
Universiti Teknikal Malaysia Melaka (UTeM),
Melaka, Malaysia
m022210014@student.utem.edu.my

Muhammad Idzdihar Idris
Micro and Nano Research Group (MINE), Centre for Telecomunication Research and Innovation (CeTRI), Faculty of Electronic and Computer Technology and Engineering.
Universiti Teknikal Malaysia Melaka (UTeM),
Melaka, Malaysia
idzdihar@utem.edu.my

Radi Husin Ramlee
Faculty of Electronic and Computer Technology and Engineering.
Universiti Teknikal Malaysia Melaka (UTeM),
Melaka, Malaysia
radihusin@utem.edu.my

Zul Atfyi Mohammed Napiah
Micro and Nano Research Group (MINE), Centre for Telecomunication Research and Innovation (CeTRI), Faculty of Electronic and Computer Technology and Engineering.
Universiti Teknikal Malaysia Melaka (UTeM),
Melaka, Malaysia
radihusin@utem.edu.my

Marziani Rashid
School of Physics,
Universiti Sains Malaysia,
Penang, Malaysia
marzaini@usm.my

Ahmad Muhajer Abdul Aziz
Faculty of Electronic and Computer Technology and Engineering.
Universiti Teknikal Malaysia Melaka (UTeM),
Melaka, Malaysia
m022210016@student.utem.edu.my

Abstract—Third-generation dye-sensitized solar cells (DSSC) provide many benefits over ordinary solar cells. DSSCs efficiently convert visible light into electrical energy and work well in low light environments. Platinum is a common counter electrode due to its electrocatalytic, conductivity, and reflection properties. Due to its drawbacks, alternative counter electrode materials have been explored. The potential counter electrode of DSSC is molybdenum Sulfide (MoS_2), an inorganic compound of molybdenum and sulphur which has good properties in terms of electrochemical properties and long-term stability. Electrodeposition is one of the deposition techniques used in thin film deposition. Due to several challenges in thin deposition techniques with control over film properties, complexity and cost, electrodeposition is a good alternative due to its cost-effectiveness, simplicity, and deposition control. In this work, we studied the effect of deposition electrolyte condition (acidic, neutral, alkaline) to the electrochemical, morphological, and optical properties of molybdenum sulfide thin film. Overall, the study reveals that MoS_2 films exhibit good hydrogen evolution reaction (HER) activity at pH 6 and pH 8, with pH 6 showing the highest performance, while also indicating minimal variation in grain size across different pH levels and highlighting the influence of pH on optical and electronic properties.

Keywords— Third Generation solar cells, thin film characterization, three electrode system, electrodeposition, pH value influence, thin film.

I. INTRODUCTION

Photovoltaic (PV) cells play a crucial role in converting sunlight into usable electricity by harnessing solar energy, primarily using silicon. As solar technology progresses, newer generations of solar cells emerge, promising improved efficiency, affordability, and scalability. Third-generation solar cells, employing innovative approaches like thin-film deposition methods and novel materials, aim to overcome existing challenges. Diverse third-generation technologies such as dye-sensitized solar cells (DSSC), Quantum Dot (QD), and Organic and Perovskite solar cells (PSC) offer exciting possibilities for enhancing solar energy utilization .

Traditionally, platinum has been a preferred material for counter electrodes in DSSCs due to its electrocatalytic properties. However, its high cost, scarcity, and susceptibility to corrosion pose significant limitations . Research endeavors have explored alternatives such as glassy carbon, copper selenide, and lead Sulfide thin films, demonstrating comparable efficiency to platinum. Molybdenum Sulfide (MoS_2) exhibit promising characteristics for DSSC counter electrodes, offering transparency, specific surface area, and catalytic activity akin to platinum. Syrrokostas et. al. [1] reported that MoS_2 exhibits extraordinary stability against corrosion and shows an unprecedented self-improving behavior with increasing cycling, indicating good long-term stability as a counter electrode of DSSC.

Electrodeposition is an electrochemical technique used for fabricating thin films and synthesizing materials, part of the bottom-up approach in material synthesis, which includes methods like PVD, CVD, ALD, and solution chemical deposition. While these methods can be complex, have temperature limitations, and may not be compatible with certain materials and substrates [2] and having low output and long reaction time for solution chemical deposition of hydrothermal [3], electrodeposition stands out for its cost-

979-8-3503-7832-0/24 $31.00 © 2024 IEEE

effectiveness, scalability, and precise control over composition and microstructure. These advantages make it suitable for thin film solar cells and semiconductors [4]. Hossain et. al. [5] presented successfully fabricated bulk MoS_2 thin films on FTO substrates via electrochemical deposition for photovoltaic applications, achieving uniform, amorphous films with indirect bandgap values between 1.3 and 1.4 eV.

In this study, the effect of pH on electrochemical, morphological and optical properties of electrodeposited MoS_2 thin film is being investigated. In addition, MoS_2 thin films were electrodeposited onto ITO-coated glass substrates using a modified setup featuring a graphite rod as the counter electrode. By adjusting the electrolyte pH, the morphological and optical properties of the deposited thin films were explored through the characterization of the film.

II. EXPERIMENTAL SETUP

A. Preparation of MoS₂ precursor solution

The precursor materials deposition electrolyte consists of 5 mM of ammonium tetrathiomolybdate $((NH_4)2MoS_4$, tracemetal basis, 99.97% Sigma Aldrich) and 0.1 M potassium chloride (KCl, 99% ACS reagent, Sigma Aldrich) in 30ml DI water. The precursor electrolyte is in base condition (pH7) after preparation. The addition of 0.1M HCl was used to adjust the pH value into pH value 6 while 0.1 M NaOH was added to adjust the pH value of 8 [6]. It is noteworthy that a pH of 6 rendered the precursor solution acidic, while a pH of 8 established an alkaline condition. The resulting aqueous precursor solution exhibited a light brown color.

B. Fabrication of MoS₂ via Electrodeposition

Fabrication of MoS_2 thin film was based on previous research [7] with some modification in terms of variation of precursor electrolyte condition of acidic (pH 6), neutral (pH 7) and alkaline (pH 8). Initially, MoS_2 was electrodeposited on ITO substrate glass. To achieve this, the ITO substrates underwent a sequential cleaning process involving multiple rinses with an aqueous soap solution, DI water, ethanol, acetone and a final rinse with IPA, each lasting 10 minutes followed by drying the cleaned ITO substrates in a oven at 100 ℃ for 10 minutes.

For the deposition of MoS_2 films, three electrode systems of electrodeposition setup were used. The setup comprised a graphite rod as counter electrode, an ITO substrate as the working electrode and Ag/AgCl (sat. KCl) served as the reference electrode. Constant potential of -1V vs Ag/AgCl reference electrode, which was chosen as the constant applied potential for the MoS_2 potentiostatic electrodeposition (Ossila Potentiostat). The MoS_2 thin was deposited in MoS_2 precursor solution for 30 min at room temperature then rinsed with DI water after depositing finished. To allow grain growth, deposited thin film was anneal in a furnace at 450℃ for 30 minutes. The process was repeating with same process but with different pH of MoS_2 precursor which manipulation of pH in precursor has been explain on section II-A. In this work, there were 3 samples which is MoS_2-pH6, MoS_2-pH7 and MoS_2-pH8.

C. Characterization of MoS₂ thin film

The morphology studies of the deposited MoS_2 thin film were analysed through field-emission scanning electron microscopy (SEM, Hitachi FE-SEM S4800). The chemical state and composition were analyzed by energy dispersive X-ray spectroscopy (EDX, Oxford Instrument). The electrocatalytic performance of thin film was examined through Cyclic Voltammetry (CV) by use of (Ossila Pontentiostat) in three electrode configurations in precursor solution with 20 mV s−1 scan rate. Optical transmission study was completed in 400–800 nm wavelength region using UV–Vis-NIR Spectrophotometer (Carry 5000, Agilent Technologies).

III. RESULT DISSCUSION

A. Electrochemical characterization of precursor solution.

Cyclic voltammetry (CV) was conducted at a scan rate of 20 mV s−1 within the potential range of -1.2V to 0.2V versus the Ag/AgCl reference electrode. Figure 2 illustrates the cyclic voltammetry analysis of a MoS_2 thin film under various pH conditions. The plot exhibits a characteristic curve featuring an anodic peak (indicating positive current) and a cathodic peak (indicating negative current). The anodic peak corresponds to oxidation processes, while the cathodic peak corresponds to reduction processes occurring at the electrode interface. Variations in pH levels of the solution lead to shifts in peak potentials, suggesting a pH dependency in the redox reactions of molybdenum Sulfide. This observed shift implies the involvement of protons (H+) in the redox mechanism of molybdenum sulfide. Additionally, alterations in pH value influence the concentration of hydrogen ions (H+) in the solution, thereby affecting both the electrochemical characteristics of the redox species and the properties of the electrode surface [8].

Typically, as pH increases, a negative shift in the anodic peak potential and a positive shift in the cathodic peak potential are observed in systems where protons directly participate in the electrochemical reaction. Current intensity at varying pH levels may indicate changes in electrochemical reaction kinetics or electron transport during the redox reaction. Notably, MoS_2 at pH 7 has lower current intensity than pH 6 and 8. Higher current intensity in MoS_2 films indicates higher hydrogen evolution reaction (HER) activity at pH 6 and pH 8 [9] MoS_2 at pH 6 had the best HER activity, which is consistent with research [6] that show acidic MoS_2 has better HER activity than neutral and alkaline.

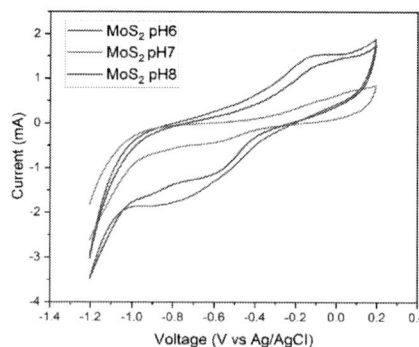

Fig. 1.   Cycle Voltammetry graph of MoS_2 thin film

B. Morphological analysis of the MoS₂ thin film

Figure 2 depicts SEM images of all samples (MoS_2-pH6, MoS_2-pH7, and MoS_2-pH8) at magnifications of 50K and 2K. It is evident from the SEM images that all samples exhibit a deposited thin film in a crystalline structure. At 50K magnification (Figure 2a, 2c, and 2e), the grain shape displays geometric shapes, indicating a well-defined grain structure. Previous studies have addressed the mechanical properties and growth of polycrystalline MoS_2, affirming its feasibility for production and characterization in such a form [10]. The grain size was determined using ImageJ software, yielding an average grain size of 0.56 μm for MoS_2-pH 6, 0.47 μm for MoS_2-pH 7, and 0.58 μm for MoS_2-pH 8. These measurements reveal a slight difference in grain size among different pH values, suggesting that pH variations have minimal impact on grains sizes.

Fig. 2.   SEM image of annealed MoS_2 thin film at different magnification. a) MoS_2 -pH 6 at 50k. b) MoS_2 -pH 6 at 2k. c) MoS_2 -pH 7 at 50k. d) MoS_2 -pH 7 at 50k. e) MoS_2 -pH 8 at 2k. d) MoS_2 -pH 8 at 50k. f) MoS_2 -pH 8 at 2k.

Figures 2b, 2d, and 2f illustrate SEM images at 2K magnification, revealing differences in grain distribution within the film. With increasing pH value, the film exhibits higher grain density, indicating a greater presence of grains. MoS_2-pH 8 demonstrates a higher density of grains compared to MoS_2-pH6, which exhibits fewer material grains within the film. It is documented that the density of material grains in thin films can significantly influence their properties, depending on the specific material and application. Thin films with fewer material grains may exhibit improved flexibility, transparency, and electrical conductivity due to higher coating density and increased mobility of thin films [11], MoS_2-pH6 demonstrates better results among the sampled pH conditions.

C. Elemental analysis of materials

EDX analysis was employed to assess the elemental composition of deposited MoS_2 thin films and to confirm the presence of MoS_2 within the films. The EDX spectrum of MoS_2 thin films across all samples is depicted in Figure 5.

In Figure 3, the EDX patterns confirm the presence of Mo and S in MoS_2 samples with pH levels of 6, 7, and 8. The spectrum shows O as the greatest peak, followed by Mo and S in Figures 4a, 4b, and 4c. Mo and S concentrations are the

second greatest peak in all samples, matching MoS_2's chemical formula, proving MoS_2 thin film deposition. Environmental conditions and a potential oxide layer on the film's surface explain the greater oxygen weight % [12].

Additionally, other materials identified in the films include Si, K, C, and Ca. Gold (Au) is present in all samples due to gold sputtering conducted on the films before SEM analysis, which is a common practice in SEM analysis. The presence of Si, K, In, and Ca can be attributed to X-ray emissions from the ITO glass substrate used in EDX analysis [13]. The presence of carbon (C) is attributed to the use of a graphite rod in the electrodeposition process.

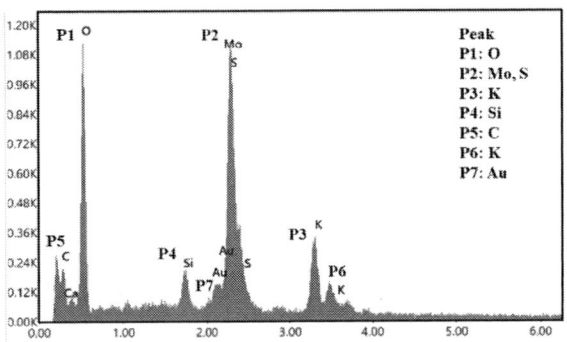

Fig. 3.   EDX spectra of MoS_2 thin film.

D. Optical Studies of MoS_2 thin film

A typical optical property of electrodeposited MoS_2 was investigated by the UV-vis spectroscopic. Based on Figure 4a, the prominent absorption peak was observed at 426.44 nm wavelength at pH 6, pH 7 and pH 8. MoS_2 pH 6 shows the higher absorption following pH 7 and pH 8 giving that increasing pH value decreasing the absorption.

Generally, the Tauc plot is used to determine the bandgap of deposited MoS_2 using the Tauc's relation (equation 1):

$$(\alpha h v)^{\frac{1}{n}} = A \left(h v - E_g \right) \qquad (1)$$

where α is absorption coefficient being a function of wavelength $\alpha(\lambda)$, h is Planck constant, E_g is an optical band gap of a semiconductor, v is frequency, A is proportionality constant, and n is Tauc exponent. The direct bandgap of MoS_2 thin films at pH 6, pH 7, and pH 8 was determined based on Figure 4b.

The bandgap values were found to be 2.25 eV for MoS_2 at pH 6, 2.16 eV for MoS_2 at pH 7, and 2.05 eV for MoS_2 at pH 8. The observed decrease in bandgap values with increasing pH implies either a slight enhancement in conductivity or modifications in the electronic structure as the pH increases. Moreover, MoS_2 at pH 6 exhibits the highest bandgap among the studied pH values. The range of obtained direct bandgap values is consistent with a previous report, which reported a direct bandgap of 2.25 eV [14].

Transmittance refers to the proportion of incident light at a specified wavelength that either passes through a sample or emerges from it. Optical transmittance measurements were conducted on the developed counter electrode within the wavelength range of 400 nm to 850 nm, and the resulting spectrum is illustrated in Figure 4c. The spectra indicates that the films display significant light transmission in the lower

979-8-3503-7832-0/24 $31.00 © 2024 IEEE   23

wavelength region, while transparency gradually increases towards the higher wavelength side. This observation aligns with findings from previous studies [7], [15]

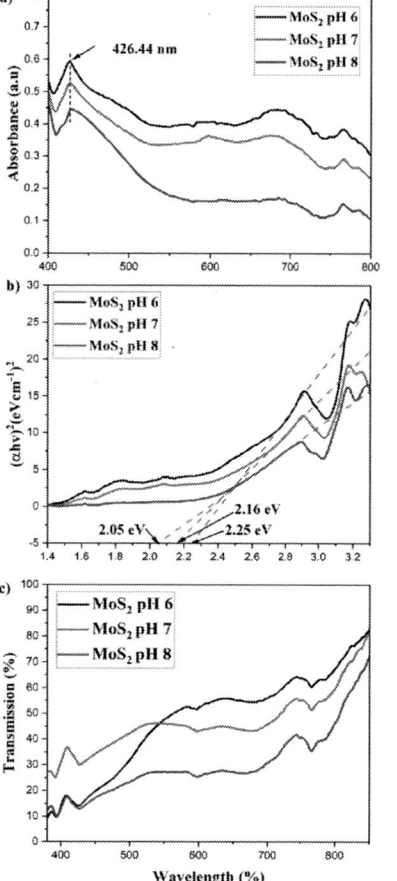

Fig. 4.   UV Vis Analysis (a) Absorbance of MoS$_2$ (b) Tauc plot showing direct band gap of MoS$_2$ thin film. c) Transmission of MoS$_2$.

IV. CONCLUSION

MoS$_2$ thin film has been successfully fabricated in different pH value via electrodeposition. The result of characterization shows that different pH values influence to the electrochemical, morphology and optical properties on deposited MoS$_2$ thin film. Cycle Voltammetry reveals that MoS$_2$ at pH 6 displays the best HER activity, as indicated by its higher current intensity. Although SEM analysis shows minimal differences in grain size among various pH levels, MoS$_2$-pH 8 exhibits a denser grain population compared to MoS$_2$-pH 6. Confirmation of MoS$_2$ presence in the films across all pH levels is provided by EDX patterns. Optical examination demonstrates that MoS$_2$-pH 6 absorbs more light than pH 7 and pH 8, with higher pH values leading to decreased absorption. Moreover, MoS$_2$ at pH 6 exhibits a direct bandgap of 2.46 eV, suggesting its potential in optoelectronic applications. Transmission spectra affirm these findings, showing substantial absorption at shorter wavelengths and increased transparency at longer wavelengths.

ACKNOWLEDGMENT

This study is funded by the Ministry of Higher Education (MOHE) of Malaysia through the Fundamental Research Grant Scheme (FRGS) No FRGS/1/2022/TK07/UTEM/02/06 .The Authors also would like to thank Universiti Teknikal Malaysia Melaka (UTeM) for all the supports.

REFERENCES

[1] G. Syrrokostas, A. Antonelou, G. Leftheriotis, and S. N. Yannopoulos, 'Electrochemical properties and long-term stability of molybdenum disulfide and platinum counter electrodes for solar cells: A comparative study', *Electrochim Acta*, vol. 267, pp. 110–121, Mar. 2018, doi: 10.1016/j.electacta.2018.02.068.

[2] B. Uzakbaiuly, A. Mukanova, Y. Zhang, and Z. Bakenov, 'Physical Vapor Deposition of Cathode Materials for All Solid-State Li Ion Batteries: A Review', *Front Energy Res*, vol. 9, May 2021, doi: 10.3389/fenrg.2021.625123.

[3] S. Yadav and A. Sharma, 'Importance and challenges of hydrothermal technique for synthesis of transition metal oxides and composites as supercapacitor electrode materials', *J Energy Storage*, vol. 44, p. 103295, Dec. 2021, doi: 10.1016/j.est.2021.103295.

[4] A. Ray, 'Electrodeposition of Thin Films for Low-cost Solar Cells', in *Electroplating of Nanostructures*, InTech, 2015. doi: 10.5772/61456.

[5] Md. A. Hossain, B. A. Merzougui, F. H. Alharbi, and N. Tabet, 'Electrochemical deposition of bulk MoS2 thin films for photovoltaic applications', *Solar Energy Materials and Solar Cells*, vol. 186, pp. 165–174, Nov. 2018, doi: 10.1016/j.solmat.2018.06.026.

[6] W.-H. Hu *et al.*, 'Effect of pH on the growth of MoS2 (002) plane and electrocatalytic activity for HER', *Int J Hydrogen Energy*, vol. 41, no. 1, pp. 294–299, Jan. 2016, doi: 10.1016/j.ijhydene.2015.09.076.

[7] M. Gurulakshmi *et al.*, 'Electrodeposited MoS2 counter electrode for flexible dye sensitized solar cell module with ionic liquid assisted photoelectrode', *Solar Energy*, vol. 199, pp. 447–452, Mar. 2020, doi: 10.1016/j.solener.2020.02.047.

[8] A. C. de Sá *et al.*, 'Electrocatalitic Detection of Hydrazine Using Chemically Modified Electrodes with Cobalt Pentacyanonitrosylferrate Adsorbed on the 3–Aminopropylsilica Surface', *Int J Chem*, vol. 9, no. 4, p. 12, Aug. 2017, doi: 10.5539/ijc.v9n4p12.

[9] X. Xu and L. Liu, 'MoS2 with Controlled Thickness for Electrocatalytic Hydrogen Evolution', *Nanoscale Res Lett*, vol. 16, no. 1, p. 137, Aug. 2021, doi: 10.1186/s11671-021-03596-x.

[10] N. Liu *et al.*, 'Growth of Multiorientated Polycrystalline MoS2 Using Plasma-Enhanced Chemical Vapor Deposition for Efficient Hydrogen Evolution Reactions', *Nanomaterials*, vol. 10, no. 8, p. 1465, Jul. 2020, doi: 10.3390/nano10081465.

[11] J. H. Lee, Y. H. Kim, S. J. Ahn, T. H. Ha, and H. S. Kim, 'Grain-size effect on the electrical properties of nanocrystalline indium tin oxide thin films', *Materials Science and Engineering: B*, vol. 199, pp. 37–41, Sep. 2015, doi: 10.1016/j.mseb.2015.04.011.

[12] T. Iqbal *et al.*, 'Comparison of optical constants of sputtered MoS2 and MoS2/Al2O3 composite thin films', *Journal of Materials Science: Materials in Electronics*, vol. 31, no. 10, pp. 7753–7759, May 2020, doi: 10.1007/s10854-020-03312-y.

[13] B. Sotillo *et al.*, 'Raman spectroscopy of femtosecond laser written low propagation loss optical waveguides in Schott N-SF8 glass', *Opt Mater (Amst)*, vol. 72, pp. 626–631, Oct. 2017, doi: 10.1016/j.optmat.2017.07.002.

[14] N. Saha *et al.*, 'Highly active spherical amorphous MoS$_2$: facile synthesis and application in photocatalytic degradation of rose bengal dye and hydrogenation of nitroarenes', *RSC Adv*, vol. 5, no. 108, pp. 88848–88856, 2015, doi: 10.1039/C5RA19442C.

[15] S. Dam, A. Thakur, and S. Hussain, 'Valence band studies of MoS2 thin films synthesised by electrodeposition method', *Mater Today Proc*, vol. 46, pp. 6127–6131, 2021, doi: 10.1016/j.matpr.2020.03.722.

Early detection of hyperuricemia using hybrid Au–ZnO biosensor and Kretschmann-based SPR at visible optical wavelengths

Yasin, M.F.H.F
IMEN, UKM
Selangor,Malaysia
mfhakim26@gmail.com

Gafor, A.H.A
HCTM, UKM
Selangor,Malaysia
halim@ukm.edu.my

Abdullah, N
UMBI, UKM
Selangor, Malaysia
noraidatulakma.abdullah@
ppukm.ukm.edu.my

Yee, L.P.
UMBI, UKM
Selangor, Malaysia
leepeyyee@ukm.edu.my

Kassim, S
FSSM, UMT
Terengganu, Malaysia
syara.kassim@umt.edu.my

Rashid; A.R.A
FST, USIM
Negeri Sembilan, Malaysia
affarozana@usim.edu.my

Shyong, S.K.
IMEN, UKM
Selangor,Malaysia
kimsiow@ukm.edu.my

Hamzah, A.A
IMEN, UKM
Selangor,Malaysia
azlanhamzah@ukm.edu.my

Mohamed, M.A
IMEN, UKM
Selangor,Malaysia
ambri@ukm.edu.my

Mustaffa, S.N.
IMEN, UKM
Selangor,Malaysia
nasuhamustaffa@gmail.
com

Jamil, N.A
IMEN, UKM
Selangor,Malaysia
nurakmarjamil83@gmail.
com

Menon, P.S.
IMEN, UKM
Selangor,Malaysia
susi@ukm.edu.my

Abstract— Hyperuricemia is a form of disease that stems from excessive amount of uric acid due to excessive purine-rich food consumption or insufficient excretion in a human body hence the need for a reliable non-invasive sensor to detect it. This study explores Surface Plasmon Resonance based on Kretchman configuration (K-SPR) as a rapid, highly sensitive and impeccably accurate biosensor. K-SPR technique was proven as the best method as it offers label-free detection of biomarkers. Experimental work was executed using Bionavis K-SPR Navi™-200 utilizing gold-zinc oxide (Au-ZnO) hybrid thin films as the sensing layer for uric acid level measurements. The minimum reflectivity, R_{min}, full-width-at-half-maximum (FWHM), sensitivity and Q-factor was calculated to determine the dependability of these layers as a sensing material. At 670 nm and 785 nm optical wavelength, respectively, the FWHM of Au-ZnO thin films produced in air was 1.271072° and 0.508751°, while their R_{min} values were 0.08747 and 0.11343. For uric acid sensing of 0.6mM, 1.2mM, 1.8mM, 2.4mM and 3.0mM, Au-ZnO has an average sensitivity of 0.05193°/mM for 670 nm and 0.04150°/mM for 785 nm. Average Q-factor obtained was 0.04085 mM^{-1} and 0.04787 mM^{-1} under 670 nm and 785 nm respectively. Lower detection limit of 0.5 mM uric acid was tested resulting in the best sensitivity of 0.0511°/mM at 670 nm optical wavelength.

Keywords—Surface plasmon resonance, SPR, Kretschmann, hyperuricemia, uric acid, biosensor, optical sensing,

I. INTRODUCTION

Elevated level of serum uric acid (UA), usually more than 6 mg/dL and 7 mg/dL for female and male respectively, could cause hyperuricemia. Hyperuricemia derives from either increased endogenous uric acid production, excessive exogenous consumption, decreased secretion or a combination of these conditions. Correlation of diseases such as cardiovascular issues, chronic kidney diseases,

hypertension, atrial fibrillation and heart failure with hyperuricemia arose significant concern. Hyperuricemia promote these diseases through controlling molecular signals that include endoplasmic reticulum stress, endothelial dysfunction, insulin resistance/diabetes, and oxidative stress[1].

Surface Plasmon Resonance (SPR) is a highly effective method for obtaining information on optical characteristics, including those of nanomaterials and biological materials for the application of food safety, drug discovery, environment protection, and medical diagnostics. Along with the quantification of many biomarkers, including proteins, deoxyribonucleic acid (DNA)(write in full), and whole cells, SPR-based platforms have shown to be among the most potent technologies for real-time molecular interaction monitoring [2]. Numerous uric acid sensor have been continuously developed throughout the years. Despite so, SPR-based sensors prevail due to its highly sensitive, rapid and label-free detection of biomarkers.

Otto, Kretschmann and Raether introduced the principle of surface plasmon excitation using prism coupling in 1698. Otto utilized the principle of total internal reflection between the prism and air interface together with evanescent field penetration, leaving slight gap between the prism and sample. As opposed to Otto, Kretschman and Raether (frequently recognized as Kretschmann configuration) applies direct contact of metal-thin-film, typically gold onto the prism. Therefore, Kretschmann based sensors are widely utilized as biomolecular sensing due to their high sensitivity, operational simplicity and real-time detection capability [3].

The application of bare gold as K-SPR sensing layer alone have several limitations such as poor adhesion of the metal layer on the substrate surface which increases the popularity of adding metal oxides on the sensing layer. Zinc Oxide (ZnO) being an n-type semiconductor with wide band gap (~3.4 eV) and exciton energy (60 meV) at room temperature equipped with resistance towards electronic degradation due to its high thermal resistance has been rigorously studied for photonics applications [4]. ZnO nanostructure can exist in plentiful morphologies comprising of nanorods, nanospheres, nanocones, nanoneedles, nanoflower, nanocomposites, nanoflakes and nanowires aside from the commonly known thin film.

Wang et al. synthesized ZrO_2 and ZnO hybrid nanocomposites for simultaneous detection of epinephrine, uric acid and folic acid. This electrochemical based sensor have low detection limit of 0.28 μM and wide detection range of 10-2400 μM for UA sensing [5]. Ahmad et .al employed high aspect ratio zinc oxide nanorods with immobilized uricase for the application of uric acid sensing and achieved low detection limit (5mM) and rapid response (~3s) on an electrochemical based sensor [6]. In the authors' knowledge, experimental testing using Au-ZnO hybrid thin film for K-SPR-based uric acid sensing have never been explored before at 670 and 785 nm optical wavelengths.

II. METHODOLOGY

A. Experimental Setup

Au-ZnO hybrid layer sensors were obtained from Bionavis Inc. Au has a thickness of 50nm while ZnO layer consist of 5nm thick as shown in Fig 1. Experimental test was conducted using Bionavis K-SPR-Navi 200™. Dual wavelengths of 670 and 785nm were utilized for the sensing purpose. Scanning angle was set to full range (38.69° to 79.4°) for air and buffered solution to act as reference. Phosphate buffered saline (PBS) with pH 7.4 was flowed into the K-SPR for 20 minutes at the rate of 100 μl/m to alleviate its reading. Throughout running the experiment, flowrate was set to 50 μl/m and injection volume was kept at 0.5ml per molarity. Buffer solution went through 2 minutes stabilization phase between each concentration.

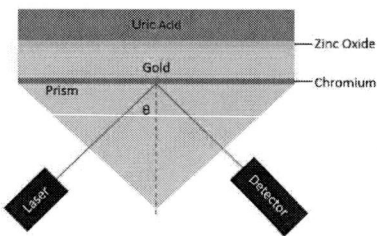

Fig. 1 Au-ZnO hybrid layer in K-SPR configuration

B. Sample Preparation

Uric acid was obtained from Sigma Aldrich in crystalline form (CAS U2625) and diluted with PBS with pH of 7.4 (CAS P4474). 3mM uric acid stock solution was made with 20ml of PBS by adding 0.0101g of uric acid and mixed with temperature of 65°C. Stock solution was left overnight to eliminate residual uric acid powder. Stock UA was diversified into 5 ml solution of 2.4 mM, 1.8 mM, 1.2 mM and 0.6 mM by diluting it with PBS with the ratio of 4:1, 3:2, 2:3, 1:4 respectively. A separate sample with 0.5mM was prepared to

test the detection limit of Au-ZnO hybrid layer thin film for UA K-SPR sensing.

C. Calculations

The K-SPR minimum reflectivity, R_{min}, full-width-at-half-maximum (FWHM), sensitivity, Q-factor and detection accuracy (D.A) are important parameters in determining a sensor's reliability. R_{min} was obtained from the lowest intensity detected on a specific wavelength. FWHM is the width of the spectrum curve measured through the half maximum amplitude from the y-axis. Sensitivity was calculated using equation (1) where Δθ is the angle shift. D.A and Q-factor were defined by equations (2) and (3). Desirable value for K-SPR- based sensor is low R_{min}, narrow FWHM with high sensitivity and Q-factor[7].

$$S = \frac{\Delta\theta(\theta_{UA} - \theta_{PBS})}{Concentration_{UA} - Concentration_{PBS}} \quad (1)$$

$$Q - factor = \frac{S}{FWHM} \quad (2)$$

$$D.A = \frac{\Delta\theta}{FWHM} \quad (3)$$

III. RESULTS AND DISCUSSIONS

Fig. 2 shows the presence of ZnO improves Au detection capabilities in PBS by improving its R_{min} and FWHM at both 670 and 785 nm optical wavelengths.

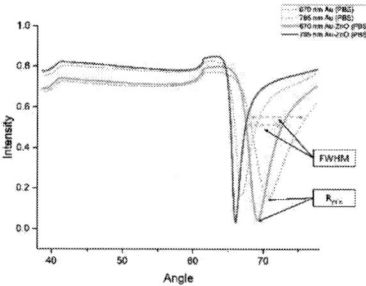

Fig. 2 K-SPR curve shift for Au and Au-ZnO sensing layers in PBS

As indicated in Table I, R_{min} value of Au-ZnO thin film obtained in PBS is 0.043601 and 0.031839 whilst their FWHM is 3.9351° and 1.7481° for 670 nm and 785 nm respectively. Meanwhile for Au, R_{min} was 0.154658 and 0.162210 for 670 nm and 785 nm. The FWHM calculated was 6.0640° at 670 nm and 2.9202° at 785 nm signifying that Au-ZnO hybrid layer significantly improved the capability of K-SPR sensor as compared to Au-only sensors in PBS analyte.

TABLE I. COMPARISON OF K-SPR VALUES OF RMIN AND FWHM USING AU AND AU-ZNO IN PBS FOR 670 AND 785 NM OPTICAL WAVELENGTH

Coating	Wavelength (nm)	R_{min}	FWHM (°)
Au-ZnO	670	0.043601	3.9351
	785	0.031839	1.7481
Au	670	0.154658	6.0640
	785	0.162210	2.9202

Due to the solubility limit of UA in distilled water, PBS is used for their neutral pH of 7.4 instead of DI water. Basic solutions such as potassium hydroxide (KOH), sodium hydroxide (NaOH) and ammonium hydroxide (NH₄OH) can also act as alternative solvents. However, it may form metallic hydroxides that affects the intensity and decline the overall performance of the sensor [8]. Consequently, results obtained in this experimental work is expected to differ with previous simulation conducted by [7].

A. Au-ZnO UA Detection

Fig 3 indicates K-SPR shift experienced in 670 nm and 785 nm optical wavelengths for the detection of 1.8 mM uric acid (UA) using the Au-ZnO hybrid sensing layer The resonance angle at 670 nm is 69° while at 785 nm optical wavelength, it is 66°.

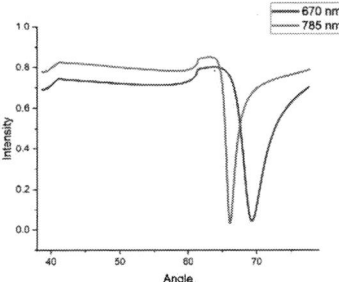

Fig. 3 K-SPR sensing of 1.8 mM UA at 670 nm and 785 nm using Au-ZnO sensing layers

Fig 4 portrays the K-SPR response from 670 nm laser in various molarities of UA, where the K-SPR R_{min} and resonance angle in PBS is 0.43 at 69.23°. Angle shift upon uric acid injection to the flow cell is 0.148224°, 0.124824°, 0.086303°, 0.066658°, 0.032829° for 3 mM, 2.4 mM, 1.8 mM, 1.2 mM and 0.6 mM respectively.

As for the 785 nm laser exposure in Fig 5, R_{min} and resonance angle in PBS is 0.03 and 66.06° respectively. Maximum shift of 0.120758° was observed during the injection of 3 mM uric acid and minimum angular shift of 0.025261° was detected upon 0.6 mM UA exposure. In comparison, 2.4 mM, 1.8 mM and 1.2 mM UA results in angle shift of 0.101517°, 0.068778° and 0.052887° respectively.

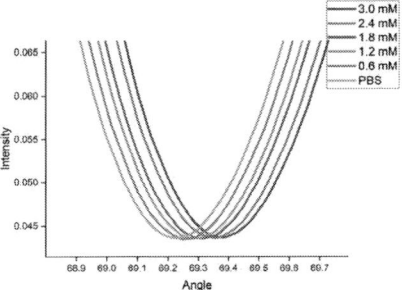

Fig. 4 Au-ZnO K-SPR response on various molarities of UA at 670 nm optical wavelength

Overall, UA exposure induce red-shifting angular response to the K-SPR as the concentration increases which correlates to an increase in refractive index of the sample [7], [9]. Table II and III designate that 785 nm has higher detection accuracy and Q-factor as compared to 670 nm as the higher value indicates better sensing capability. Despite the higher sensitivity on 670 nm, narrow FWHM in 785 nm greatly contributes to its performance, signifying the

superiority of 785 nm wavelength in sensing low limits of UA.

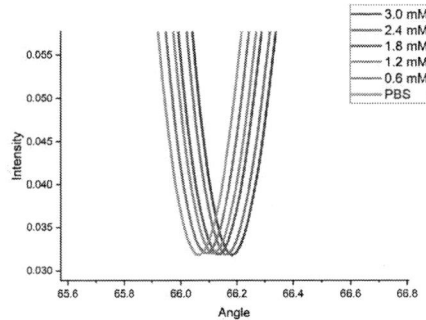

Fig. 5 Au-ZnO K-SPR response on various malorities of UA at 785 nm optical wavelength

TABLE II.   COMPARISON OF K-SPR VALUES OF RMIN AND FWHM USING AU AND AU-ZNO IN PBS FOR 670 AND 785 NM OPTICAL WAVELENGTH

Wavelength (nm)	Concentration (mM)	Resonance Angle Shift (°)	Sensitivity (°/mM)
670	0.6	0.03283	0.05473
	1.2	0.06666	0.05555
	1.8	0.08630	0.04795
	2.4	0.12482	0.05201
	3.0	0.14822	0.04941
785	0.6	0.02526	0.04210
	1.2	0.05289	0.04407
	1.8	0.06978	0.03877
	2.4	0.10152	0.04230
	3.0	0.12076	0.04025

TABLE III.   CALCULATION OF Q-FACTOR AND D.A OF AU-ZNO K-SPR SENSOR FOR VARIOUS MOLARITIES OF UA

Wavelength (nm)	Concentration (mM)	FWHM (°)	Q-Factor (mM⁻¹)	Detection Accuracy
670	0.6	3.962861	0.04305	0.00828
	1.2	3.948388	0.04370	0.01688
	1.8	4.004799	0.03772	0.02155
	2.4	3.932953	0.04092	0.03174
	3.0	4.003807	0.03887	0.03702
785	0.6	1.748893	0.08275	0.01444
	1.2	1.748764	0.08663	0.03024
	1.8	1.748696	0.07620	0.03990
	2.4	1.748572	0.08314	0.05806
	3.0	1.763031	0.07915	0.06849

B. Sensitivity and Q-Factor Comparison between Experimental and Simulation work.

The data used in the experimental calculation is the average of 3 different runs with deviation of less than 10%. Table 4 shows that the sensitivity of UA detection using Au-ZnO hybrid thin film in this work, is 85.46% and 340% greater compared to previous simulative experiments conducted for both 670 nm and 785 nm wavelength. Coherently, Q-factor in this study improved by 670% for 670 nm and 457% for 785 nm. Actual average values were summarized in Table IV 4. The difference is associated with the actual experimental parameters that are mostly neglected in simulation such as temperature, humidity, pH stability and actual refractive index of UA under those conditions.

979-8-3503-7832-0/24 $31.00 © 2024 IEEE   27

TABLE IV. EXPERIMENTAL AND SIMULATION DATA COMPARISON FOR AU-ZNO K-SPR SENSOR FOR UA DETECTION

Parameter	Experimental [this work]		Simulation [7]	
Wavelength (nm)	670	785	670	785
R_{mm}	0.0875	0.1134	0.0218	0.1330
FWHM	1.2711	0.5087	1.4606	0.7116
Average Sensitivity (°/mM)	0.05193	0.04150	0.02800	0.00943
Average Q-factor (mM^{-1})	0.04085	0.04787	0.00530	0.00858

IV. DETECTION LIMIT

Another contribution in this work as compared to previous research by [7] is in the lowest detection limit capability of the Au-ZnO K-SPR sensors for the UA detection. Previous work tested minimum of 0.6 mM of UA whilst this study test the limit up to 0.5mM which is the borderline healthy level of UA in an adult as portrayed in Fig 6 for both 670 and 785 nm optical wavelengths [10].

Three separate UA injections were administered hence resulting in an average resonance angle of 0.02555° for 670 nm and 0.02085° for 785 nm. The sensitivity and Q-factor for 0.5 mM of UA obtained for 670 nm and 785 nm were 0.0511 °/mM, 0.0417 °/mM, 0.04021 mM^{-1} and 0.08195 mM^{-1} correspondingly. The low detection limit thereby proved that Au-ZnO sensor is capable of boosting the sensing properties of Au for K-SPR application.

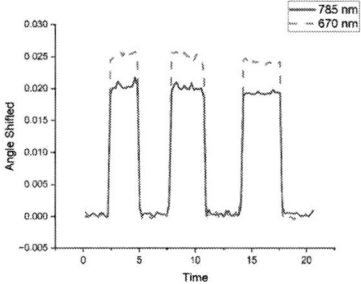

Fig. 6 Sensorgram for the detection of 0.5 mM of UA using Au-ZnO K-SPR sensors at 670 and 785 nm optical wavelengths

V. CONCLUSION

In conclusion, Au-ZnO hybrid layer was tested as uric acid K-SPR sensor experimentaly and the result was compared to existing simulation data. The sensitivity of Au-ZnO layer utilized is 0.05193°/mM and 0.04150°/mM for 670 nm and 785 nm. This value indicates 85.46% and 340% increase for 670 nm and 785 nm sensing as compared to previous simulation work due to the influence of temperature, humidity, pH stability and RI. Lower detection limit of 0.5 mM was tested and achieved sensitivity of 0.0511 °/mM and 0.0417 °/mM for 670 nm and 785 nm indicating the reliability of ZnO as a sensitivity enhancing layer for Au thin film. Therefore, it is believed that ZnO serves as relevant performance enhancer that could sense hyperuricemia in its early stages and further experimental works utilizing uricase to fabricate enzymatic K-SPR sensor will be executed to further increase its accuracy.

ACKNOWLEDGMENT

This project was funded by the National University of Malaysia (UKM) using grants UKM-TR2023-09. Institute of Microengineering and Nanoelectronics (IMEN) is also acknowledged for the financial support.

REFERENCES

[1] C. George, S. W. Leslie, and D. A. Minter, "Hyperuricemia," *Urology at a Glance*, pp. 107–109, Oct. 2023, doi: 10.1007/978-3-642-54859-8_23.

[2] J. H. Qu, A. Dillen, W. Saeys, J. Lammertyn, and D. Spasic, "Advancements in SPR biosensing technology: An overview of recent trends in smart layers design, multiplexing concepts, continuous monitoring and in vivo sensing," *Anal Chim Acta*, vol. 1104, pp. 10–27, Apr. 2020, doi: 10.1016/J.ACA.2019.12.067.

[3] N. Shukla, P. Chetri, R. Boruah, A. Gogoi, and G. A. Ahmed, "Surface plasmon resonance biosensors based on Kretschmann configuration: basic instrumentation and applications."

[4] G. S. Mei, P. Susthitha Menon, and G. Hegde, "ZnO for performance enhancement of surface plasmon resonance biosensor: A review," *Materials Research Express*, vol. 7, no. 1. Institute of Physics Publishing, 2020. doi: 10.1088/2053-1591/ab66a7.

[5] Q. Wang, H. Si, L. Zhang, L. Li, X. Wang, and S. Wang, "A fast and facile electrochemical method for the simultaneous detection of epinephrine, uric acid and folic acid based on ZrO2/ZnO nanocomposites as sensing material," *Anal Chim Acta*, vol. 1104, pp. 69–77, Apr. 2020, doi: 10.1016/J.ACA.2020.01.012.

[6] R. Ahmad, N. Tripathy, M. S. Ahn, and Y. B. Hahn, "Solution Process Synthesis of High Aspect Ratio ZnO Nanorods on Electrode Surface for Sensitive Electrochemical Detection of Uric Acid," *Scientific Reports 2017 7:1*, vol. 7, no. 1, pp. 1–8, Apr. 2017, doi: 10.1038/srep46475.

[7] S. N. Mustaffa *et al.*, "Visible and angular interrogation of Kretschmann-based SPR using hybrid Au–ZnO optical sensor for hyperuricemia detection," *Heliyon*, vol. 9, no. 12, Dec. 2023, doi: 10.1016/j.heliyon.2023.e22926.

[8] L. Singh *et al.*, "Gold Nanoparticles and Uricase Functionalized Tapered Fiber Sensor for Uric Acid Detection," *IEEE Sens J*, vol. 20, no. 1, pp. 219–226, Jan. 2020, doi: 10.1109/JSEN.2019.2942388.

[9] V. Sharma, D. Verma, and G. S. Okram, "Influence of surfactant, particle size and dispersion medium on surface plasmon resonance of silver nanoparticles," *Journal of Physics Condensed Matter*, vol. 32, no. 14, Apr. 2020, doi: 10.1088/1361-648X/ab601a.

[10] "Uric acid - blood." Accessed: May 14, 2024. [Online]. Available: https://www.ucsfhealth.org/medical-tests/uric-acid----blood-

Synthesis of Multi-Spike Gold Nanostar using the Surfactant-Free Method as LSPR Substrate Material

Tita Oktavia Cahya Rahayu
Institute of Microengineering
and Nanoelectronics (IMEN)
UKM Bangi, Malaysia
titaoktaviacr@gmail.com

Izzah Hanaanah Ab Aziz
Institute of Microengineering
and Nanoelectronics (IMEN)
UKM Bangi, Malaysia
p138569@siswa.ukm.edu.my

Nur Hidayah Azeman
Institute of Microengineering
and Nanoelectronics (IMEN)
UKM Bangi, Malaysia
nhidayah.az@ukm.edu.my

Tengku Hasnan Tengku Abdul
Aziz
Institute of Microengineering
and Nanoelectronics (IMEN)
UKM Bangi, Malaysia
hasnanaziz@gmail.com

Mohd Suzeren Md Jamil
Faculty Science and Technology
(FST)
UKM Bangi, Malaysia
suzeren@ukm.edu.my

Ahmad Rifqi Md Zain
Institute of Microengineering
and Nanoelectronics (IMEN)
UKM Bangi, Malaysia
rifqi@ukm.edu.my

Abstract—**Gold nanostars (AuNNs) were synthesized by a surfactant-free wet-chemistry technique. The UV-Vis spectra and surface morphology showed the formation of AuNNs with average diameters of 20-30 nm and 100 nm, including those spikes. From the EDX and AFM analysis, the sample has high purity (above 79 wt%) and was successfully deposited on ITO glass with a 100-300 nm thickness. The current methods for synthesizing AuNNs often involve surfactants, which can interfere with their purity and application in sensitive optical sensors. There is a need for a reliable, surfactant-free synthesis method that produces high-purity AuNNs with consistent morphology, which can be effectively utilized in biosensing applications, particularly for detecting biomolecules such as creatinine. This study aims to synthesize AuNNs using a surfactant-free wet-chemistry method and to characterize their UV-Vis spectra, surface morphology, and purity. These findings are important for various optical sensor applications. As an initial measurement, AuNNs were promising to be used as an LSPR substrate material and applied to biosensors. Synthesized AuNNs and the initial optical performance using LSPR for creatinine detection are highlighted in this study.**

Keywords—gold nanostar, LSPR, spikes, creatinine, surfactant-free.

I. INTRODUCTION

Nowadays, noble metals (such as silver, Ag and gold, Au) nanostructures hold great promise for disease diagnostics and ultrasensitive detection because of the advent of nanotechnology in a wide spectrum of biotechnology fields [1, 2]. Due to their tunable qualities (including sizes, forms, and optoelectronic properties), superior biocompatibility, and great stability, gold nanoparticles (AuNPs) have become the material of choice for researchers. To date, AuNPs with diverse structural shapes, including nanospheres [3], nanorods [4], nanostars [5, 6], and nanoflowers [7], have been produced and studied for medicinal purposes. In contrast, their localized surface plasmon resonance (LSPRs) are located in the near-infrared (NIR) and visible parts of the light spectrum. Although silver

NPs show higher enhancement, their LSPR is blue-shifted compared to the AuNPs [8]. As mentioned before, AuNPs exhibit unique and significant optical properties, including the LSPR, surface-enhanced fluorescence (SEF), which is also known as metal-enhanced fluorescence (MEF) and surface-enhanced Raman scattering (SERS) [9].

LSPR is an optical device that has the potential to be further developed. LSPR refers to the potential of certain substances to transfer energy, into the collective oscillation of conduction band electrons from the photons. The noble metal nanostructure or nanoparticle confines the plasmonic field at the core of LSPR. The light extinction profile in this LSPR confinement completely determined by the nanoparticle shape, material, aspect ratio, size and the interspaces between the nanoparticles or nanostructures surrounding it [10]. The sizes, surface functionalization of AuNPs, and shapes can be adjusted to achieve the intended surface plasmonic properties.

In this research, we proposed gold nanostars (AuNNs) as a material to improve their surface plasmon properties for LSPR measurement. The biocompatible, surfactant-free production of AuNNs in water-based solution is particularly relevant for their bio application. Several seed-mediated synthesis techniques or "one-pot" employing mostly cetyltrimethylammonium bromide (CTAB) or poly(N-vinylpyrrolidone) (PVP) as the surfactant since 2003 [11, 12, 13]. Ag ions were added to enhance the shape and production as a result of the underpotential deposition of Ag ions on specific crystal surfaces of Au seeds. Many branches were believed to form because certain facets were blocked by the silver ion or surfactant. [14]. Due to (1) CTAB's possible toxicity, (2) the formation of aggregation after several washes, and (3) the difficulty of substituting the surfactants, CTAB or PVP, in biofunctionalization, the usage of nanostars has been limited. Therefore, creating surfactant-free nanostar method of synthesis may overcome the problems and bring important progress for developing nanostars in future uses. Surfactant-free gold nanostar

979-8-3503-7832-0/24 $31.00 © 2024 IEEE

synthesis is a method for generating gold nanoparticles with a star-shaped morphology without using surfactants. This method is based on the seed-mediated growth process, in which small gold nanoparticles (seeds) are developed into bigger nanostars using a reducing agent such as ascorbic acid in the presence of gold and, in some cases, silver ions. The procedure is normally carried out in an aqueous medium and is intended to produce biocompatible and stable gold nanostars ideal for a variety of biomedical and sensor applications [15].

AuNPs that has star-shaped exhibit plasmon bands that are discovered to be adjustable and tunable in NIR range. These plasmon bands can be characterised as hybrid resonances related to the individual protecting spikes and spherical core, resulting in dark and bright modes. Furthermore, they showed that the placement of the multipolar modes relies on the sharpness and length of the spikes, as well as their relative alignment with the polarization of the incoming field. Sharper and longer spikes demonstrate more red-shifted resonances that effectively enhance the local electromagnetic (EM-field). [14, 16]. AuNSs have been extensively utilised in sensing, LSPR, biological imaging, nanotherapeutics, and nanomedicine due to these essential features. The electric charge density is focused on the sharp ends of AuNSs, leading to a significant enhancement of the electronic coupling effect and local electromagnetic field due to the anisotropic structure [17]. In this study, we offer a novel and straightforward technique for synthesizing AuNNs for biosensing applications, particularly the detection of creatinine by LSPR measurement, employing a surfactant-free wet-chemistry technique.

II. METHODOLOGY

A. Materials and Reagents

Tetrachloroaurate trihydrate ($HAuCl_4 \cdot 3H_2O$, $\geq$ 99.9%), Silver nitrate ($AgNO_3$, $\geq$ 99.0%), Hydrochloric acid (HCl, 36.5% − 38%) and sodium citrate ($Na_3C_6H_5O_7$) were all purchased from Sigma-Aldrich. Ascorbic acid (Vitamin C) as reducing agent was purchased from the real C. In experiments the deionized water (18 MΩ) was used.

B. Gold Nanostars (AuNSs) Synthesis

In the synthesis of AuNSs, three main ingredients are needed: chloroauric acid ($HauCl_4$), a reducing agent, and a shape-directing or capping agent. The reducing agent used in this synthesis was ascorbic acid. Meanwhile, the shape used is silver ion (Ag^+) [2, 4, 5, 6]. In this research, a surfactant-free technique was utilized, which only involves gold seeds, gold ions, sodium citrate, and small quantities of silver ions. Nanostars were produced through a seed-mediated growth process, in which 15 ml of 1% citrate solution was mixed into 100 ml of boiling 1 mM $HAuCl4$ solution and vigorously agitated for 15 minutes. This approach yields a solution of brick-red gold nanoparticles. The process for synthesizing nanostars involved adding 100 μl of the citrate-stabilized seed solution to a 0.25 mM HAuCl4 solution (with 10 μl of 1M HCl) in a 20 ml container, to a total volume of 10 ml while stirring moderately. Simultaneously, 33 μl of silver nitrate ($AgNO_3$) and rapidly, 50 μl of ascorbic acid were added, causing the solution to be agitated as its color swiftly changed from light red to blue or bluish-black. Immediately thereafter, a 15-minute centrifugal wash was done to stop the nucleation.

C. Characterizations of the Gold Nanostars (AuNNs)

The prepared AuNNs were characterized by using Ultraviolet-visible spectroscopy (Uv-vis). Transmission electron microscopy (TEM) and field emission scanning electron microscopy (FESEM) with machine model ZEISS SUPRA 55VP operated at an acceleration voltage of 3 kV and atomic force microscopy (AFM) were used to analyze the AuNNs surface morphology. Lastly, the AuNNs will be characterized using Energy-Dispersive X-ray spectroscopy (EDX) and localized surface plasmon resonance (LSPR).

III. RESULT AND DISCUSSION

Fig. 1(A) shows the results for the UV-Vis characterization. The black line is gold nanoparticles (spheres). Meanwhile, the other lines are gold nanostars, AuNNs with the concentration variation of Ag^+ ions ranging from 1 − 5 μL. Based on the spectra, the highest intensity of gold nanoparticles was shown at a wavelength (λ_{max}) of 525 nm. This result is similar to the research of G.H. Jeong et al. (2009), which produced maximum intensity at a 520 − 530 nm wavelength. On the other hand, based on the UV-Vis spectrum of AuNNs shown by the red, blue, green, purple, and yellow lines. The maximum absorbance of the material is indicated at 620 - 800 nm wavelength. Such an outcome aligns with the research done by Supriya and Philip (2018), which produced a AuNNs with maximum absorbance intensity observed at a wavelength ranging from 650 to 917 nm [16, 17]. The successful synthesis of AuNNs is shown by the morphological image from Fig. 1(B), which results from the FESEM test using 50.52 K magnification. Based on the figure, gold nanostar particles (AuNNs) have an average diameter of 30 - 50 nm, and when added to the spike length on both sides, it reaches up to 134 nm. The presence of Ag^+ ions direct spike development, as they selectively adsorb onto specific crystal facets of the gold seeds, facilitating anisotropic growth. This produces the distinctive spiky morphology of AuNNs, with the number and length of spikes regulated by the concentration of Ag^+ ions.

The average particle size of AuNNs in this study was still smaller than the previous study by Salinas et al. (2014), which resulted in an average particle size of 180 nm [19]. Based on the FESEM image, each particle has many spikes, with an average of 12-15 spikes for each particle. The number of spikes depends on the concentration of Ag^+. As shown in Fig. 1(A) previously, the plasmon band showed a red shift as the concentration of Ag^+ increased, creating longer, sharper, and more numerous branches with minimal overall size variation. Here, Ag^+ acts as a shape-directing agent [14].

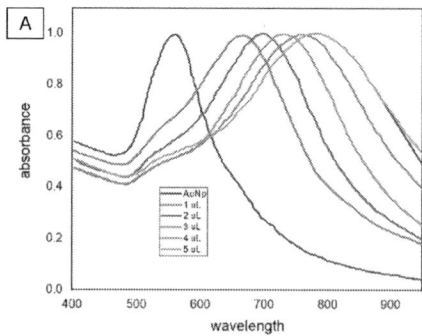

979-8-3503-7832-0/24 $31.00 © 2024 IEEE

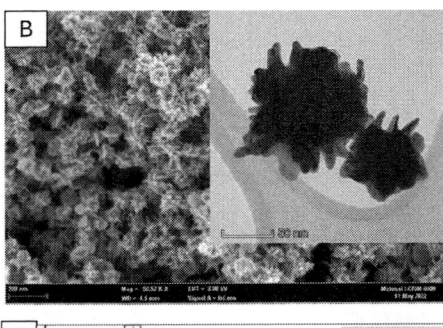

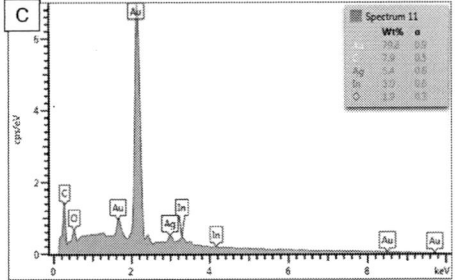

Fig. 1. (A) UV-Vis spectra of gold nanoparticle and gold nanostars, (B) FESEM and TEM image, and (C) EDX spectra of synthesized gold nanostar.

The energy dispersive X-ray transmission (EDX) in Fig. 1(C) shows that gold (Au) was the element with the highest composition (79.8 wt%). This indicates that the synthesized sample has high purity. This result is still higher than the gold composition in the research conducted by Salinas et al. (2014) and Mayoral et al. (2009) [19,20]. The gold element is followed by carbon (C), silver (Ag), indium (In), and oxygen (O). Carbon, silver, and oxygen were residues from the synthesis stage, while indium represents the coating on the ITO glass substrate. The presence of indium in the EDX spectra confirms that the sample has been successfully deposited onto the ITO glass substrate.

The results of atomic force microscopy (AFM) are shown in Figure 2 (A) and (B). Figure 2(A) is a two-dimensional image of the surface of the ITO substrate deposited AuNNs, while Figure 2(B) is three-dimensional. Based on the AFM image, AuNNs form a rough and pointed surface on ITO glass. The roughness value is 12.5 nm. The roughness value affects the plasmonic response generated by the optical device [21]. Based on research conducted by Byun et al. (2008), samples with roughness >5nm, which means surfaces that are extremely rough can generate highly intensified localized plasmons, resulting in a notable decline in the performance of LSPR biosensors. [22]. From the illustration of AuNNs film in Figure 2(B), there is a confirmation of 100-300 nm thick film. In some areas, there are 2-3 particles of gold nanostars stacked on top of each other so that they have a greater thickness than their surroundings.

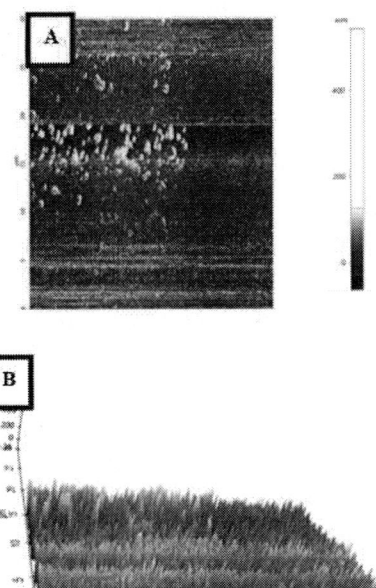

Fig. 2. Atomic Force Microscope of Gold Nanostars on ITO glass, (A) 2-Dimensional and (B) 3-Dimensional.

The synthesized sample was then tested for potential use as LSPR biosensors with 0.1 mM creatinine solution. The plasmonic resonance's intensity changes were observed and analysed when transitioning from the base solution to the creatinine solution.

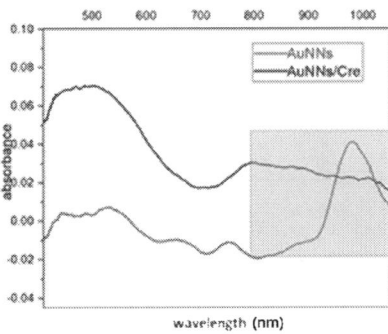

Fig. 3. Absorbance Spectra of Gold Nanostars (AuNNs) and Gold Nanostars with Cre (AuNNs/Cre).

Based on Figure 3, it can be seen that there is a difference between gold nanoparticle samples before and after the addition of creatinine. The absorbance peaks are attributed to the LSPR of the gold nanostars. LSPR occurs when conduction electrons on the surface of gold nanostars resonate with incident light, which leads to strong absorbance at specific wavelengths [24]. Figure 3 shows that the primary peak for AuNNs alone is around 950-1000 nm; upon the addition of creatinine, a broader peak appears around 550-600 nm. This shift suggests that creatinine had altered the local dielectric environment of the gold nanostars or induced changes in their aggregation state. It also proves that an interaction occurs between AuNNs and creatinine. Creatinine molecules had adsorbed onto the surface of the

AuNNs, by modifying their surface properties and thereby their plasmonic behavior is affected. The adsorption has the potential to change the refractive index in the vicinity of the nanostars, which can result in modifications to both the position and intensity of the LSPR peak. Pure AuNNs exhibit a distinct absorbance peak around 950-1000 nm, characteristic of their sharp tips and high aspect ratios, which enhance the plasmonic effect. The absorbance peak undergoes a significant shift and widening upon the introduction of creatinine, signifying robust interactions between the AuNNs and creatinine molecules. Hence, there is a significant shift and broadening of the absorbance peak when creatinine is introduced, suggesting that AuNNs can be effectively used in biosensing applications. The LSPR peak is sensitive to variations in the nearby environment makes gold nanostars an excellent candidate for detecting biomolecules such as creatinine.

IV. CONCLUSION

Gold nanostars were effectively synthesized through a surfactant-free wet-chemistry process. The UV-Vis spectra and FESEM showed the emergence of gold nanostars measuring 20-30 nm on average, with some reaching up to 100 nm, including spikes. The high presence of gold elements in EDX spectra showed the high purity of gold nanostars (79.8 wt%). Based on the AFM characterization, gold nanostars form a rough and pointed surface on ITO glass with an average thickness of 100-300 nm. The samples were tested for performance using LSPR, a buffer solution, and creatinine as the analyte. The change in absorbance intensity and the formation of a curve peak at a wavelength of 950-1000 nm indicate that the nanostar is promising to be further developed as a sensor for creatinine and other biomolecules using LSPR measurement.

ACKNOWLEDGMENT

This research was funded by the Malaysia Ministry of Higher Education (MOHE), Malaysia, through Fundamental Research Grant Scheme (FRGS), grant number FRGS/1/2022/STG05/UKM/02/4 and the Universiti Kebangsaan Malaysia (UKM) through the Strategic Research Fund, grant number KRA-2023-001.

REFERENCES

[1] T. O. C. Rahayu, N. L. W. Septiani, B. Yuliarto, D. R. Adhika, and I. Anshori, "Modification of gold substrate with magnetite (Fe3O4) and graphene (Gr) composite as a surface plasmon resonance (SPR) biosensor based material," *4Th Int. Conf. Mater. Metall. Eng. Technol. 2020*, vol. 2384, p. 100001, 2021, doi: 10.1063/5.0077602.

[2] C. Pothipor et al., "Transmission Surface Plasmon Resonance Imaging of Gold Grating and Silver Nanoparticle for Creatinine Detection," *Conf. Proc. ISEIM 2017*, pp. 461–463, 2017.

[3] S. Li et al., "Talanta A sensitive SPR biosensor based on hollow gold nanospheres and improved sandwich assay with PDA-Ag @ Fe 3 O 4 / rGO," *Talanta*, vol. 180, no. September 2017, pp. 156–161, 2018, doi: 10.1016/j.talanta.2017.12.051.

[4] C. R. Rekha, V. U. Nayar, and K. G. Gopchandran, "Synthesis of highly stable silver nanorods and their application as SERS substrates," *J. Sci. Adv. Mater. Devices*, vol. 3, no. 2, pp. 196–205, 2018, doi: 10.1016/j.jsamd.2018.03.003.

[5] F. Liebig et al., "A simple one-step procedure to synthesise gold nanostars in concentrated aqueous surfactant solutions," *RSC Adv.*, vol. 9, no. 41, pp. 23633–23641, 2019, doi: 10.1039/c9ra02384d.

[6] J. Lee et al., "Tailoring surface plasmons of high-density gold nanostar assemblies on metal films for surface-enhanced Raman spectroscopy," *Nanoscale*, vol. 6, no. 1, pp. 616–623, 2014, doi: 10.1039/c3nr04752k.

[7] F. Yu, H. Huang, J. Shi, A. Liang, and Z. Jiang, "A new gold nanoflower sol SERS method for trace iodine ion based on catalytic amplification," *Spectrochim. Acta Part A Mol. Biomol. Spectrosc.*, vol. 255, p. 119738, Jul. 2021, doi: 10.1016/J.SAA.2021.119738.

[8] B. Andreiuk, F. Nicolson, L. M. Clark, and S. R. Panikkanvalappil, "Nanotheranostics Design and synthesis of gold nanostars-based SERS nanotags for bioimaging applications," vol. 6, 2022, doi: 10.7150/ntno.61244.

[9] J. Ou, Z. Zhou, Z. Chen, and H. Tan, "Optical diagnostic based on functionalized gold nanoparticles," *Int. J. Mol. Sci.*, vol. 20, no. 18, 2019, doi: 10.3390/ijms20184346.

[10] B. A. Prabowo, A. Purwidyantri, and K. C. Liu, "Surface Plasmon Resonance Optical Sensor: A Review on Light Source Technology," *Biosensors*, vol. 8, no. 3, 2018, doi: 10.3390/bios8030080.

[11] S. Barbosa et al., "Tuning size and sensing properties in colloidal gold nanostars.," *Langmuir*, vol. 26, no. 18, pp. 14943–14950, Sep. 2010, doi: 10.1021/la102559e.

[12] S. Trigari, A. Rindi, G. Margheri, S. Sottini, G. Dellepiane, and E. Giorgetti, "Synthesis and modelling of gold nanostars with tunable morphology and extinction spectrum," *J. Mater. Chem.*, vol. 21, no. 18, pp. 6531–6540, 2011, doi: 10.1039/C0JM04519E.

[13] G. Chirico, M. Borzenkov, and P. Pallavicini, Gold nanostars: Synthesis, properties and biomedical application. 2015.

[14] H. Yuan, C. G. Khoury, and H. Hwang, "Gold nanostars : surfactant-free synthesis , 3D modelling , and two-photon photoluminescence imaging," vol. 075102, 2012, doi: 10.1088/0957-4484/23/7/075102.

[15] P.M. Moses, M.D. Wingrove and V.B. Christiaan. "Seedless gold nanostars with seed-like advantages for biosensing applications". *R. Soc. Open Sci* vol. 6181971, 2019, doi:10.1098/rsos.181971.

[16] T. V Tsoulos, S. Atta, M. J. Lagos, P. E. Batson, G. Tsilomelekis, and L. Fabris, "Rational Design of Gold Nanostars with Tailorable Plasmonic Properties," *ChemRivx*, no. January 2018, 2018, doi: 10.26434/chemrxiv.6552743.

[17] Q. Wu et al., "Gold nanostar-enhanced surface plasmon resonance biosensor based on carboxyl-functionalized graphene oxide," *Anal. Chim. Acta*, vol. 913, pp. 137–144, 2016, doi: 10.1016/j.aca.2016.01.063.

[18] G. H. Jeong, Y. W. Lee, M. Kim, and S. W. Han, "High-yield synthesis of multi-branched gold nanoparticles and their surface-enhanced Raman scattering properties," *J. Colloid Interface Sci.*, vol. 329, no. 1, pp. 97–102, 2009, doi: 10.1016/j.jcis.2008.10.004.

[19] K. Salinas, Z. Kereselidze, F. DeLuna, X. G. Peralta, and F. Santamaria, "Transient extracellular application of gold nanostars increases hippocampal neuronal activity," *J. Nanobiotechnology*, vol. 12, no. 1, pp. 1–7, 2014, doi: 10.1186/s12951-014-0031-y.

[20] A. Mayoral, A. Vazquez-Duran, D. Ferrer, J. M. Montejano-Carrizales, and M. Jose-Yacaman, "On the structure of stellated single crystal Au/Ag nanoparticles," *CrystEngComm*, vol. 12, no. 4, pp. 1090–1095, 2010, doi: 10.1039/b914749g.

[21] H. T. Phan, C. Vinson, and A. J. Haes, "Gold Nanostar Spatial Distribution Impacts the Surface-Enhanced Raman Scattering Detection of Uranyl on Amidoximated Polymers," *Langmuir*, vol. 37, no. 16, pp. 4891–4899, 2021, doi: 10.1021/acs.langmuir.1c00132.

[22] K. M. Byun, S. J. Yoon, and D. Kim, "Effect of surface roughness on the extinction-based localized surface plasmon resonance biosensors," *Appl. Opt.*, vol. 47, no. 31, pp. 5886–5892, 2008, doi: 10.1364/AO.47.005886.

[23] T. O. C. Rahayu, N. L. W. Septiani, G. Gumilar, D. R. Adhika, Suyatman, and B. Yuliarto, "Modification of Gold Substrate with Fe3O4-Graphene Nanocomposite to Increase Resolution of Surface Plasmon Resonance (SPR) Glucose Sensor," *IEEE Sens. J.*, vol. 21, no. 18, pp. 19959–19966, 2021, doi: 10.1109/JSEN.2021.3093107.

[24] S. Pranveer. "LSPR Biosensing: Recent Advances and Approaches." In: Geddes, C. (eds). *Reviews in Plasmonics*, vol 2016, pp. 211-238, 2016.https://doi.org/10.1007/978-3-319-48081-7_10.

Reliability Analysis of Multibridge Channel Field Effect Transistor

Huang Ganlin
Faculty of Electrical
Engineering,
Universiti Teknologi Malaysia,
81310 Johor Bahru, Malaysia
47826650@qq.com

N. Ezaila Alias*
Faculty of Electrical
Engineering,
Universiti Teknologi Malaysia,
81310 Johor Bahru, Malaysia
ezaila@fke.utm.my

M. L. Peng Tan
Faculty of Electrical
Engineering,
Universiti Teknologi Malaysia,
81310 Johor Bahru, Malaysia
michael@utm.my

M. N. Hamzah
Faculty of Electrical
Engineering,
Universiti Teknologi Malaysia,
81310 Johor Bahru, Malaysia
muhammadnaziiruddin@gradua
te.utm.my

Yasmin Abdul Wahab
Nanotechnology & Catalysis
Research Centre,
Universiti Malaya,
50603 Kuala Lumpur, Malaysia
yasminaw@um.edu.my

Hanim Hussin
School of Electrical
Engineering,
College of Engineering,
Universiti Teknologi MARA,
40450 Shah Alam, Malaysia
hanimh@uitm.edu.my

Abstract—In this paper, reliability studies of Positive Bias Temperature Instability (PBTI) characteristics in n-channel MBCFETs and Negative Bias Temperature Instability (NBTI) in p-channel MBCFETs are conducted. Similar to NBTI, PBTI is also a significant reliability issue in transistors. An analytical study of MBCFET concerning the degradation/shifting and recovery of threshold voltage (ΔV_{th}) and on-current (ΔI_{on}) by varying different device parameters such as channel length, stress voltage, and stress time was carried out before and after stress application. This project aims to provide extensive data on the degradation mechanism of PBTI and NBTI in MBCFETs. This is achieved by simulating the 12nm MBCFET's device structure and applying stress tests on the proposed device. Several sets of stress voltages ranging from -2 to -5V are applied to the gate terminal of p-MBCFETs for a stress time up to 900 seconds to observe NBTI degradation, and stress voltages ranging from +2 to +5V are applied to the gate terminal of n-MBCFETs for a stress time up to 900 seconds to observe PBTI degradation. NBTI degradation results in a notable V_{th} shift, ranging from 19.5mV to 31.6mV, attributed to a temporary trap charge, which is comparatively larger than PBTI. It exhibits a significant recovery effect over time, with a V_{th} shift due to a permanent trap charge ranging from 0.2mV to 0.6mV. In contrast, PBTI degradation induces a smaller V_{th} shift of about 4.4mV, with limited and prolonged recovery.

Keywords— NBTI, PBTI, Multi-bridge FET, reliability, transistor.

I. INTRODUCTION

The performance of a MOSFET is influenced by both the channel length and width. Decreasing the channel length may lead to short channel effects that can compromise device performance. To address this challenge, non-planar structures like FinFETs and gate-all-around nanosheet FETs have been introduced. These structures aim to improve electrostatic control over the channel, enabling more effective gate length scaling and mitigating the impact of short channel effects.

The structure of Multibridge Channel Field Effect Transistor (MBCFET) is a type of transistor that can be classified as a Gate-All-Around (GAA) technology [1]. However, it deviates from the industry standard of using nanowires and instead uses wider sheet-like structures. By using nanosheets instead of nanowires, MBCFET can provide several key advantages such as continuously adjusting the channel width by controlling the sheet width. In addition, MBCFET can provide structural variations that allow all four sides of the device to act as channels. With these innovations, MBCFET can provide lower operating voltage, higher current efficiency (i.e., drive current capability), and a high degree of design flexibility.

As the gate length shortens, there is an observed increase in the shift of the threshold voltage [2]. This phenomenon can be attributed to the reduction in the interface area between the oxide and silicon when the gate length is reduced. The diminished interface area results in a more concentrated electric field due to the shorter length, causing the electric field to be more dispersed and less concentrated. The reliability of MOSFETs can be affected by various factors such as their geometry, material properties, and operating conditions. Manufacturers conduct reliability and qualification tests to ensure that their MOSFETs meet certain standards and can perform reliably under different conditions [3-6].

Reliability in the context of multi-channel field-effect transistors refers to the ability of these devices to perform their intended functions consistently and predictably over an extended period of time under various operating conditions [7-10]. It involves assessing the durability, stability, and resistance to failure mechanism. Some of the key aspects related to the reliability of the MBCFET are Aging and Degradation, Material and Interface Reliability, Stress Testing, Failure Analysis and Statistical Analysis. Table 1 summarizes the impact of failure mode in MOSFET.

Authors would like to acknowledge the financial support under the UTM Fundamental Research Grant Project No.Q.J130000.3823.22H52. Also, thanks to the Research Management Center (RMC) of Universiti Teknologi Malaysia (UTM) for providing an excellent research environment in which to complete this work.

TABLE I. SUMMARY ON IMPACT OF FAILURE MODE IN MOSFET

Failure Mode		Impact
HCI	Hot Carrier Injection	Transistor
BTI	Bias Temperature Instability	
TDDB	Time dependent dielectric breakdown	Gate-oxide
EM	Electro migration	Gate (metal)
SM	Stress migration	
Low k TDDB	Low k time dependent dielectric breakdown	
PID	Plasma induced trap charge	Processing

In this work, to address this gap by specifically investigating PBTI and NBTI on MBCFETs with channel lengths ranging from 10 to 20nm, aiming to contribute valuable insights into the reliability of MBCFETs. Several sets of stress voltage ranging from -2 to -5V are applied on the gate terminal of p-channel device for a stress time up to 900s to observe NBTI degradation, and stress voltage ranging from +2 to +5V are applied on the gate terminal of n-channel device for a stress time up to 900s to observe PBTI degradation. The recovery time was set to 1000s to observe the post-recovery behaviour for both devices. The work concerns on the degradation/shift and recovery of threshold voltage (ΔV_{th}) and on-current (ΔI_{on})

II. DEVICE SIMULATION

A. Device structure modeling and parameter determination

To begin with, the structure of the MBCFET is developed using Silvaco TCAD tool called SDE. In this work, a three-channel nanosheet stacking is performed, while the parameters and doping type of the device are referenced from [16]. The gate material used is Titanium Nitride (TiN), and Silicon Nitride (Si$_3$N$_4$) serves as the spacers. The metal work function of the gate is 4.52 eV for NMOS and 5.01 eV for PMOS. The thickness of the Silicon Oxide (SiO$_2$) and Hafnium Oxide (HfNO$_2$) are 0.7 nm. Each nanosheet has a thickness of 5 nm (NSH$_{TH}$) and a width of 50 nm (NSH$_W$). Source and drain doping concentration is 1×10^{17} cm^{-3}. Channel doping concentration is 3×10^{20} cm^{-3}.

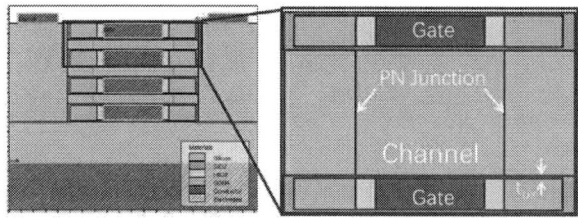

Fig. 1. Device structure of MBCFET with 3-channels. The right figure is the cross-sectional area showing 1-channel region.

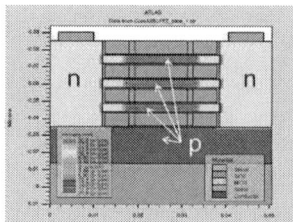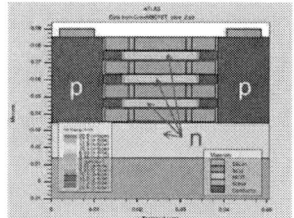

Fig. 2. Device structure of MBCFET with 3-channel (Left: N-channel, right: P-channel).

Fig. 1 and 2 illustrate the composition of the simulated structure. In the image, the left side represents the source terminal, the right side represents the drain terminal, and the middle region corresponds to the main channel area of the MBCFET. For N-Channel, the junction is NPN type. While for P-Channel, the junction is PNP type. The main channel is surrounded by gate material, which is responsible for controlling the flow of electrons within the channel. Between the gate material and the channel, there is a layer of gate oxide and a spacer region composed of Silicon Oxide (SiO$_2$) and Hafnium Oxide (HfNO$_2$), serving as insulation and isolation barriers to prevent electron leakage from the gate into the channel region.

B. Implementation of PBTI and NBTI stress test on MBCFET

The stress conditions applied to the device structure are shown in Table 2, in the exploration of Negative Bias Temperature Instability (NBTI), the stress voltage range is meticulously set between -2V and -5V. To ensure the precision and visibility of results, a nominal bias voltage of -0.1V is judiciously introduced at the drain terminal. This supplementary bias voltage not only aids in the accurate measurement of effects but also enhances the reliability of the observed outcomes, contributing to a comprehensive assessment of NBTI-induced impacts on the device. On the other hand, in the investigation of Positive Bias Temperature Instability (PBTI), the stress voltage is within the range of 2V to 5V. This carefully chosen bias voltage ensures effective observation and contributes to the precision of measurements, facilitating a thorough evaluation of the device's response to PBTI.

TABLE II. THE STRESS CONDITIONS APPLIED TO THE MBCFET DEVICE STRUCTURE

Stress	V_{DS}	V_{GS}	Temperature
NBTI	-0.1V	-2 to -5 V	100°C
PBTI	0.1V	+2 to +5 V	100°C

III. RESULT AND DISCUSSION

The Id-Vg characteristics for p-channel and n-channel of MBCFET are shows in Fig. 3. Table 3 tabulates the electrical paramters of both devices, it shows that the subthreshold swing (SS) for both devices are almost the same which are 70.02 mV/dec and 75.79 mV/dec for n-channel and p-channel, respectively. However, the on-off current ratio for n-channel device has a significant value with 1.39×10^6 as compared to the p-channel device with 1.65×10^3. From this

979-8-3503-7832-0/24 $31.00 © 2024 IEEE

electrical performance, it will affect the reliability performances which is discussed in the next subsection.

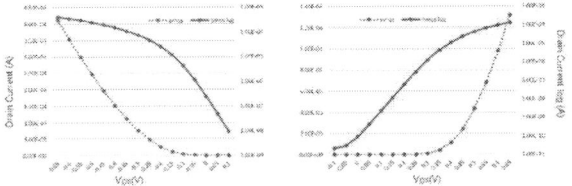

Fig. 3. I_d-V_g graph of MBCFET for p-channel (left) and n-channel (right).

TABLE III.   ELECTRICAL CHARACTERISTIC OF THE SIMULATED MBCFET

Electrical Characteristics	V_{th} (V)	SS (mV/dec)	I_{on}/I_{off}
N-channel	0.46	70.02	1.39×10^6
P-channel	-0.29	75.79	1.65×10^3

The initial performance of the measured structure was assessed with V_{ds} set at -0.65V and V_{gs} ranging from -3V to 3V, resulting in the output I_d-V_{gs} curve, as depicted by the black line (before stress application) as shown in Fig. 4. Subsequently, a stress voltage of -2V was applied to the gate terminal for a duration of 1s, under the same measurement conditions, yielding the I_d-V_{gs} curve shifted to the left represented by the red line (after stress). Following the removal of stress and a recovery period of 1000s, the I_d-V_{gs} curve was shifted to the right towards the initial state represented by the blue line (after recovery). Under this NBTI stress application, despite the removal of the stress voltage and a recovery period of 1000s, there was a persistent leftward deviation compared to the initial state, indicating incomplete restoration to the same state. This suggests that under NBTI stress, the p-channel MBCFET is not fully recovered after 1000s when the stress is removed.

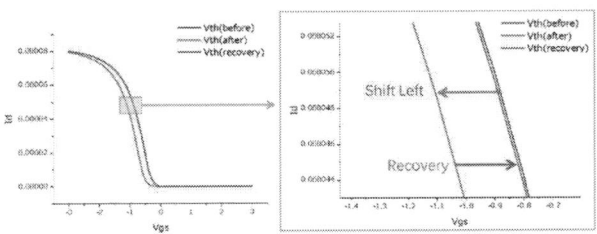

Fig. 4. I_d-V_{gs} graph of 2D single p-channel structure under NBTI stress. The right figure is the I-V shifted to the left.

Fig. 5 and 6 show the V_{th} and I_d shifts versus stress time up to 900s due to NBTI stress, respectively for different stress voltages. The stress conditions are -2V, -3V, and -5V. The figures show that the V_{th} and I_d shifted significantly for the 1s and slowly increased up to 900s. It exhibits the same trend for all stress voltage applications. At the highest V_{stress} = -5V, NBTI degradation results in a prominent V_{th} shift, ranging from 30.5mV (1s) to 31.6mV (900s). However, after 1000s of recovery, the V_{th} is fully recovered almost 100%. This is attributed to a temporary trap charge under NBTI. It should be noted that, over time, the device tends to recover a

significant portion of the initially induced threshold voltage shift, demonstrating resilience to the stress-induced effects.

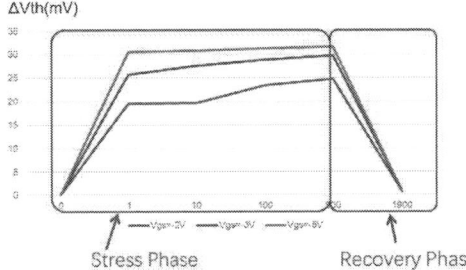

Fig. 5. ΔV_{th} under different stress voltage and stress time due to NBTI.

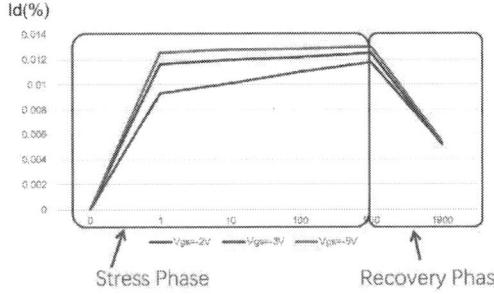

Fig. 6. ΔI_d(%) under different stress voltage and stress time due to NBTI.

The initial performance of the measured structure was assessed with V_{ds} set at 0.65V and V_{gs} ranging from -3V to 3V, resulting in the output I_d-V_g curve, as depicted by the black line (before stress) in Fig. 7. Subsequently, a stress voltage of +2V was applied to the gate terminal of the indicated structure for a duration of 1s, under the same measurement conditions, yielding the I_d-V_{gs} curve shifted to the right side represented by the red line (after stress) in Fig. 7. Following the removal of stress and a recovery period of 1000s, the I_d-V_{gs} curve shows almost not shifted, represented by the blue line (after recovery). Under PBTI stress application, there was almost no deviation compared to the initial state, indicating a significant incomplete restoration to the same state. This suggests that under PBTI stress, the n-channel MBCFET is not fully recovered after 1000s when the stress is removed. It may need a longer relaxation time to recover.

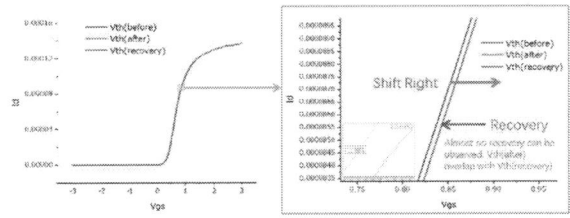

Fig. 7. I_d-V_g graph of 2D single n-channel structure under PBTI stress. The right figure is the I-V shifted to the right.

Fig. 8 and 9 show the V_{th} and I_d shifts versus stress time up to 900s due to PBTI stress, respectively for different stress voltages. The stress conditions are +2V, +3V and +5V. The figures show that the V_{th} and I_d shifted significantly for the 1s and almost no shift up to 900s. It looks like the lines overlap with each other for all stress voltage applications. At the highest V_{stress} of 5V, PBTI degradation results in an unnoticeable V_{th} shift, ranging from 4.4mV (1s) to 4.4mV (900s). Additionally, after 1000s of recovery, the V_{th} slightly recovers, as shown in the graph. This is attributed to a permanent trap charge and prolonged recovery under PBTI. Insensitive V_{th} shift during stress phase where the shift in V_{th} appears to be less sensitive to both stress voltage and stress time when compared to NBTI. Unlike NBTI, where the V_{th} shift is more pronounced with increasing stress duration, PBTI exhibits a less responsive V_{th} shift to variations in stress voltage and time during the stress phase.

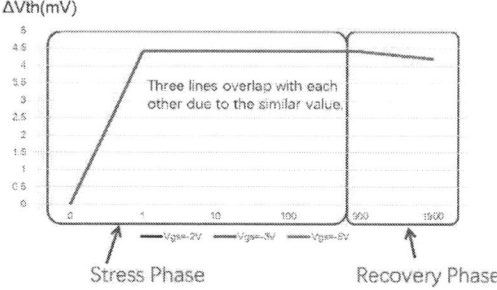

Fig. 8. ΔV_{th} under different stress voltage and stress time due to PBTI.

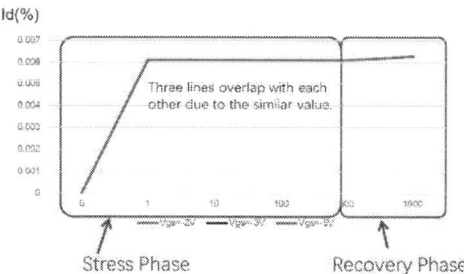

Fig. 9. $\Delta Id(\%)$ under different stress voltage and stress time due to PBTI.

In the post-recovery phase under PBTI, it shows very insignificant recovery can be observed in comparison to NBTI. Unlike NBTI, where a substantial recovery is noted over time, PBTI exhibits limited recovery post-stress. This suggests that the device's V_{th} and I_d shifts induced by PBTI are less reversible, pointing to a potentially greater long-term impact on device performance.

IV. CONCLUSION

In conclusion, the simulation of the Negative Bias Temperature Instability (NBTI) and Positive Bias Temperature Instability (PBTI) reliability issue on a simplified 2D single-channel structure was done and the effects of the reliability issues on the MBCFET in terms of its electrical characteristic degradations such as the threshold voltage shift and on current was analyzed. From the findings, it is crucial to note that the n-channel MBCFET experiences

degradation primarily due to PBTI, while the p-channel is susceptible to NBTI. In comparing the recovery properties between NBTI and PBTI, the recovery of NBTI-induced degradation was observed to be significant. In contrast, PBTI exhibits very little recovery, possibly due to the extended recovery time required for electron trap charges. The significant recovery of NBTI suggests potential possibilities to develop elastic mechanisms and optimize device architecture to enhance the overall robustness of p-channel MBCFETs. In contrast, limited recovery from PBTI requires further investigation. In summary, this work helps to understand the performance of MBCFET under different stress conditions and helps to understand the characteristics of MBCFET with respect to reliability.

ACKNOWLEDGMENT

Authors would like to acknowledge the financial support under the UTM Fundamental Research Grant Project No.Q.J130000.3823.22H52. Also, thanks to the Research Management Center (RMC) of Universiti Teknologi Malaysia (UTM) for providing an excellent research environment in which to complete this work.

REFERENCES

[1] S. Afidah Affandi, N. Ezaila Alias, Afiq Hamzah, M.L Peng Tan and Hanim Hussin "Performance Analysis of Junctionless Multi-Bridge Channel FET with Strained SiGe Application," Proceedings in 2022 IEEE International Conference on Semiconductor Electronics (ICSE), pp. 17-20, 2022.

[2] M. Naziiruddin Hamzah, N. Ezaila Alias and M. L. Peng Tan. "Negative Bias Temperature Instability Analysis of a 15 nm p-channel Junctionless Fin Field Effect Transistor (pJLFinFET)," Proceedings in 2022 IEEE 20th Student Conference on Research and Development (SCOReD), pp. 73-76, 2022.

[3] Hao Chang et al., "Degradation Mechanism of Short Channel p-FinFETs under Hot Carrier Stress and Constant Voltage Stress," Proceedings in 2020 IEEE International Symposium on the Physical and Failure Analysis of Integrated Circuits (IPFA), pp. 1-4, 2020.

[4] O. Prakash, S. Maheshwaram, S. Beniwal, N. Gupta, N. Singh, and S. Manhas, "Impact of time zero variability and BTI reliability on SiNW FET-based circuits," IEEE Transactions on Device and Materials Reliability, vol. 19, no. 4, pp. 741-750, 2019.

[5] T. Cho, R. Liang, G. Yu, and J. Xu, "Reliability analysis of P-type SOI FinFETs with multiple SiGe channels on the degradation of NBTI," Proceedings in 2020 IEEE Silicon Nanoelectronics Workshop (SNW), pp. 101-102, 2020.

[6] Miaomiao Wang et al., "Bias Temperature Instability Reliability in Stacked Gate-All-Around Nanosheet Transistor," Proceedings in 2019 IEEE International Reliability Physics Symposium (IRPS), pp. 1-4, 2019.

[7] Changze Liu et al., "New insights into 10nm FinFET BTI and its variation considering the local layout effects," Proceedings in 2017 IEEE International Reliability Physics Symposium (IRPS), pp. 1-4, 2017.

[8] W. Mizubayashi et al., "PBTI for N-type tunnel FinFETs," 2015 International Conference on IC Design & Technology (ICICDT), Leuven, 2015

[9] L. Zhou et al., "A comparative study of TiN thickness scaling impact on DC and AC NBTI kinetics in replacement metal gate pMOSFETs," Proceedings in 2019 IEEE International Integrated Reliability Workshop (IIRW), pp. 1-4, 2019.

[10] L. Zhou et al., "Physical mechanism underlying the time exponent shift in the ultra-fast NBTI of high-k/metal gated p-CMOSFETs," Proceedings in 2018 IEEE International Symposium on the Physical and Failure Analysis of Integrated Circuits (IPFA), pp. 1-4, 2018.

An investigation on the interplay between strain and defects on the thermal conductivity of monolayer graphene

Dharma Darren Ram
Institute of Microengineering and Nanoelectronics
Universiti Kebangsaan Malaysia
43600, Bangi, Selangor
P116360@siswa.ukm.edu.my

Muhammad Aniq Shazni Mohammad Haniff
Institute of Microengineering and Nanoelectronics
Universiti Kebangsaan Malaysia
43600, Bangi, Selangor
aniqshazni@ukm.edu.my

Abdul Manaf bin Hashim
Malaysia−Japan International Institute Of Technology
Universiti Teknologi Malaysia
54100, Kuala Lumpur
abdmanaf@utm.my

Mohd Ambri Mohamed
Institute of Microengineering and Nanoelectronics
Universiti Kebangsaan Malaysia
43600, Bangi, Selangor
ambri@ukm.edu.my

Abstract— Strain engineering is rapidly gaining momentum as a method to tuning the thermal properties of materials. We have performed extensive molecular dynamics studies on the thermal properties of defective single layer graphene (d-SLG) to understand its effects on its thermal conductivity. We have created a model of d-SLG, and we have then applied a varying strain amount onto the graphene model to determine if it is possible to rectify the defect-induced thermal conductivity reduction. Using a Tersoff manybody potential, we have calculated the thermal conductivity of our model. Our results show a quantitative ability of strain to rectify thermal conductivity reductions up to a point. Our simulation found that there are 2 points of interest, 5% ε_x and, 10% ε_x and ε_y. Compared to their unstrained counterparts, 5% ε_x on 0.3% d-SLG shows a 192% increase in λ_{pvr}. We also found that 10% ε_x and ε_y on 0.6% d-SLG shows a 246% and 323% increase in λ_{ovr}. We have then explained this phenomenon using the models of phonon-phonon and phonon-defect scattering. We believe this research can be applied to different 2D materials and advance our understanding of thermal transport in these low-dimensional materials in heat management applications.

Keywords— Thermal conductivity, molecular dynamics, Strain engineering, Defective graphene, Two dimensional materials, Green-Kubo method

I. INTRODUCTION

As the fields of micro- and nanotechnology grow, and the materials used in these fields require significant downsizing to accommodate the field's changing requirements. The increased heat density from this size reduction makes thermal transport a critical issue. Two-dimensional (2D) materials, whose properties vary significantly from that of their bulk counterparts, are gaining momentum in this application. Graphene, the most well-known among these materials, has been shown to be an excellent prospect in the field of nanoscale heat dissipation due to its exceedingly high thermal conductivity.

The thermal conductivity of pristine single layer graphene (p-SLG) has been shown to be between 1800 W/m.K and 5300 W/m.K, with results usually around 2500 W/m.K [1]. Compared to silver (429 W/m. K) and copper (398 W/m. K), the thermal conductivity of graphene is exceptional. Producing wafer scale p-SLG remains a challenge with current methods (i.e., CVD and epitaxial

growth) invariably introducing defects, such as vacancies, dislocations, and multiple grain boundaries.

Defect and strain engineering is a common approach taken to tailor material properties to specific applications. The introduction of defects affects the mean free path of the main heat carriers of 2D materials, significantly reducing thermal conductivity. In graphene, the compressive strain forms ripples and wrinkles in the sheet reducing thermal conductivity due, in part, to the increased phonon scattering from the corrugation effect. The impact of tensile strain remains inconsistent, with some studies suggesting an increase and some a decrease.

Given these complexities and disputed results, this study aims to investigate the combined effects of strain and defects on the thermal conductivity of suspended single layer graphene (SLG). In this study, we use equilibrium molecular dynamics (EMD) to examine the thermal transport properties of SLG with different topological defects (single and double vacancy) and the impact of various tensile strains on said properties. The results are then analysed, considering phonon states and lattice vibrations. We aim to determine if the thermal conductivity reduction caused by defects can be reliably corrected using strain engineering.

II. METHODOLOGY

The simulations are all performed on a square "sheet" of monolayer graphene measuring 10nm × 10nm as shown in Fig. 1(a). The square samples are more ideal for EMD simulations due to the identical dimensions in both chiral directions. The topographical defects studied in this paper are limited to single and double vacancy defects only illustrated by Fig. 1(b, c). The exclusion of the Stone-Wales 5-7-7-5 defects is due to hardware limitations. The locations of the defects were randomized by our simulation code, which uses a pseudo-random generator, which removes atoms at random locations while saving the configuration of the defective sheet facilitating easy replicability for multiple simulation runs. The "random" deletion effectively includes the in-sheet coordinates of the defects as well as the orientation of the defects (the effects of defect orientation are outside the scope of this paper). We believed this randomness of defect location and orientation accurately represents defects present in real-world scenarios, whether

979-8-3503-7832-0/24 $31.00 © 2024 IEEE

the monolayer is obtained by chemical vapour deposition (CVD) or mechanical exfoliation (ME).

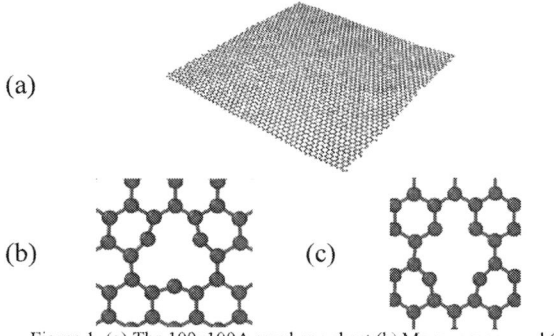

Figure 1. (a) The 100x100A graphene sheet (b) Monovacancy and (c) Divacancy defects

A. Molecular Dynamics

Calculation on the thermal conductivity is performed by classical molecular dynamics. This method calculates the position of atoms at progressing timesteps based on Newton's 2nd Law and predefined interatomic potentials. This paper uses the Green-Kubo method, which calculates the thermal conductivity, λ, by integrating the time dependent heat flux vectors [2, 3]

$$\lambda = \frac{V}{k_B T^2} \int_{-\infty}^{\infty} \{J(t) \cdot J(0)\} dt \qquad (1)$$

The in-plane thermal conductivity of single-layer graphene (SLG) is determined by averaging the thermal conductivity in the x and y directions (λxx and λyy) using periodic boundary conditions. This calculation involves the system volume (V), the Boltzmann constant (kB), and the heat flux (J(t)) in the x, y, or z direction. Periodic boundary conditions are imposed in the x and y directions. To ensure sufficient convergence this study employs a simulation time of 100 ps.

The simulations were conducted using LAMMPS [4] with input files modified from examples included in the LAMMPS package. Equilibrium molecular dynamics (EMD) simulations were performed to calculate the thermal conductivity of the samples. The equations of motion are integrated using the Verlet velocity algorithm. Initial temperatures were randomized with a starting temperature of 300 K. Data files were created using the Open Visualizer Tool (OVITO) [5] and Visual Molecular Dynamics (VMD). Intermolecular interactions were described using the Tersoff potential.

The x and y directions were treated as the in-plane directions with periodic boundary conditions, while the z direction was treated as normal with fixed boundary conditions. The timestep was set to 0.1 fs. The system was equilibrated twice: first, in the isothermal-isobaric ensemble for 1.25 ns, then in the canonical ensemble for 150 ns, both using a Nosé–Hoover thermostat. Finally, the system was run in the NVE ensemble to compute the thermal conductivity at 300 K.

B. Density Functional Theorem

DFT calculations were performed using CASTEP using a norm-conserving pseudopotential in the framework of the General Gradient Approximation (GGA) with the Perdew-Burke-Ernzerhof exchange correlation functional [7] The optimized in-plane lattice constants for graphene were calculated to be $a_{SLG} = b_{SLG} = 2.4599$ Å and $c_{SLG} = 8.3048$ Å, which is consistent with the work of Lynch et al [8]. The kinetic energy cutoff was set to 770 eV and a $7\times7\times2$ Monkhorst-Pack grid was used to optimize each sheet to ensure accurate and repeatable results. The system convergence tolerance was set to 5×10^{-7} eV/atom

III. RESULTS AND DISCUSSIONS

The graphene sheet in this study has been oriented in a way such that the zigzag and armchair chiral directions correspond to the x- and y-coordinates respectively, as shown in Fig 2.

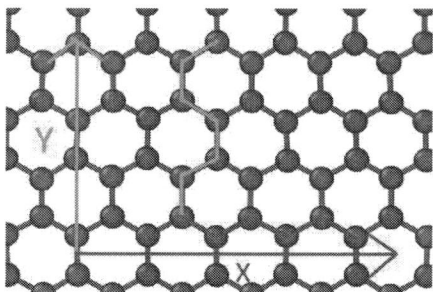

Figure 2. The directional configuration of the graphene sheet

In order to validate our methodology, we first ran the simulation on a sheet of p-SLG (200 Å × 200 Å, 16064 atoms). This was run 5 times at 300K and with different initial atomic velocities. The result was then averaged to achieve our benchmark thermal conductivity value. The size of the simulation cell was adjusted through trial and error seeking to optimize the simulation run time while maintaining accuracy. In our previous work, it was found that reducing the cell size beyond a certain point would result in artificially high thermal conductivity, which is due to the improper modelling of phonon scattering [9]. We chose a cell size of 100x100 Å² due to its reasonable simulation run time and its thermal conductivity result of 2563 W/m.K, which is in range of previously published results [1, 12]. The anisotropic properties of graphene's thermal conductivity are clearly shown in Table 1. where the calculated λ values vary significantly depending on the direction of measurement.

TABLE I.   STRAIN EFFECTS ON THE THERMAL CONDUCTIVITY, Λ (W/M.K) OF P-SLG

ε (%)	ε_x			ε_y		
	λ_{ovr}	λ_x	λ_y	λ_{ovr}	λ_x	λ_y
0	2563.45	2458.36	2668.54	2563.45	2458.36	2668.54
2.5	971.23	853.68	1088.79	438.74	402.63	474.86
5	618.01	495.32	740.7	690.1	775.57	604.63
7.5	583.83	772.88	394.78	570.45	800.74	340.15
10	265.16	269.72	260.62	591.64	633.89	549.39
12.5	224.85	180.06	269.65	448.11	454.38	441.85

979-8-3503-7832-0/24 $31.00 © 2024 IEEE

A. Defect Effects

Next, we introduced single and double vacancy defects at concentrations of 0.3% and 0.6%. These values were chosen for their large, sharp effect on the thermal conductivity of the monolayer. Three models with randomly distributed defects were simulated independently, and their thermal conductivities were averaged. As defect concentration increased, thermal conductivity decreased significantly, with a 0.3% and 0.6% defect concentration reducing the calculated thermal conductivity by 75.7% and 91.6%, respectively as shown in Table 2.

TABLE II. DEFECT EFFECTS ON THE THERMAL CONDUCTIVITY, Λ (W/M.K) OF UNSTRAINED GRAPHENE

Defect Density (%)	λ_{ovr} (W/m. K)	% reduction compared to unstrained PG
0	2563.447	0
0.3	369.0325	75.7
0.6	241.3965	91.6

To understand the mechanisms behind these extreme thermal conductivity reductions the role of phonons was studied. The phonon density of states (DOS) was calculated by performing a Fourier transform on the heat current autocorrelation function:

$$DOS(\omega) = \int_{-\infty}^{\infty} \langle v_a(t)v_a(0)\rangle e^{-2\pi i \omega t} dt \quad (2)$$

Where v_a is a contraction of the directions, ω is the frequency, t is the correlation time, and $\langle v_a(t)\, v_a(0)\rangle$ is the velocity autocorrelation function.

$$\Lambda = c_{qs}\tau_{qs} \quad (3)$$

Our DOS calculation shows that there are three obvious peaks at 50 THz, ~26 THz and ~13 THz. The 50THz. As the defect density increases from 0% to 0.6%, there is an obvious reduction in the peak's amplitude. These changes signify that the phonon lifetimes are reduced and, according to Eq. 3 showing the proportionality of phonon mean free path to the thermal conductivity, so has the thermal conductivity.

B. Strain Effects on d-SLG sheets

Strain affects the vibrational properties of materials by modifying phonon dispersion relations, thereby altering speed of sound, frequency range, scattering rates, and thermal conductivity. For strained graphene, the ZA mode linearizes along the strain axis but remains unchanged perpendicular to it, resulting in non-vanishing group velocity for the ZA modes propagating along the strain axis.

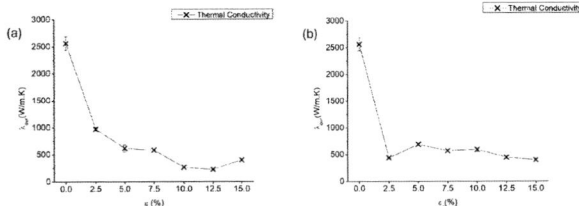

Figure 3. The effect of strain in the (a) x- and (b) y-direction on the thermal conductivity of p-SLG

In-plane acoustic modes soften slightly in both directions, while high-frequency optical phonon degeneracy at the zone center breaks.

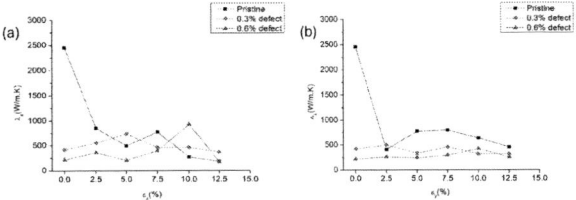

Figure 4. Effects of strain (a) ε_x (b) ε_y on the thermal conductivity of graphene in the zigzag (x-axis) compared to that of p-SLG

The thermal conductivity (λ) in the x- and y-direction computed as 2458.356 W/m. K and 2668.537 W/m. K, respectively. These values are lower than experimental data due to the higher mean free path (MFP) in graphene (775 nm) compared to the study's graphene length [10]. When the MFP of phonons exceeds the system's characteristic length, heat transport is ballistic; otherwise, it is diffusive. In p-SLG, λ decreases significantly with tensile strain. Experimental studies also show that tensile strain reduces graphene's λ [11]. Figs. 4 and 5 show λ variation with tensile strain at different strain values and directions. For strain in the x-direction (ε_x), the thermal conductivity of p-SLG reduces by 65-92% (λ_x), 59-89% (λ_y) and 60%-90%(λ_{ovr}). For strain in the Y direction (ε_y),the thermal conductivity of graphene reduces by 67-83%% (λ_x), 77-87% (λ_y) and 73-82% (λ_{ovr}). Overall, strain affects the thermal conductivity in the direction perpendicular to it more than it affects the thermal conductivity in the direction parallel to it.

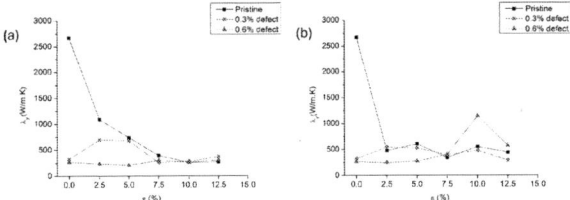

Figure 5. Effects of strain (a) ε_x (b) ε_y on the thermal conductivity of graphene in the armchair (y-axis) compared to that of p-SLG

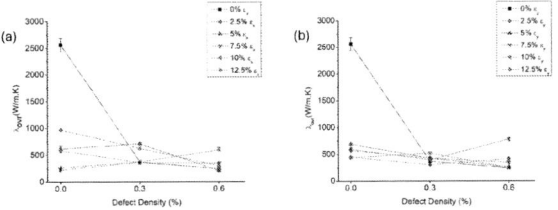

Figure 6. Effect of (a) ε_x and (b) ε_y on the thermal conductivity of d-SLG sheets

DOS analysis shows that unstrained graphene's G peak at 52.8 THz undergoes a redshift with strain, softening phonon modes and reducing λ. DOS computation along different directions reveals the impact of strain on phonon scattering, affecting λ which is influenced by phonon MFP, with the LA mode contributing the most in unstrained graphene. Under strain, low-frequency phonons' heat-carrying capacity diminishes, and ZA mode MFP increases due to higher group velocity and relaxation time according to Eq. (3). Mode-wise thermal contributions shift under strain, with in-plane acoustic modes dominating in p-SLG and ZA mode contributions increasing under strain.

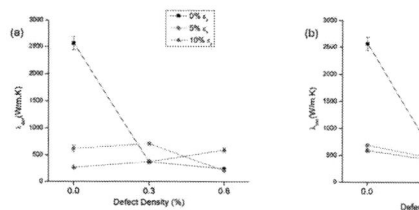

Figure 7. Notable points of interest of the effect of (a) ε_x and (b) ε_y on d-SLG sheets

Fig. 7 shows the effects of varying strain on d-SLG, which clearly demonstrates a performance increase when imposing a strain onto d-SLG.

IV. CONCLUSION

In this study, we have verified that our model of SLG accurately represents accepted thermal conductivity values, if not real-world performance. We have also demonstrated that defects severely affect the thermal conductivity of the SLG model and that the difference between single and double vacancy defects are negligible. The density of defects is the main correlating factor to this reduction. We have shown that imposing a strain on this d-SLG model increases the thermal conductivity, particularly for 10% ε_x. The complex interplay of these factors necessitates careful consideration in applications involving graphene's thermal properties. Our DOS analysis revealed that differences in the attenuation of out-of-plane low-frequency phonons were the primary drivers of variations in thermal conductivity.

In conclusion, our work showed that the thermal conductivity of d-SLG can be increased through strain engineering. More research is needed to explain the mechanisms behind this anomalous effects. We also anticipate that experimental verification of our predictions could be feasible, offering insights into the behaviour of strained graphene in real-world applications.

ACKNOWLEDGMENT

Research reported in this publication was supported by the Ministry of Higher Education under the Fundamental Research Grant Scheme (FRGS/1/2020/STG05/UKM/02/9).

REFERENCES

[1] A. A. Balandin *et al*. Superior thermal conductivity of single-layer graphene. *Nano Letters*, 8(3), 902-907, 2008.

[2] M. S. Green. Markoff random processes and the statistical mechanics of time-dependent phenomena. II. Irreversible processes in fluids. *The Journal of chemical physics*, 22(3), 398-413,1954.I. S. Jacobs and C. P. Bean, "Fine particles, thin films and exchange anisotropy," in Magnetism, vol. III, G. T. Rado and H. Suhl, Eds. New York: Academic, 1963, pp. 271–350.

[3] R. Kubo. Statistical-mechanical theory of irreversible processes. I. General theory and simple applications to magnetic and conduction problems. *Journal of the physical society of Japan, 12*(6), 570-586,1957.

[4] S. Plimpton. Fast parallel algorithms for short-range molecular dynamics. *Journal of computational physics, 117*(1), 1-19,1995.

[5] A. Stukowski. Visualization and analysis of atomistic simulation data with OVITO–the Open Visualization Tool. *Modelling and simulation in materials science and engineering, 18*(1), 015012,2009.

[6] S. J. Clark *et al*. First principles methods using CASTEP. *Zeitschrift für kristallographie-crystalline materials, 220*(5-6), 567-570,2005.

[7] J. P. Perdew, K. Burke, and M. Ernzerhof. Generalized gradient approximation made simple. Physical review letters, 77(18), 3865,1996.

[8] R. Lynch and H. Drickamer. Effect of high pressure on the lattice parameters of diamond, graphite, and hexagonal boron nitride. The Journal of Chemical Physics, 44(1), 181-184,1966.

[9] Ram, D.D., Haniff, M.A.S.M., Hashim, A.M.B. and Mohamed, M.A., 2023, August. Thermal Conductivity of Stacked Hexagonal Boron Nitride (hBN) and Graphene–A Molecular Dynamics Approach. In *2023 IEEE Regional Symposium on Micro and Nanoelectronics (RSM)* (pp. 5-8). IEEE.

[10] H. T. Nguyen-Truong, B. Da, L. Yang, Z. Ding, H. Yoshikawa, and S. Tanuma. Low-energy electron inelastic mean free path for monolayer graphene. Applied Physics Letters, 117(3), 2020.

[11] K. Nakagawa, K. Satoh, S. Murakami, K. Takei, S. Akita, and T. Arie. Controlling the thermal conductivity of multilayer graphene by strain. Scientific Reports, 11(1), 19533,2021

[12] Pop, E., Varshney, V. and Roy, A.K., 2012. Thermal properties of graphene: Fundamentals and applications. *MRS bulletin, 37*(12), pp.1273-1281.

Biomimetic Microstructure Design for Superhydrophobic

Structure in Triboelectric Nanogenerator Device

Firdaus Jamal Rashid
Institute Kejuruteraan Mikro dan Nanoelektronik
Universiti Kebangsaan Malaysia
Bangi, Malaysia
P118428@siswa.ukm.edu.my

Abang Annuar Ehsan
Institute Kejuruteraan Mikro dan Nanoelektronik
Universiti Kebangsaan Malaysia
Bangi, Malaysia
aaehsan@ukm.edu.my

Muhammad Aniq Shazni Mohammad Haniff
Institute Kejuruteraan Mikro dan Nanoelektronik
Universiti Kebangsaan Malaysia
Bangi, Malaysia
aniqshazni@ukm.edu.my

Abstract— **Triboelectric nanogenerators (TENGs) convert kinetic energy into electrical charge, with liquid-solid TENGs utilizing the movement of water droplets for energy generation. To optimize efficiency, the surface must be superhydrophobic, inspired by the natural properties of taro leaves. These leaves exhibit contact angles exceeding 150°, indicating superhydrophobicity. By mimicking the unique surface structure of taro leaves, biomimetic technology has been developed to achieve similar properties. Using Field Emission Scanning Electron Microscopy (FESEM), the specific design responsible for these characteristics was analysed. An equation derived from the Cassie-Baxter theory confirmed high contact angles, with the maximum observed angle of 168.45° achieved using a 10 μm cylindrical pillar diameter (S_{cy}) and a 30 μm distance between pillars (d_{cy}). This study demonstrates that replicating the microstructure of taro leaves can effectively create superhydrophobic surfaces.**

Keywords—Biomimetic, Superhydrophobic, Contact angle, Taro leave.

I. INTRODUCTION

Energy harvesting is a method of converting energy from the environment, such as nature, into usable electrical energy [1]. There have been many studies on energy harvesting, including solar energy technology, thermal dan Radio Frequency [2]. Among these technologies, the nanogenerator is a recent innovative discovery in the field of energy harvesting that includes three types of nanogenerators namely piezoelectric, pyroelectric and triboelectric [3].

The triboelectric nanogenerators (TENGs) is a new nanogenerator that can effectively generate electricity from the effect of motion through contact, producing an electric charge [4]. In particular, liquid-solid contact TENGs, which use the liquid itself as a triboelectric material, have gained significant attention because they can overcome the inevitable frictional wear between two solid materials in conventional TENGs. Water has also been identified as one of the triboelectric materials that can generate electricity through contact electrification [5]. Liquid-solid contact TENGs can produce stable energy output. However, solid surfaces must be hydrophobic or superhydrophobic to repel liquids after they are dropped [5]. There are several methods to produce hydrophobic or superhydrophobic properties on a surface, such as vapor deposition, plasma treatment, or monolayer coating, to create micro/nanostructures and lower the energy or adhesion force on the surface. Micromachining is one of the interesting approaches that can be used because it can produce precise geometry and small tolerances [6]. Additionally, this method can also generate a rougher surface and can contribute to increasing the superhydrophobic properties of the surface [7]. However, the structure and design need to be studied for this surface to be effectively produced.

Biomimetic technology has been identified as an approach to create a hydrophobic and superhydrophobic surface. Inspired by the biomimetic technology, surfaces in nature need to be studied to identify suitable surface structures and shapes that can induce superhydrophobic properties. Additionally, properties that affect hydrophobicity also need to be investigated. The aim of this study is to evaluate the hydrophobic and superhydrophobic properties using the Cassie-Baxter model. Furthermore, the surface microstructure was designed and analysed to mimic hydrophobic surfaces.

II. METHODOLOGY

Hydrophobic and superhydrophobic taro plants were subjected to testing using syringes filled with tap water. The procedure involved applying water droplets onto the surface and observing the resultant surface characteristics, where a superhydrophobic surface would exhibit a contact angle above 150°. Taro leaves exhibiting superhydrophobic properties were then selected for FESEM characterization to elucidate the underlying design and structural factors influencing these properties. The acquired surface knowledge was subsequently compared to a theoretical model of hydrophobicity, and the resulting equations were employed to assess these characteristics. These equations were derived from established models, including the Young, Wenzel, and Cassie-Baxter models [8, 9]. From this derivation process, the final equation, as shown in equation (1), was obtained:

$$cos\theta_{CB} = \frac{\pi d_{cy}^{2}}{4(d_{cy} + s_{cy})^{2}} cos\theta + \frac{\pi d_{cy}^{2}}{4(d_{cy} + s_{cy})^{2}} - 1.$$

where θ_{CB} represents the contact angle for the Cassie-Baxter model, and θ represents the contact angle for the Young model. Additionally, d_{cy} denotes the diameter parameter of the circular pillar, and S_{cy} indicates the separation distance between two cylindrical pillars. Finally, a design capable of mimicking the surface features of a taro leaf was developed, and the parameters were rigorously evaluated to ascertain the optimal design configuration.

III. RESULTS AND DISCUSSION

Superhydrophobic surfaces can be identified based on the contact angle of water droplets on the surface. The contact angle of water on superhydrophobic surfaces is greater than 150° where water droplets will form a sphere on the surface [10]. Figure 1(a) shows the hydrophobicity test that has been conducted on the surface of the taro plant. Based on this test, the surface of this plant was found to be superhydrophobic. This is proven by the water drop test performed on it, obtaining a contact angle of more than 150° when the water drop is on the surface. [11].

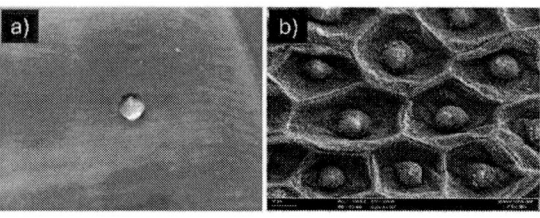

Figure 1. Water drop test on taro leafs to test zero hydrophobicity on the surface and FESEM test on the surface of taro leafs at 1000X magnification

The surface of the taro leaf is analysed using FESEM, as shown in Figure 1(b). The FESEM diagram reveals a unique structure on the surface of the taro leaf, where the formation of a homogenous circular distribution of bumps around the surface can be observed. The development of this microstructure significantly influences the roughness of the surface and, consequently, affects the superhydrophobic nature of the leaf surface [12]. This phenomenon occurs because the microstructure has reduced the interaction of water molecules on the leaf surface which results in surface tension and the formation of spheres on the water [13].

Figure 2. Schematic design diagram to imitate the surface of taro leafs.

In order to mimic the structure of taro leaves surface. The structure which is shown in figure 2 has been design to create the surface roughness of the taro leaves to mimic the superhydrophobic surface of taro leaves. The design were selected base on the tendency and possibility of the structure to be fabricated using the CNC machine method. Based on the design, the prediction of the surperhydrophobic where calculated by using the equation (1). *Polytetrafluoroethylene* (PTFE) is assumed as the surface because this material will be used to produce TENG devices. Without any structure, Polytetrafluoroethylene has a contact angle of 126°. There are

two key parameters considered in designing the surface morphology of the sample: the diameter of the circular pillar (d_{cy}) and the distance between the two cylindrical pillars (S_{cy}).

The effect of variation in design parameters was evaluated and optimized by modifying these parameters. The study focused on investigating the effect of pillar diameter (d_{cy}), as illustrated in Figure 3, which was determined by varying the pillar's distances from 10 μm, 50 μm, and 100 μm (S_{cy}). The size of the pillar was selected based on considerations of production feasibility. Various values of the pillar diameter (d_{cy}) were considered, including 10 μm, 15 μm, 20 μm, 25 μm, and 30 μm. These values of d_{cy} and S_{cy} were incorporated into Equation 1 to examine the anticipated changes in the contact angle when altering the — pillar diameter. From the graph, it can be observed that the contact angle of water exponentially increased as the pillar diameter increased from 10 μm to 30 μm. For the 10 μm S_{cy}, the contact angle increased from 156.79° to 168.45° as d_{cy} increased from 10 μm to 30 μm. In the case of the 50 μm S_{cy}, the contact angle increased from 140.82° to 150.87°, while for the 100 μm S_{cy}, it increased between 137.09° and 143.94° as d_{cy} increased from 10 μm to 30 μm.

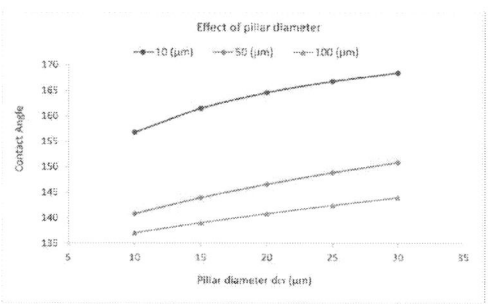

Figure 3. The effect of pillar diameter (d_{cy}) on the contact angle varies with different distances of cylindrical pillars (S_{cy}): 10 μm, 50 μm, and 100 μm.

To achieve superhydrophobic properties, the contact angle must exceed 150°. From the results, it can be observed that a circular pillar separation distance (S_{cy}) of 100 μm resulted in a contact angle below 150°. Therefore, this parameter was not suitable for achieving superhydrophobicity. However, for separation distances (S_{cy}) of 10 μm and 50 μm, it can be observed that some of the parameters can achieve superhydrophobicity. In the case of a 50 μm separation distance (S_{cy}), it is observed that only a 30 μm diameter achieved a contact angle surpassing the targeted threshold of 150.87°. Conversely, for a separation distance of 10 μm, all pillar diameters (d_{cy}) surpassed the superhydrophobic contact angle of 150°.

In addition to the pillar diameter, the effect of the separation distance between two cylindrical pillars (S_{cy}) on the contact angle was also studied. Figure 4 illustrates the impact of the distance between two pillars, ranging from 10 μm to 200 μm, while varying the size of the pillars. The size of the pillars was set at 10 μm, 15 μm and 20 μm, as this average diameter was anticipated to yield a higher contact angle. From the graph, it can be observed that all pillar diameter sizes (d_{cy}) exhibit the same pattern, wherein the

979-8-3503-7832-0/24 $31.00 © 2024 IEEE   42

contact angle decreases as the separation distance between two cylindrical pillars (S_{cy}) increases. For pillar separation distance (S_{cy}) above 50 µm, none of the parameters achieved superhydrophobicity, with contact angles falling below 150°. However, at a separation distance of 20 µm, it is evident that both 15 µm and 20 µm diameters (d_{cy}) attained superhydrophobicity, measuring at 153.41° and 156.78°, respectively. Conversely, the combination of a 10 µm pillar diameter with a 20 µm separation distance did not achieve superhydrophobicity, reaching 148.88°. For a separation distance of 10 µm, it can be observ ed that all parameters attained superhydrophobicity, with 10 µm, 15 µm, and 20 µm having contact angles of 156.79°, 151.48°, and 164.59°, respectively.

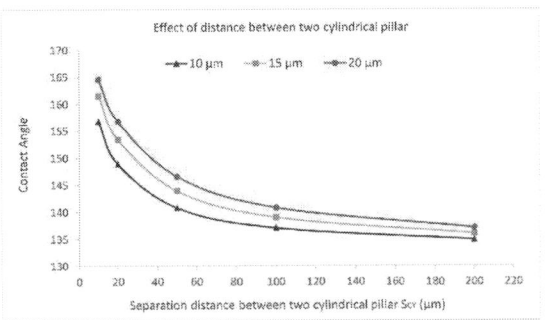

Figure 4. The effect of different distances of cylindrical pillars (S_{cy}) on the contact angle varies with circular pillar diameter (d_{cy}): 10 µm, 15 µm, and 20 µm.

Both graphs demonstrate that the pillar diameter and distances between pillars play crucial roles in influencing the water contact angle. A very high separation distance and a very low circular pillar diameter would result in a reduction of the contact angle, as the droplet would swiftly move onto the surface, be drawn into the intricate microstructure, and quickly spread, ultimately covering the entire surface. Overall, the results indicate that the highest contact angle was achieved with a 30 µm distance between cylindrical pillars (d_{cy}) and a 10 µm cylindrical pillar diameter (S_{cy}), resulting in a contact angle of 168.45°. This illustrates that the biomimetic surface design can effectively confer superhydrophobic properties, achieving a contact angle exceeding 150°. Therefore, this structure modeling design on the PTFE surface is suitable for application in TENG devices to enhance energy harvesting efficiency. The superhydrophobic properties would help to effectively repel liquid after it falls, creating better mechanical motion for contact electrification [14].

IV. CONCLUSION

In conclusion, biomimetic technology inspired by the environment can be developed. The superhydrophobic phenomenon is evident on the surfaces of plants such as taro leaves, as confirmed by the water drop test, which yields a contact angle of more than 150° and the formation of a rolling sphere on its surface. FESEM characterization reveals that the surface of taro leaves has a unique design structure, rendering it superhydrophobic. The structural theory of this design is studied based on the Cassie-Baxter model, which provides the equation for the interfacial energy between the surface of taro leaves and water droplets. The equation demonstrates the feasibility of obtaining superhydrophobic properties based on the designed structure. The highest contact angle obtained was 168.45° by using the parameters of a 30 µm distance between cylindrical pillars (S_{cy}) and a 10 µm cylindrical pillar diameter (d_{cy}). Therefore, changing the structure of the surface design by mimicking the microstructure of taro leaves can produce a superhydrophobic surface, which could be effectively applied to PTFE to enhance the surfaces of TENG devices and increase energy harvesting performance.

ACKNOWLEDGMENT

This work was supported by Research University Fund (GUP-2023-013). Firdaus Jamal Rashid would like to thanks to Universiti Kebangsaan Malaysia for fundings this research.

REFERENCES

[1] Y. Zhou, W. Deng, J. Xu, and J. Chen, "Engineering materials at the nanoscale for triboelectric nanogenerators," *Cell Reports Physical Science*, vol. 1, no. 8, 2020.

[2] J. W. Matiko, N. J. Grabham, S. P. Beeby, and M. J. Tudor, "Review of the application of energy harvesting in buildings," *Measurement Science and Technology*, vol. 25, no. 1, p. 012002, 2013.

[3] A. A. Mathew, A. Chandrasekhar, and S. Vivekanandan, "A review on real-time implantable and wearable health monitoring sensors based on triboelectric nanogenerator approach," *Nano Energy*, vol. 80, p. 105566, 2021.

[4] Q.-T. Nguyen and K.-K. K. Ahn, "Fluid-based triboelectric nanogenerators: A review of current status and applications," *International Journal of Precision Engineering and Manufacturing-Green Technology*, vol. 8, pp. 1043-1060, 2021.

[5] J. Chung, D. Heo, B. Kim, and S. Lee, "Superhydrophobic water-solid contact triboelectric generator by simple spray-on fabrication method," *Micromachines*, vol. 9, no. 11, p. 593, 2018.

[6] B. Z. Balázs, N. Geier, M. Takács, and J. P. Davim, "A review on micro-milling: recent advances and future trends," *The International Journal of Advanced Manufacturing Technology*, vol. 112, pp. 655-684, 2021.

[7] H. Yu, X. Zhang, Y. Wan, J. Xu, Z. Yu, and Y. Li, "Superhydrophobic surface prepared by micromilling and grinding on aluminium alloy," *Surface Engineering*, vol. 32, no. 2, pp. 108-113, 2016.

[8] H. Y. Erbil and C. E. Cansoy, "Range of applicability of the Wenzel and Cassie– Baxter equations for superhydrophobic surfaces," *Langmuir*, vol. 25, no. 24, pp. 14135-14145, 2009.

[9] S. Parvate, P. Dixit, and S. Chattopadhyay, "Superhydrophobic surfaces: insights from theory and experiment," *The Journal of Physical Chemistry B*, vol. 124, no. 8, pp. 1323-1360, 2020.

[10] B. Zhang and W. Xu, "Superhydrophobic, superamphiphobic and SLIPS materials as anti-corrosion and anti-biofouling barriers," *New Journal of Chemistry*, vol. 45, no. 34, pp. 15170-15179, 2021.

[11] S. Li, J. Huang, Z. Chen, G. Chen, and Y. Lai, "A review on special wettability textiles: theoretical models, fabrication technologies and multifunctional applications," *Journal of Materials Chemistry A*, vol. 5, no. 1, pp. 31-55, 2017.

[12] Y. Chen *et al.*, "Biomimetic taro leaf-like films decorated on wood surfaces using soft lithography for superparamagnetic and superhydrophobic performance," *Journal of Materials Science*, vol. 52, pp. 7428-7438, 2017.

[13] K. Negara, I. Wardana, D. Widhiyanuriyawan, and N. Hamidi, "The role of the slope on taro leaf surface to produce electrical energy," in *IOP Conference Series: Materials Science and Engineering*, 2019, vol. 494: IOP Publishing, p. 012084.

[14] D. Choi, D. Yoo, K. J. Cha, M. La, and D. S. Kim, "Spontaneous occurrence of liquid-solid contact electrification in nature: Toward a robust triboelectric nanogenerator inspired by the natural lotus leaf," *Nano Energy*, vol. 36, pp. 250-259, 2017.

Unveiling the photovoltaic properties of ETL-free, lead-free, and graphene-based PSC

Nabilah A. Jalaludin
Faculty of Electronics & Computer Technology and Engineering (FTKEK), Universiti Teknikal Malaysia Melaka (UTeM),
Melaka, Malaysia
p022220006@student.utem.edu.my

F. Salehuddin
Faculty of Electronics & Computer Technology and Engineering (FTKEK), Universiti Teknikal Malaysia Melaka (UTeM),
Melaka, Malaysia
fauziyah@utem.edu.my

F. Arith
Faculty of Electronics & Computer Technology and Engineering (FTKEK), Universiti Teknikal Malaysia Melaka (UTeM),
Melaka, Malaysia
faiz.arith@utem.edu.my

A. S. Mohd Zain
Faculty of Electronics & Computer Technology and Engineering (FTKEK), Universiti Teknikal Malaysia Melaka (UTeM),
Melaka, Malaysia
anissuhaila@utem.edu.my

K. E. Kaharudin
Faculty of Engineering and Built Environment, Lincoln University College (Main Campus),
Petaling Jaya, Selangor, Malaysia
khairilezwan@yahoo.com.my

S. A. Mat Junos
Faculty of Electronics & Computer Technology and Engineering (FTKEK), Universiti Teknikal Malaysia Melaka (UTeM),
Melaka, Malaysia
aisah@utem.edu.my

I. Ahmad
College of Engineering (CoE), Universiti Tenaga Nasional (UNITEN),
Kajang, Selangor, Malaysia
aibrahim@uniten.edu.

Abstract—The entire photovoltaic field has recently experienced significant achievement in perovskite solar cell (PSC) research with remarkable progress in device performance and power conversion efficiency (PCE). High-performance perovskite solar cells typically utilize n-i-p and p-i-n device structures, with separate electron transport layer (ETL) and hole transport layer (HTL), which are considered as necessary components for effective photogenerated carrier extraction. However, conventional PSCs are sensitive to heat, humidity, and light leading to prolonged instability. Herein, an ETL-free approach was investigated, combining lead-free methylammonium tin triiodide (MASnI$_3$) as the absorber with graphene oxide (GO) as the HTL. The study yielded notable results with a PCE of 17.83%. The insights revealed in this research have the potential to be applied in the development of cost-effective ETL-free PSCs with enhanced performance.

Keywords—ETL-free, pb-free, graphene-oxide, PSC

I. INTRODUCTION

The field of photovoltaics has witnessed substantial breakthroughs in perovskite solar cell (PSC) research, marked by notable advancements in device performance and power conversion efficiency (PCE). Various device structures have been developed, with optimal PCEs achieved through modifications in material combinations and deposition sequences. High-performance PSCs typically employ n-i-p and p-i-n structures, featuring separate electron transport layers (ETLs) and hole transport layers (HTLs), which are crucial for efficient photogenerated carrier extraction [1]–[3]. However, PSCs are susceptible to degradation when exposed to light, humidity, and heat, which can result in long-term instability issues. Furthermore, traditional PSC architectures often require high-temperature processing when using inorganic ETLs. In response, ETL-free PSCs have been developed, enabling low-cost and large-area applications while also facilitating low-temperature processes that can mitigate instability.

ETL-free PSCs are perovskite solar cells with the absence of ETL. ETL-free PSCs are highly advantageous for low-temperature processes because they decrease the lack of stability induced by the ETL [4]. Metal oxides such as ZnO and TiO$_2$ are commonly used n-type semiconductor materials or ETLs in PSCs to achieve high efficiency [5], [6]. Both are accessible, are basic in composition, have high stability, and are inexpensive [7]. However, the deposition of metal oxide films often necessitates a high-temperature sintering procedure resulting in prolonged instability [8]. Given these factors, removing the ETL layer is considered a favourable approach to enhance the photovoltaic performance and minimize the instability of PSCs. In addition, a study by D. Liu et al. [9] found an increase in device stability by eliminating the ETL layer.

In another study, L. Huang et al. [10] successfully designed and simulated ETL-free devices using spiro-MeOTAD and MAPbI$_3$-xClx as the HTL and the absorber layer. Their study found that the interface at the FTO/absorber has a more significant influence on PCE than the absorber/HTL interface, implying that more focus should be devoted to the interface at the FTO/absorber to improve the efficiency of ETL-free PSC further. They obtained an efficiency of 21%.

In addition, T. Wang et al. [11] have also simulated the ETL-free and ETL-based structures of PSC in 2022. Their work reported that ETL-free efficiency might surpass the ETL-based in a thick absorber layer. The highest efficiency obtained for ETL-free is 26.64% during the doping optimization. L. Hao et al. [12] designed and simulated an ETL-free and lead-free device structure solar cells using SCAPS simulation software. Their work found that an increase in detrimental recombination will occur if the absorber layer is too thick, and the defect density is the primary factor impacting device performance. They also stated that the interface defect layer (IDL) is a crucial constituent influencing device performance. The result shows that the highest PCE achieved is 15.41%.

979-8-3503-7832-0/24 $31.00 © 2024 IEEE

II. DEVICE SIMULATION

SCAPS-1D is a solar cell modelling or simulation program created at Gent University. This software enables the simulation of the behaviour and performance of multi-layer solar cell structures.

The device structure depicted in Fig. 1 consists of the following layers: Fluorine-doped Tin Oxide (FTO), MASnI₃ as absorber layer, GO as HTL and Gold (Au). The performance of PSC is assessed by analysing the power conversion efficiency (PCE) under diverse conditions, such as varying layer thicknesses and doping concentrations. The numerical analysis was conducted using the SCAPS-1D software under AM1.5G illumination conditions. Previous studies have provided an overview of the SCAPS input parameters and fundamental properties for each material [13], [14] which are presented in Table I. The metal work functions of the FTO and Au are specified as 4.4 eV and 5.1 eV respectively [15].

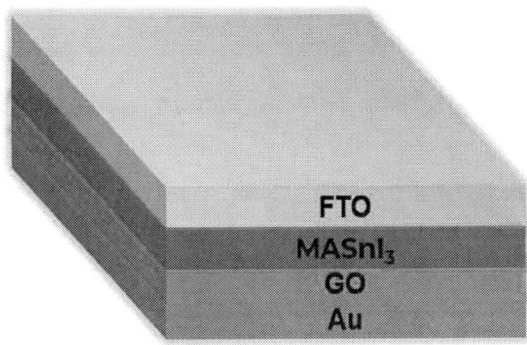

Fig. 1.   The structure of PSC

TABLE I.   INPUT PARAMATERS OF THE ETL-FREE GRAPHENE OXIDE-BASED PSC

Parameters	FTO	MASnI₃	GO
d (nm)	500	800	100
E_g (eV)	3.5	1.41	2.48
χ, (eV)	4	4.17	2.3
εr	9	8.2	10
N_C (cm⁻³)	2.2×10^{18}	1.0×10^{18}	2.2×10^{18}
N_V (cm⁻³)	1.8×10^{19}	1.0×10^{18}	1.8×10^{19}
V_e (cm/s)	1×10^{7}	1×10^{7}	5.2×10^{7}
V_h (cm/s)	1×10^{7}	1×10^{7}	5.0×10^{7}
μ_e (cm²/Vs)	20	1.6	26
μ_p (cm²/Vs)	10	1.6	123
N_D (cm⁻³)	2×10^{19}	NA	NA
N_A (cm⁻³)	NA	3×10^{15}	2×10^{18}

III. RESULT AND DISCUSSION

A. Analysis of absorber layer thickness

The perovskite absorber layer is a critical component of PSCs due to its exceptional light absorption capabilities, directly influencing the efficiency of PSCs. Lead-based perovskites are commonly utilized in PSCs as absorber layers, renowned for their wide bandgap, high carrier mobility, high absorption coefficient, and low exciton binding energy [16]. The potential for commercializing PSCs is largely driven by the integration of perovskite materials, attributed to their cost-effectiveness and versatile deposition methods, facilitating the production of high-quality thin films during manufacturing. To achieve high PCE in PSCs, the perovskite layer is systematically optimized by considering two key factors: thickness and doping acceptor density.

The thickness of the absorber layer is a crucial parameter that plays a pivotal role in determining the PSC performance. Initially, the absorber layer thickness was changed between 100 nm and 1000 nm to observe the efficiency trend of PSCs. Subsequent analysis aimed to determine the maximum solar cell efficiency achievable with an optimal absorber layer thickness. Fig. 2 demonstrates a large rise in both open-circuit voltage (V_{OC}) and short-circuit current density (J_{SC}) with increased perovskite layer thickness. Decreasing the perovskite layer thickness typically reduces light absorption and exciton formation, impacting J_{SC}. Excessive thickness can lead to charge carrier recombination before reaching the surface [17].

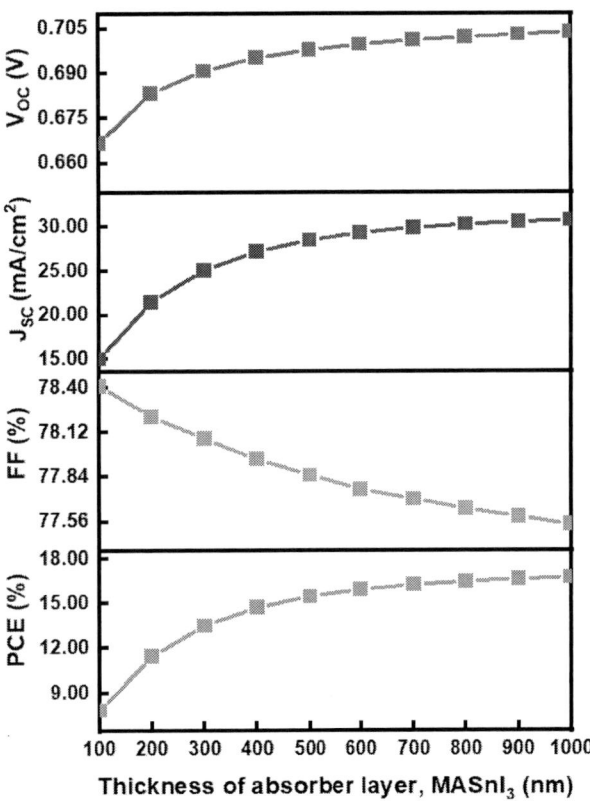

Fig. 2.   The variation of absorber thickness on PV parameters of simulated PSC

The thicker perovskite layer resulted in higher JSC, attributed to improved absorption of longer wavelength photons within the thicker layer. Fig. 2 also shows a decrease in fill factor (FF) from 78.41% to 77.54% as the perovskite layer thickness increased. The findings indicate that perovskite layer thickness significantly affects PSC efficiency. PCE increased from 7.87% to 16.74% with a perovskite layer thickness extension to 700 nm, with

979-8-3503-7832-0/24 $31.00 © 2024 IEEE   45

marginal improvement beyond 800 nm. Increased absorber layer thickness raises the risk of charge recombination due to longer diffusion paths for charge carriers [18]. Insufficient light absorption in thinner absorber layers may reduce electron-hole pair production, leading to decreased PCE [18]. Enhanced PCE results from charge carriers' diffusion length exceeding the absorber thickness, allowing most carriers to reach the electrode. Excessive thickness, however, can increase recombination by creating longer paths for charge carriers, hindering performance [12].

B. Analysis of absorber layer doping acceptor density

The performance of PSC is influenced by various factors, including the thickness of the perovskite layer and the doping acceptor concentration within the layer. In this study, the doping acceptor density was varied from 10^{15} cm^{-3} to 10^{20} cm^{-3}. Fig. 3 shows that as the acceptor density increases, the V_{OC} exhibits an upward trend and begin to decrease at 10^{19} cm^{-3}, while the J_{SC} decreases. The reduced conductivity of the PSC is due to the existence of a smaller depletion area, which results in less carrier collection [19], [20].

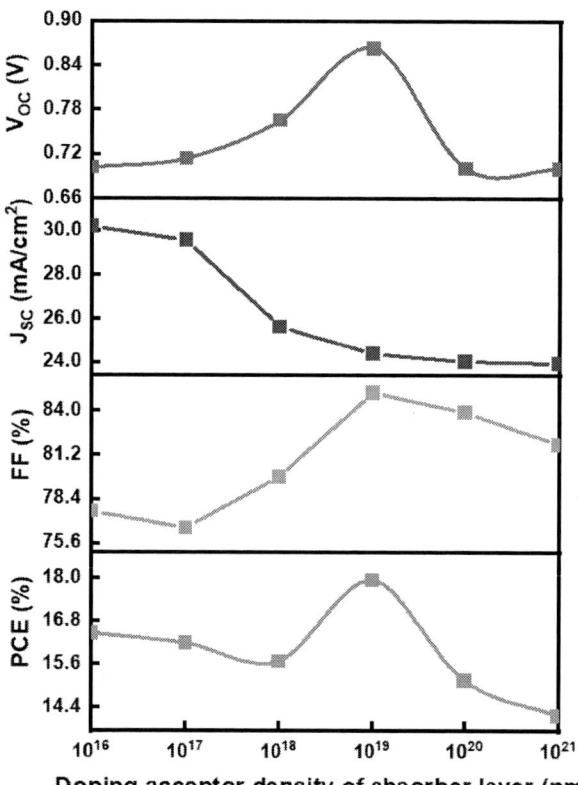

Fig. 3. The variation of absorber doping acceptor density on PV parameters of simulated PSC

The graph also demonstrates that the V_{OC} is proportional to the doping density, as the saturation current increases with the doping density. Subsequently, the fill factor and PCE follow a closely related trajectory, with the PV performance improving as the acceptor density rose to 10^{18} cm^{-3}, followed by a sharp fall at higher doping concentrations. This is due to the fact that as the acceptor density rises, the resistance lowers, resulting in an increase

in charge mobility, hence improving the performance of the PSC. [19].

C. Optimized device performance

The optimal values for the thickness and doping density of the device layers were obtained after modeling the solar cell structure with the SCAPS-1D software. The optimal thickness of the absorber layer was found to be 800 nm. Additionally, the optimal doping concentration was determined to be 3×10^{19} cm^{-3}. Furthermore, the findings from this study were compared to the simulation results presented by L. Hao et al., [12] from the J-V graph shown in Fig. 4. The comparison reveals that this work has achieved an improved power conversion efficiency of approximately 31.48% compared to the previous study.

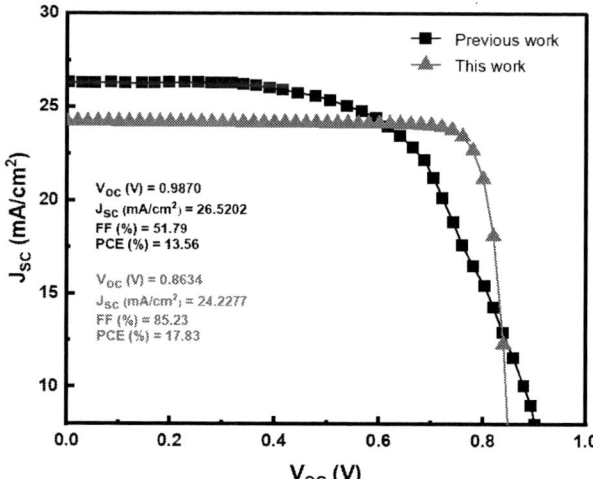

Fig. 4. J-V graph of ETL-free PSC device for previous work and this work

IV. CONCLUSION

In this study, an enhanced cell efficiency for an ETL-free perovskite solar cell (PSC) with an FTO/MASnI$_3$/GO/Au device structure was simulated. The absorber layer was thoroughly investigated and optimized in terms of thickness and doping density. The results indicate that both parameters significantly influence the performance of the PSC. The increase in power conversion efficiency (PCE) of the PSCs is attributed to the charge carriers' diffusion length being larger than the absorber layer's thickness, allowing most charge carriers to reach the electrode, contributing to the improved performance. Further analysis of the doping density in the absorber layer revealed that the photovoltaic (PV) performance enhanced as the acceptor density increased to 10^{18} cm^{-3}. However, a dramatic decline in performance was observed at higher doping concentrations due to the fact that excessive acceptor concentration leads to an increase in recombination rates, which deteriorates the overall performance of the PSC.

After conducting optimizations, it was found that utilizing GO as the HTL resulted in the attainment of optimal performance. The achieved values were a V_{OC} of 0.8634 V, a J_{SC} of 24.2277 mA/cm^2, a fill factor of 85.23%, and a PCE of 17.83%. Besides, this work enhanced efficiency by 31.48% over the earlier study by L. Hao et al., [12]. The findings demonstrated in this study can potentially be used in

979-8-3503-7832-0/24 $31.00 © 2024 IEEE

developing more affordable PSCs with improved performance.

ACKNOWLEDGMENT

The authors express their gratitude to the Ministry of Higher Education (MOHE) for providing funding for this work under project (FRGS/1/2022/TK07/UTEM/02/47). The authors also acknowledge the support and encouragement received from MiNE, CeTRI, and the Faculty of Electronics & Computer Technology and Engineering (FTKEK) at Universiti Teknikal Malaysia Melaka (UTeM) throughout the project. Additionally, the authors would like to thank Dr. Marc Burgelman of the University of Gent in Belgium for providing the SCAPS 1D simulation program used in this research.

REFERENCES

[1] K. E. Kaharudin, N. Ahmad Jalaludin, F. Salehuddin, F. Arith, A. S. M. Zain, I. Ahmad, S. A. M. Junos, and A. M. Abdul Hamid "Optimal modeling of perovskite solar cell with graphene oxide as hole transport layer using L32 (28) Taguchi design," *Int. J. Nanoelectron. Mater.*, vol. 17, no. 1, pp. 20–27, 2024.

[2] N. A. Johari, N. Ahmad Jalaludin, F. Salehuddin, F. Arith, K. E. Kaharudin, A. S. Mohd Zain, S. A. Mat Junos, and I. Ahmad, "Analysis of Inverted Planar Perovskite Solar Cells with Graphene Oxide as HTL using L9 OA Taguchi Method," *J. Adv. Res. Micro Nano Eng.*, vol. 16, no. 1, pp. 48–60, 2024.

[3] N. S. Noorasid, F. Arith, O. V. Aliyaselvam, F. Salehuddin, A. N. Mustafa, P. Chelvanathan, M. A. Azam, and N. Amin, "Low-temperature sol-gel synthesized TiO2 with different titanium tetraisopropoxide (TTIP) molarity for flexible emerging solar cell," *J. Sol-Gel Sci. Technol.*, vol. 109, no. 3, pp. 826–834, 2024.

[4] C. W. Jang, D. H. Shin, and S. H. Choi, "Photostable electron-transport-layer-free flexible graphene quantum dots/perovskite solar cells by employing bathocuproine interlayer," *J. Alloys Compd.*, vol. 886, p. 161355, 2021.

[5] J. Han, H. Kwon, E. Kim, D. W. Kim, H. J. Son, and D. H. Kim, "Interfacial engineering of a ZnO electron transporting layer using self-Assembled monolayers for high performance and stable perovskite solar cells," *J. Mater. Chem. A*, vol. 8, no. 4, pp. 2105–2113, 2020.

[6] M. Ebrahimi, A. Kermanpur, M. Atapour, S. Adhami, R. Haji Heidari, E. Khorshidi, N. Irannejad, and B. Rezaie, "Performance enhancement of mesoscopic perovskite solar cells with GQDs-doped TiO2 electron transport layer," *Sol. Energy Mater. Sol. Cells*, vol. 208, no. November 2019, 2020.

[7] R. L. Z. Hoye, K. P. Musselman, and J. L. Macmanus-Driscoll, "Research update: Doping ZnO and TiO2 for solar cells," *APL Mater.*, vol. 1, no. 6, 2013.

[8] R. Sandoval-Torrientes, J. Pascual, I. García-Benito, S. Collavini, I. Kosta, R. Tena-Zaera, N. Martín, and J. L. Delgado, "Modified Fullerenes for Efficient Electron Transport Layer-Free Perovskite/Fullerene Blend-Based Solar Cells," *ChemSusChem*, vol. 10, no. 9, pp. 2023–2029, 2017.

[9] D. Liu, J. Yang, and T. L. Kelly, "Compact layer free perovskite solar cells with 13.5% efficiency," *J. Am. Chem. Soc.*, vol. 136, no. 49, pp. 17116–17122, 2014.

[10] L. Huang, X. Sun, C. Li, R. Xu, J. Xu, Y. Du, Y. Wu, J. Ni, H. Cai, J. Li, Z. Hu , and J. Zhang, "Electron transport layer-free planar perovskite solar cells: Further performance enhancement perspective from device simulation," *Sol. Energy Mater. Sol. Cells*, vol. 157, pp. 1038–1047, 2016.

[11] T. Wang, G. J. Xiao, R. Sun, L. B. Luo, and M. X. Yi, "High efficiency ETM-free perovskite cell composed of CuSCN and increasing gradient CH3NH3PbI3," *Chinese Phys. B*, vol. 31, no. 1, pp. 0–12, 2022.

[12] L. Hao, M. Zhou, Y. Song, X. Ma, J. Wua, Q. Zhu, Z. Fu, Y. Liu, G. Hou, and T. Li, "Tin-based perovskite solar cells: Further improve the performance of the electron transport layer-free structure by device simulation," *Sol. Energy*, vol. 230, no. 2103, pp. 345–354, 2021.

[13] F. Jahantigh and M. J. Safikhani, "The effect of HTM on the performance of solid-state dye-sanitized solar cells (SDSSCs): a SCAPS-1D simulation study," *Appl. Phys. A Mater. Sci. Process.*, vol. 125, no. 4, pp. 1–7, 2019.

[14] E. Widianto, Shobih, E. S. Rosa, K. Triyana, N. M. Nursam, and I. Santoso, "Performance analysis of carbon-based perovskite solar cells by graphene oxide as hole transport layer: Experimental and numerical simulation," *Opt. Mater. (Amst).*, vol. 121, no. September, p. 111584, 2021.

[15] N. D. Lam, "Modelling and numerical analysis of ZnO/CuO/Cu2O heterojunction solar cell using SCAPS," *Eng. Res. Express*, vol. 2, no. 2, 2020.

[16] P. Mahajan, R. Datt, W. Chung Tsoi, V. Gupta, A. Tomar, and S. Arya, "Recent progress, fabrication challenges and stability issues of lead-free tin-based perovskite thin films in the field of photovoltaics," *Coord. Chem. Rev.*, vol. 429, p. 213633, 2021.

[17] I. M. D. L. Santos, H. J. Cortina-Marrero, M.A. Ruíz-Sánchez, L. Hechavarria-Difur, F.J. Sánchez-Rodríguez, M. Courel, and H. Hu "Optimization of CH3NH3PbI3 perovskite solar cells : A theoretical and experimental study," *Sol. Energy*, vol. 199, no. January, pp. 198–205, 2020.

[18] F. Anwar, R. Mahbub, S. S. Satter, and S. M. Ullah, "Effect of Different HTM Layers and Electrical Parameters on ZnO Nanorod-Based Lead-Free Perovskite Solar Cell for High-Efficiency Performance," *Int. J. Photoenergy*, vol. 2017, 2017.

[19] S. Srivastava, A. K. Singh, P. Kumar, and B. Pradhan, "Comparative performance analysis of lead-free perovskites solar cells by numerical simulation," *J. Appl. Phys.*, vol. 131, no. 17, 2022.

[20] N. Syamimi, N. Faiz, A. Ahmad, N. Mustafa, and P. Chelvanathan, "Improved performance of lead - free Perovskite solar cell incorporated with TiO 2 ETL and CuI HTL using SCAPs," *Appl. Phys. A*, pp. 1–16, 2023.

Kretschmann-based Surface Plasmon Resonance sensor using Au/Graphene for Glucose Detection

Nur Akmar Jamil
Institute of Microengineering and Nanoelectronics
Universiti Kebangsaan Malaysia
Selangor,Malaysia
nurakmarjamil83 @gmail.com

Abdul Halim Abdul Gafor
Hospital Canselor Tuanku Muhriz
Universiti Kebangsaan Malaysia
Selangor,Malaysia
halim@ukm.edu.my

Noraidatulakma Abdullah
UKM Medical Molecular Biology Institute
Universiti Kebangsaan Malaysia
Selangor,Malaysia
noraidatulakma.abdullah @ppukm.ukm.edu.my

Lee Pey Yee
UKM Medical Molecular Biology Institute
Universiti Kebangsaan Malaysia
Selangor,Malaysia
leepeyyee @ ukm.edu.my

Syara Kassim
Faculty of Science and Marine Enviroment
Universiti Malaysia Terengganu
Terengganu, Malaysia
syara.kassim@umt.edu. my

Siow Kim Shyong
Institute of Microengineering and Nanoelectronics
Universiti Kebangsaan Malaysia
Selangor,Malaysia
kimsiow@ukm.edu.my.

Azrul Azlan Hamzah
Institute of Microengineering and Nanoelectronics
Universiti Kebangsaan Malaysia
Selangor,Malaysia
azlanhamzah@ukm.edu. my

Mohd Ambri Mohamed
Institute of Microengineering and Nanoelectronics
Universiti Kebangsaan Malaysia
Selangor,Malaysia
ambri@ukm.edu.my

Muhammad Feidhul Hakim Fatah Yasin
Institute of Microengineering and Nanoelectronics
Universiti Kebangsaan Malaysia
Selangor,Malaysia
mfhakim26@gmail.com

P. Susthitha Menon
Institute of Microengineering and Nanoelectronics
Universiti Kebangsaan Malaysia
Selangor,Malaysia
susi@ukm.edu.my

Abstract— Diabetes is a chronic condition characterized by high blood sugar (glucose) levels. Early and accurate detection of diabetes is crucial to prevent serious complications. Surface Plasmon Resonance (SPR) sensors offer a label-free, real-time approach to glucose detection. A hybrid layer Kretschmann-based SPR (K-SPR) sensor for detecting glucose is presented using gold and graphene thin films (Au-Gr) over a BK7 prism at visible-light angular interrogation. Our findings demonstrate an average sensitivity of 5.0 °/M and 3.9 °/M for detecting glucose levels ranging from 4 mM to 20 mM at 670 nm and 785 nm wavelengths respectively. This study highlights the need for an optical biosensing method that is convenient and patient-friendly for diabetes management, prompting the development of non-invasive biosensors for early detection.

Keywords— *surface plasmon resonance, Kretschmann, glucose, graphene, optical biosensor, sensitivity*

I. INTRODUCTION

Non-communicable diseases (NCD) such as diabetes mellitus has become a global pandemic, posing a major threat to public health and economies worldwide. The 2019 Malaysian National Health and Morbidity Survey (NHMS) revealed a marked increase in diabetes prevalence among adults aged ≥18 years, rising from 11.2% in 2011 to 18.3%. This translates to approximately 3.9 million adults living with diabetes in Malaysia [1]. The often-asymptomatic nature of early-stage diabetes leads to delayed diagnosis, significantly increasing the risk of complications like kidney failure, blindness, heart disease, and stroke. This, coupled with the substantial economic burden diabetes places on healthcare systems globally, underscores the critical need for prioritizing early detection and diagnosis [2]. Pre-diabetes is characterized by impaired fasting glucose (IFG) and/or impaired glucose tolerance (IGT). Individuals with pre-diabetes exhibit fasting, blood glucose levels between 5.6 –7.0 mmol/L, while two hours after a glucose challenge, levels range from 7.8–11.1

mmol/L. In comparison, healthy individuals typically have fasting blood glucose levels below 6.1 mmol/L and post-challenge levels under 8.0 mmol/L. Untreated pre-diabetes can progress to diabetes, where fasting glucose levels surpass 7.0 mmol/L and post-challenge levels exceed 11.1 mmol/L [3].

Recent research has shown a growing interest in surface plasmon resonance (SPR) sensors for their potential applications in biomedicine. SPR sensors based on the Kretschmann configuration (K-SPR) have emerged as a promising technique for non-invasive and label-free glucose detection. These sensors exploit the phenomenon of surface plasmon resonance, where incident light excites collective electron oscillations on a thin metal film's surface. The resonance angle, which is highly sensitive to the surrounding medium's refractive index, experiences a shift upon binding of biomolecules such as glucose. This characteristic makes K-SPR sensors sensitive to biomolecule adsorption on the metal surface.

Gold (Au) is most used as the plasmon active material for excitation due to its well-understood properties. However, its sensitivity for biomolecule detection can be improved by incorporating other materials. Recent research has focused on two-dimensional (2D) materials such as graphene (Gr) and transition metal dichalcogenides (TMDCs) to improve the capabilities of SPR sensors. These materials offer a distinct advantage over traditional SPR sensors that rely mainly on Au. Gr exhibits the unique ability to support surface plasmon polaritons (SPPs) across a much broader spectral range, extending from the mid-infrared regime to the terahertz domain. This expanded range is attributed to Gr's exceptional optoelectronic properties, including its tunable conductivity and high carrier mobility. Karki et. al demonstrated the sensor utilizes a combination of thin layers of silver, MXene, ZnO, and Gr for glucose detection resulted in a significant improvement and enhancement at the metal-Gr interface, leading to stronger interaction with the target analyte[4]. Similarly, Rahayu et. al reported a novel non-enzymatic glucose sensor utilizing SPR composite material formed by combining magnetite (Fe_3O_4) nanoparticles with Gr. This sensor demonstrated a significant shift in resonance intensity (18.2 units) when detecting a 5 mM glucose solution making it a potential candidate for future glucose monitoring applications[5].

II. METHODOLGY

A. Experimental Setup

Fig. 1 schematically illustrates the proposed biosensor, a five-layer Kretschmann configuration. The first layer is borosilicate (BK7) glass prism chosen for its refractive index (n_{prism} = 1.5202 and 1.5162) at the respective wavelengths, to ensure proper light momentum matching within the sensor structure. Following the prism, a thin layer of chromium and Au is deposited. The 50 nm-thick Au sensor slides were purchased from BioNavis (Tampere, FI). Atop the Au layer lies multilayer chemical vapor deposited (CVD) -graphene (Gr) layer, incorporated for its unique ability to enhance the surface plasmon resonance (SPR) effect [6]. Finally, the topmost layer serves as the biomolecule interaction site, where the target analyte (e.g., glucose) binds.

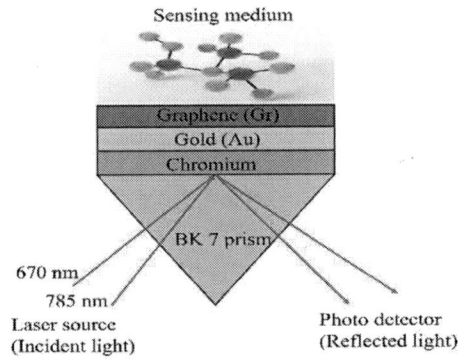

Fig. 1 Schematic view of Kretschmann-based SPR sensor with hybrid Au-Gr thin films

An experimental test was conducted using Bionavis SPR-Navi 200™ (Tampere, FI) with a prism that had an elastomeric coating with a matching refractive index (RI) to run the angular interrogation SPR analysis. The instrument employs the Kretschmann configuration, where the incident light is coupled via the BK7 prism to enhance the wave number of the incident light to match that of the excited surface plasmon in air and with water in the fluidic channels. Monochromatic p-polarized light (TM Mode) at 670 and 785 nm wavelengths were directed on the film to excite the plasmonic effect at the sensor surface. The changes in the SPR curve can provide information on the exciton-plasmon coupling strength over a wide angular range of 38° to 78° to establish a baseline response using air and DI water as reference points. To minimize initial noise in the readings, DI water was flowed through the K-SPR sensor for 20 minutes at a rate of 100 µl per minute. During the main experiment, the flow rate was maintained at 50 µl per minute. For each tested concentration, an injection volume of 0.5 ml was used. To ensure stable readings between concentration changes, a 2-minute stabilization phase with buffer solution was implemented. The initial stage of the experiment involved testing the sensor with a range of glucose analyte solutions at varying molarities. Molarity, expressed as moles of solute per liter of solution, reflects the concentration of a dissolved substance. In this study, the chosen molarity range (4 mM to 20 mM) corresponds to the physiological levels of glucose found in human blood.

B. Sample Preparation

Glucose was obtained from Sigma Aldrich in powder form (CAS 50-99-7) and diluted with DI water. The preparation process involved creating a stock solution. Here, 0.18016 grams of glucose powder was precisely weighed and dissolved in 50 ml of DI water. This resulted in a concentrated stock solution with a molarity of 20 mM. Subsequently, a serial dilution technique was employed. 4 ml of the stock solution were diluted with 1 ml of DI water and mixed thoroughly. This process was repeated six more times, yielding additional solutions with progressively lower molarities of 16 mM, 14 mM, 12 mM, 10 mM, 8 mM, and finally, 4 mM.

C. Performances

Surface plasmon resonance (SPR) sensors rely not only on the minimum reflectance (Rmin) but also on the Full-Widthat-Half-Maximum (FWHM) of the resonance curve for optimal performance. Sensitivity was an important parameter in determining a sensor's reliability. Rmin ¬is obtained from the lowest intensity detected on a specific wavelength. FWHM is the width of the spectrum curve measured through the half maximum amplitude from the y-axis. Sensitivity is calculated using Equation (1).

$$S = \frac{\Delta\theta_{glucose} - \Delta\theta_{water}}{Concentration_{glucose} - Concentration_{water}} \quad (1)$$

III. RESULTS AND DISCUSSIONS

A. Au-Gr sensor for Glucose Detection

K-SPR was first tested in water (refractive index, RI = 1.33) using Au-only samples, before Gr was deposited on Au. Sharp and narrow resonance in the reflective intensity of the K-SPR curves is observed as in Fig. 2. Nevertheless, the performance of K-SPR biosensors and the resonance of the laser incident angle depend largely on the materials constructing the sensor structure. In the Au only sensing layers, the K-SPR peaks exhibited resonance angles of 65.8496° to 69.0188° at 785 nm and 670 nm respectively. We obtained R_{min} of 0.0784 and 0.0955 and the FWHM calculated was 1.4815° and 3.5262°, respectively.

In Fig. 3, the Au-Gr K-SPR sensing peaks resonated at the angles of approximately 66.8988° to 70.6846° at 785 nm and 670 nm, respectively. The observed increase in K-SPR resonance angles can be attributed to the requirement for a larger wavevector component parallel to the plasmonic interface. The study revealed that incorporating Gr broadens the resonance curve, as evidenced by an increased FWHM. FWHM of Au-Gr thin film obtained in water is 2.5212° and 6.2640° whilst the R_{min} value is 0.1267 and 0.1101 for 785 nm and 670 nm wavelengths respectively. While our observations indicate an increase in both R_{min} and FWHM, indicating a broadening effect due to damping of surface plasmons in the Gr layer, there's a crucial advantage. The rightward shift in the K-SPR curve's incidence angle signifies a more pronounced change in refractive index (RI) in the vicinity of the sensor surface that makes Gr a promising material for analyte detection.

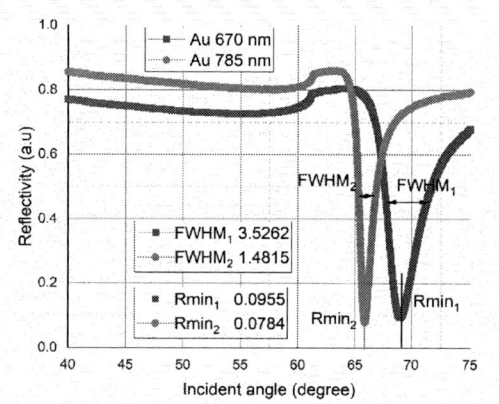

Fig. 2 K-SPR curve shift of Au-only sensor in DI water at 670 and 785 nm optical wavelengths

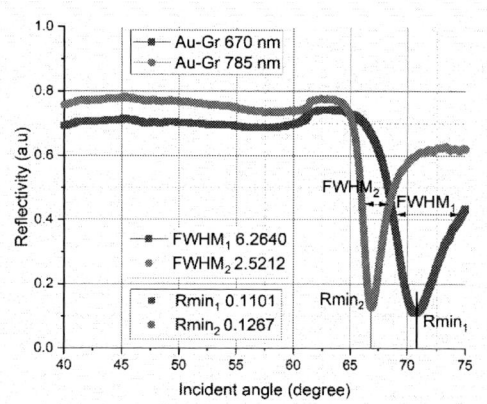

Fig. 3 K-SPR curve shift of Au-Gr sensor in DI water at 670 and 785 nm optical wavelengths

This investigation explores the potential of the Au-Gr sensor for detecting varying glucose concentrations, potentially differentiating between healthy and unhealthy levels. Fig. 4 and Fig. 5 depict the K-SPR reflectance curves obtained at 670 nm and 785 nm for a range of glucose concentrations (4 mM to 20 mM). As observed in the graphs, the resonance angle exhibits a red shift with increasing glucose concentration. This phenomenon can be attributed to a rise in the RI of the surrounding medium due to the interaction of glucose molecules with the sensing layer. This interaction likely alters the absorption properties of the sensing layer, consequently affecting the reflectance value. The observed angle shift in the K-SPR curve highlights the potential of the Au-Gr layer at this specific wavelength for glucose sensor applications utilizing the angular interrogation method.

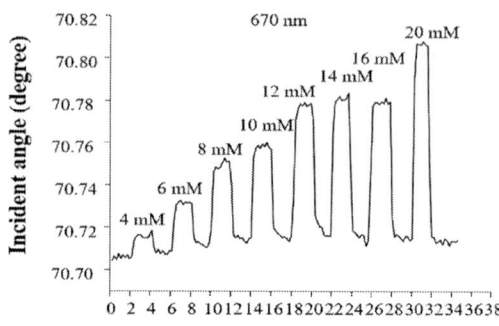

Fig. 4 Sensorgram of glucose detection at 670 nm using Au-Gr sensing layers

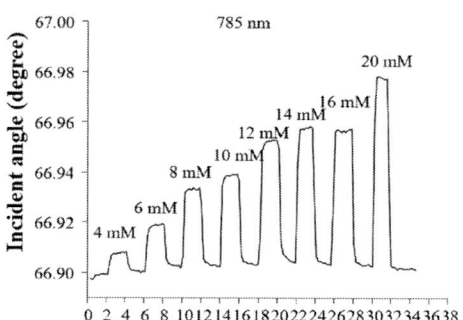

Fig. 5 Sensorgram of glucose detection at 785 nm using Au-Gr sensing layers

B. Sensitivity

From the graph in Fig. 4 and 5, the shifted resonance angle versus concentration of glucose is plotted and shown in Fig. 6. The linear fit line of the data is presented in Equation (2) and (3)

$$\Delta\theta_{SPR(670\,nm)} = 0.0118x + 0.005 \tag{2}$$

$$\Delta\theta_{SPR\,(785nm)} = 0.0091x + 0.0038 \tag{3}$$

where $\Delta\theta_{SPR}$ represents the shift in the resonance angle and x denotes the glucose concentration. These equations reflect the sensor's sensitivity, which is 0.0118° and 0.0091° per 1 mM change in glucose concentration for wavelengths 670 nm and 785 nm, respectively. The high R^2 values (0.9704 and 0.9666) at both wavelengths indicate excellent sensor performance in terms of linearity. This linearity signifies a direct correlation between the resonance angle shift and the glucose concentration in the solution. The shifted resonance angle and sensitivity data are tabulated in Table I.

Fig. 7 shows the sensor's sensitivity to varying glucose concentrations. The sensitivity initially increases at lower molarities but reaches a saturation point at higher concentrations, evident from the diminishing shift in the resonant angle. This saturation occurs because glucose molecules can proportionally change the surrounding medium's refractive index based on their concentration. However, this effect has a limit, and the sensor response plateaus once a certain refractive index is reached.

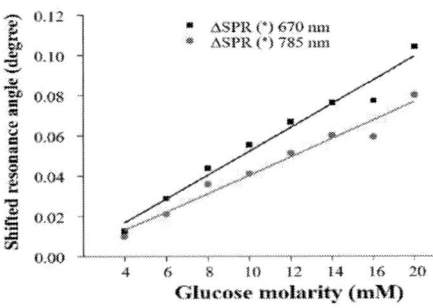

Fig. 6 Au-Gr response on glucose at 670 nm and 785 nm

TABLE I.   SHIFTED RESONANCE ANGLE AND SENSITIVITY OF AU-GR SENSOR AT 670 NM AND 785 NM

Wavelength (nm)	Concentration (mM)	Resonance Angle Shift $\Delta\theta_{SPR}$ ($°$)	Sensitivity ($°$/mM)
670	4	0.012	3.125
	6	0.029	4.817
	8	0.044	5.475
	10	0.055	5.530
	12	0.067	5.575
	14	0.076	5.443
	16	0.077	4.825
	20	0.104	5.205
785	4	0.010	2.450
	6	0.021	3.517
	8	0.036	4.488
	10	0.041	4.110
	12	0.051	4.275
	14	0.060	4.300
	16	0.059	3.706
	20	0.080	4.005

As the analyte's RI increases, the effective refractive index experienced by the surface plasmons changes as well. The Gr layer modulates this change, leading to a more pronounced shift in the resonant wavelength. This larger shift ultimately translates to enhanced sensitivity and resolution of the Au-Gr hybrid biosensor for biomolucles detection.

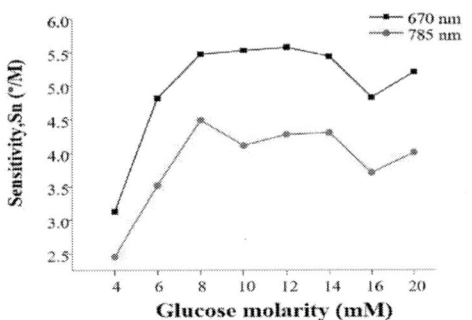

Fig. 7 Sensitivity for glucose detection using Au-Gr sensor at 670 and 785 nm optical wavelengths

Previous research has demonstrated similar trends with the sensitivity of Au-based sensors only [7]. They achieved sensitivities of 3.41°/M and 2.73°/M at 670 nm and 785 nm, respectively. In contrast, the Au-Gr sensor developed in this study demonstrates significantly enhanced sensitivity, reaching an average of 5.0°/M and 3.9°/M at the same wavelengths.

IV. CONCLUSION

This study explored a Kretschmann-based SPR approach for glucose detection at 670 nm and 785 nm using an Au-Gr sensor. Sensitivity towards high and low glucose concentrations was evaluated via K-SPR response analysis. Achieved sensitivity falls within the range suitable for diagnosing healthy and diabetic glucose levels. Enzyme immobilization holds promise for further sensitivity enhancement. The sensor's inherent advantages of stable sensitivity and real-time monitoring highlight its potential for glucose detection.

ACKNOWLEDGMENT

This project was funded by National University of Malaysia (UKM) using grants UKM-TR2023-09. Institute of Microengineering and Nanoelectronics (IMEN) is also acknowledged for the financial support

REFERENCES

[1] L.-L. Lim, Z. Hussein, N. M. Noor, A. S. A. Raof, N. Mustafa, M. B. L. Bidin, R. A. Ghani, S. Samsuddin, S.-L. Yong, and S.-H. Foo, "Real-world evaluation of care for type 2 diabetes in Malaysia: A cross-sectional analysis of the treatment adherence to guideline evaluation in type 2 diabetes (TARGET-T2D) study," Plos one, vol. 19, no. 1, pp. e0296298, 2024..

[2] K. K. Sadasivuni, J.-J. Cabibihan, A. K. A. Al-Ali, and R. A. Malik, Advanced Bioscience and Biosystems for Detection and Management of Diabetes: Springer, 2022.

[3] T. Sitasuwan, and R. J. B. o. Lertwattanarak, "Prediction of type 2 diabetes mellitus using fasting plasma glucose and HbA1c levels among individuals with impaired fasting plasma glucose: a cross-sectional study in Thailand," BMJ open, vol. 10, no. 11, pp. e041269, 2020.

[4] B. Karki, A. Jha, A. Pal, and V. Srivastava, "Sensitivity enhancement of refractive index-based surface plasmon resonance sensor for glucose detection," Optical Quantum Electronics, vol. 54, no. 9, pp. 595, 2022.

[5] T. O. C. Rahayu, N. L. W. Septiani, G. Gumilar, D. R. Adhika, and B. J. I. S. J. Yuliarto, "Modification of Gold Substrate With Fe 3 O 4-Graphene Nanocomposite to Increase Resolution of Surface Plasmon Resonance (SPR) Glucose Sensor," IEEE Sensors Journal, vol. 21, no. 18, pp. 19959-19966, 2021.

[6] P. S. Menon, N. A. Jamil, G. S. Mei, A. R. M. Zain, D. W. Hewak, C.-C. Huang, M. A. Mohamed, B. Y. Majlis, R. K. Mishra, and S. J. I. j. o. t. e. d. s. Raghavan, "Multilayer CVD-Graphene and MoS₂ Ethanol Sensing and Characterization Using Kretschmann-Based SPR," IEEE journal of the electron devices society, vol. 8, pp. 1227-1235, 2020.

[7] P. S. Menon, B. Mulyanti, N. A. Jamil, C. Wulandari, H. S. Nugroho, G. S. Mei, N. F. Z. Abidin, L. Hasanah, R. E. Pawinanto, and D. D. Berhanuddin, "Refractive index and sensing of glucose molarities determined using Au-Cr K-SPR at 670/785 nm wavelength," Sains Malaysiana, vol. 48, no. 6, pp. 1259-1265, 2019.

VCSELs-based Optoelectronics Transceiver for Free Space Photonics Packet Switching Network

Clarence Augustine TH Tee
College of Physics & Electrical Information Engineering,
Zhejiang Normal University
321004 Zhejiang
People's Republic of China
catht@zjnu.edu.cn

Burhanuddin Yeop Majlis
Institute of Microengineering and Nanoelectronics (IMEN)
Universiti Kebangsaan Malaysia
43600 UKM Bangi
Selangor Malaysia
burhan@ukm.edu.my

Sheng Li
Zhejiang Institute of Optoelectronics,
321004 Zhejiang
People's Republic of China
shengli@zjnu.edu.cn

Han Bin Sun
Photonics & Sensors Key Research Laboratory,
Zhejiang Normal University
321004 Zhejiang
People's Republic of China
hanbinsun@zjnu.edu.cn

Muhamad Ramdzan Buyong
Institute of Microengineering and Nanoelectronics (IMEN)
Universiti Kebangsaan Malaysia
43600 UKM Bangi
Selangor Malaysia
muhdramdzan@ukm.edu.my

Le Song
State Key Laboratory of Precision Measurement Technology and Instruments,
Tianjin University
300072 Tianjin
People's Republic of China
songle@tju.edu.cn

W P Yeo
Photonics & Sensors Key Research Laboratory,
Zhejiang Normal University
321004 Zhejiang
People's Republic of China
svcat15@gmail.com

Ahmad Rifqi Md Zain
Institute of Microengineering and Nanoelectronics (IMEN)
Universiti Kebangsaan Malaysia
43600 UKM Bangi
Selangor Malaysia
rifqi@ukm.edu.my

Yelong Zheng
State Key Laboratory of Precision Measurement Technology and Instruments,
Tianjin University
300072 Tianjin
People's Republic of China
zhengyelongby@tju.edu.cn

Abstract— For the realization of a large packet switch with throughput in the range of Tera-to-Petabits/s, an optoelectronics transceiver is one of the crucial modules in the free space photonics link. Here, a novel high speed and capacity free space optoelectronics communication transceiver of 850nm, single-moded and polarization-stable VCSELs, has been designed. The VCSELs transceiver with its driving interface system contain a dual operational amplifier with constant current source module and direct digital synthesizer (DDS) enabling high frequency signal generation module up to 500MHz and beyond, for the realization of the specification for an optoelectronic switching module suitable for a packet router. The designed module could be scaled up for a throughput of 500 Gb/s and act as the basic building block for a larger switch with a throughput of 131 Tb/s.

Keywords— *Optical Packet Switch; Spatial Light Modulator; Direct Digital Synthesizer, VCSELs, Free Space.*

I. INTRODUCTION

Free space chip-to-chip optical interconnections has been demonstrated as a viable means of increasing the scalability of large electronic IP switches i.e., hardware implementation of channel selection in the optical interconnect via optical fan-out, followed by optical fan in of channel selection [1-4]. Our version of the architecture is of liquid crystal over silicon devices were used as input transducers to modulate the optical channels and to reconfigure the interconnect [1-5]. The motivation of the work here is in demand due to the requirement of the new information age of electronics and photonics convergence which necessitates large scale, monolithic integration of photonics circuits with electronic circuits, the threshold commercial displacement of electronic interconnects by optical ones and the superior optical interconnect performance in terms of switching speed, cost effectiveness, energy saving, reduced electromagnetic interference and a smaller volume footprint [1-7]. The Cisco's report [8] has predicted that the surge in IP rate would directly lead to the rapid bandwidth demand in the future network, a bottleneck in the legacy telecommunication network for transmission speed and bandwidth. In tandem, the optoelectronics switch employing the photonics packet switching architecture and network is the core for the realization of large switch with throughput from terabytes to petabytes, which is made possible with the high speed and capacity free space optoelectronics communication of a single mode and polarization stable VCSEL laser source [1-5]. In the coming sections, full designs, electrical, optical characterization of VCSELs and driving interface, implementation as well as commissioning the transmitter would be detailed. It also shows the paramount importance of the role offered by the VCSELs and driving system in the realization of the large packet switch network

In the high frequency domain, the traditional cable interconnection system has the problems of large loss, serious crosstalk and high power consumption. Meanwhile, the cable interconnection needs strict impedance matching, which is difficult to design, and the interconnection of different systems is difficult, which is not flexible enough [9]. Compared with the traditional electrical interconnection, optical interconnection has the advantages of high speed, low cost, low power consumption, and no electromagnetic interference, so optical communication is becoming the mainstream choice of high-speed transmission system [10-11]. The transmitting module with interfacing driver is an important link in the optical interconnect system, using the modulated signal to modulate the laser to complete the conversion of electrical signal to optical signal [12-13] In this paper, a commercially available high speed VCSELs has been commissioned and designed with the interfacing driver and circuitry consisting of the VCSELs laser constant current control module and DDS module for high frequency signal generation.

979-8-3503-7832-0/24 $31.00 © 2024 IEEE

II. METHODOLOGY

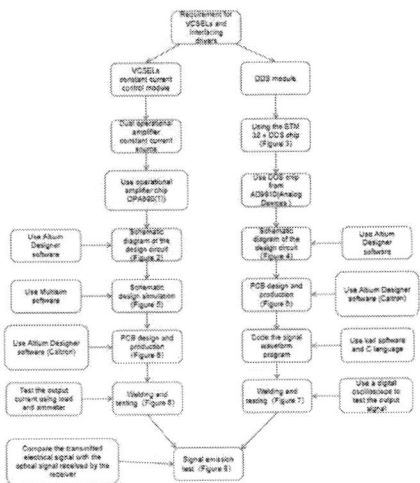

Fig. 1. Framework of Research, Design & Process Flowchart

Figure 1. shows the overall research framework and methodology. The flowchart shows the design, process, testing and commissioning of VCSELs laser constant current control and DDS modules. Constant current control module transforms the input voltage value into the modulated current of the laser to control the output light intensity i.e. a key step to convert the electrical signal into the optical signal. While DDS module directly and accurately generates the signals needed for communication, enabling the switching network flexibly applied in various scenarios.

LASING MODULE

A. Design of the constant current source circuit

The basic principle of the constant current source circuit is to use a stable current as the reference current and then connect with the load resistance, and realize the stable output current under the condition of load resistance change. An operational amplifier is an electronic device that has the function of amplifying a voltage signal or a current signal. Here, the design uses a feedback mechanism, which could amplify the input signal, so that the output signal has a greater gain than the input signal [13].

In a dual operational amplifier constant current source, one operational amplifier acts as the main amplifier which is responsible for amplifying signals; the other operational amplifier is used to provide a stable working environment. The main amplifier is usually in the feedback mode, and the magnification and gain can be controlled by adjusting the ratio of the feedback resistance and the input resistance. The auxiliary amplifier is responsible for providing a stable reference voltage or current to ensure the proper operation of the main amplifier.

The dual operational amplifier constant current source circuit has been chosen to drive the single-mode 850nm VCSELs laser. Compared with the traditional constant current source circuit, it can provide the higher output impedance, thus reducing the impact on the load and having better stability and accuracy. In this experiment, the OPA690 high-speed operational amplifier designed by Texas Instruments has been used with a transport bandwidth gain product of 500MHz.

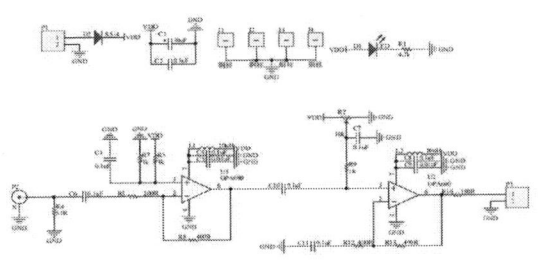

Fig. 2. Schematic of constant current source circuit of dual operational amplifier

The schematic diagram of the constant current source circuit of the dual operational amplifier is shown in Figure 2. The 12V single power supply is used to reduce the design complexity of the power supply module.

As the main amplifier, the operational amplifier U1 is set to double amplification to amplify the input medium and high frequency signal. U2 serves as the auxiliary amplifier to ensure the normal operation of the main amplifier. By adjusting the sliding rheostat R2 to increase or decrease the reference current, the output current of the output can be adjustable. The maximum constant output constant current is 190mA, and the maximum signal input frequency is 200MHz.

B. DDS circuit design

The integration of DDS generating module into the transmitting module of the laser makes the VCSELs transmitter module more flexible while saving cost. The DDS module consists of the following functional modules and communication interface i.e., DDS generator; DDS control unit; LCD display controller; keyboard for controlling waveform; SAM interface for output signals.

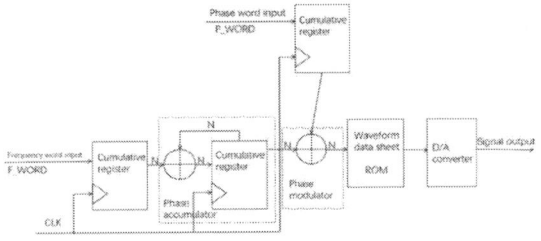

Fig. 3. Basic structure of the DDS

As shown in Figure 3, the basic structure of DDS is mainly composed of four major structures i.e., phase accumulator, phase modulator, waveform data table ROM, and D / A converter [14-15]. If the accumulator length is N, 2N storage units can store the sampling data. If the sampling of 2N points of a periodic waveform is performed, then m= 2N. The relationship between the output frequency, FOUT, the system clock frequency, FCLK, the phase accumulator length N and the frequency control word F_WORD is as:

$$F\_WORD = 2^N * FOUT/FCLK \qquad (1)$$

Compared to the traditional use of DDS which consist of phase accumulators, phase modulators, waveform data sheet and D/A converter, the main advantages of using digital frequency synthesizer chips are fast frequency jump speed, extremely fast switching speed with almost no establishment time and having more refined tuning resolution and channel flexibility [16-17]. Under the premise of getting a higher main frequency, there is a lower power consumption, higher stability and lower cost.

The direct digital frequency synthesizer chip used in this experiment is the AD9910 DDS chip from Analog Devices. The chip has the maximum of 1 GSPS clock frequency and 14bits data conversion accuracy. The highest resolution of the chip can reach the frequency tuning word resolution of 32bits, allowing the DDS chip to receive a 16-bit write frequency tuning word at every 4ns, thus fully supporting our demanding output of 500MHz signal. STM32F104 has been used to write the chip frequency word F _ WORD to let the module output corrects the frequency waveform. And OLED display screen and button module are equipped with better control the frequency and waveform of the output signals. The schematic diagram of the AD9910 module is shown in Figure 4.

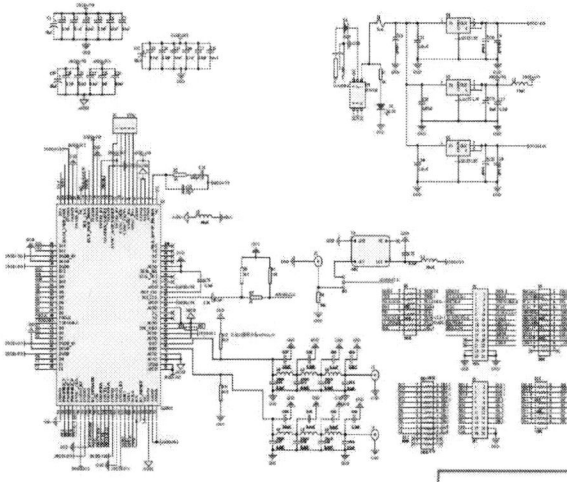

Fig. 4. AD9910 Schematic diagram

III. SIMULATION AND EXPERIMENTAL RESULTS

The dual operational amplifier constant current source circuit is simulated and the result is shown in Figure 5. The output circuit is 10mA when the sliding rheostat is at 13%.

The system experimental setup is shown in Figure 6 (a). From left to right, there are the AD9910 digital signal generator board, dual operational amplifier constant current source circuit and STM 32 controller module. Figure 6 (b) shows the system in operational mode.

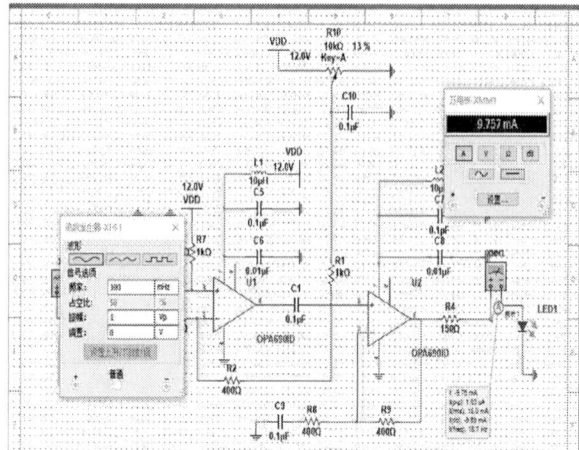

Fig. 5. Simulation results of the constant-current source circuit

（a）

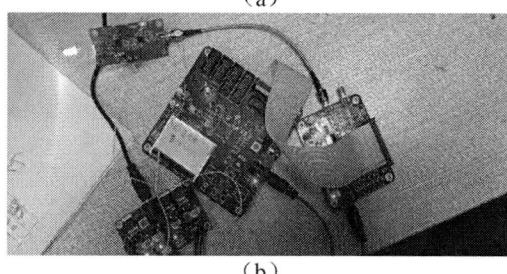

（b）

Fig. 6. Module and system experimental setup

Fig. 7. Testing the DDS signal transmission

TABLE I.   DDS OUTPUT RESULTS

Theoretical frequency(HZ)	Actual frequency(HZ)
50	50
500	500
5K	4.99K
50K	50.1K
500K	501.1K
5M	5M
50M	50M
500M	499.9M

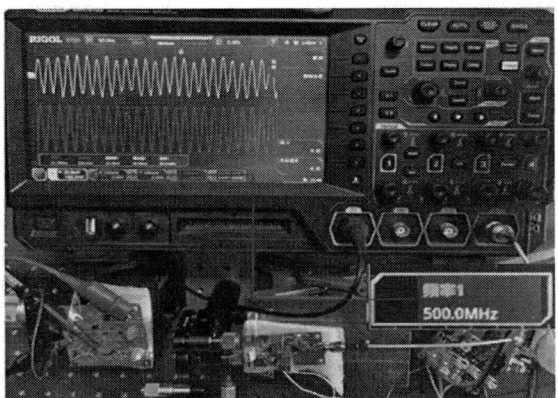

Fig. 8. Signal emission test

As shown in Figure 7 and Table I, the output tests were performed for the signals of 50 HZ to 500 MHZ. The output error tolerance is less than 0.02% for 500MHz (Fig. 8). Thus, the DDS signal generator can fulfill the experimental requirements as stated. Figure 8 shows the VCSELs transmitter with DDS module performed the interconnect test at 500MHz signal with the supporting receiver terminal.

IV. CONCLUSION

The high speed VCSELs-based transmitter system for 850nm single-mode laser source has been designed for the free-space photonics packet switch network application. The feasibility of a free space optical link with high data rate of 256 Mbits/sec for the transceiver system will be demonstrated with the scalability and simplicity of the transceiver system, acting as the basic building block for the petabytes throughputs system.

ACKNOWLEDGMENT

The research work here has been supported by the research team (and international collaborators) within the Photonics & Sensors Key Research Laboratory as part of the cross-disciplined Research Project "Multi-Dimensional Nano-Microelectronics Functional Devices & System Engineering (MINDS)", an interdisciplinary and foresight collaborative key research initiative. The work is funded and supported by the China National High Talent Foreign Expert Grant under Grant G2021016010L, Zhejiang Normal University's Distinguished (Eminent) Professorship Grant and Zhejiang's "Pioneering and Leading Goose" R&D Programme Grant 2022C01030.

REFERENCES

[1] C. A. T. H. Tee, W. A. Crossland, N. Collings, T. D. Wilkinson, "A Reconfigurable VCSEL Array Link Suitable for Use in Large Electronic IP Switches", Fifth Optoelectronics and Communications Conference (OECC 2000) Technical Digest (IEICE, IEEE, LEOS, OSA), July 2000, Makuhari Messe, Chiba, Japan.

[2] C. A. T. H. Tee, W. A. Crossland, N. Collings, T. D. Wilkinson, P. Robertson, "Design of a free space optical switch demonstrator for a VCSEL-based photonic packet switch", Optics in Computing (Invited paper), OSA, IEEE/LEOS, Lake Tahoe, Nevada, January 10-12, 2001.

[3] C. A. T. H. Tee, W. A. Crossland, T. D. Wilkinson, A. B. Davey, "Binary Phase Modulation Using Electrically Addressed Transmissive and Silicon Backplane Spatial Light Modulators", Opt. Eng. 39(9), pp. 2527-2534, Sept 2000, Optical Engineering Journal, USA.

[4] VIVALDI was a UK EPSRC project under the Optical System Integration (OSI) initiative, Reference: GR/M29283, "VCSEL based very high-capacity photonic packet switches, VIVALDI", 1997-2000.

[5] Crossland et al., TELECOMMUNICATIONS SWITCH ARCHITECTURE, US Patent Number: 5,576,873.

[6] Kimerling, "THE ECONOMICS OF SCIENCE: FROM PHOTONS TO PRODUCTS," Optics & Photonics News 9(10), 19- (1998), Opt. 16(1) 10-11 (1977).

[7] R Kirchain, L Kimerling, "A roadmap for nanophotonics", Nature photonics, VOL 1, JUNE 2007.

[8] P. J. Delfyett et al., 'Recent Advances in Stabilized Ultrafast Modelocked Semiconductor Diode Lasers for High-Speed Information Based Applications', in LEOS 2006 - 19th Annual Meeting of the IEEE Lasers and Electro-Optics Society, pp. 791–791, Oct. 2006.

[9] T. Tapen and A. Apsel, 'Ultra wideband Frequency Synthesis Using the Compact Tunable Transmission Line (CTTL)', IEEE Trans. Microwave Theory Technology, vol. 70, no. 7, pp. 3374–3384, Jul. 2022.

[10] I. I. Kim, B. McArthur, E. J. Korevaar, 'Comparison of laser beam propagation at 785nm and 1550nm in fog and haze for optical wireless communications', , presented at the Information Technologies 2000, E. J. Korevaar, Ed., Boston, MA, pp. 26–37, Feb. 2001.

[11] R. V. Chkalov, N. S. Pokryshkin, M. N. Gerke, K. S. Khorkov, D. A. Kochuev, V. G. Prokoshev, 'High-precision medium power laser diode driver with microprocessor-based control system', in 2018 International Conference Laser Optics (ICLO), pp. 174–174, Jun. 2018.

[12] R. Tao, M. Berroth, Z. G. Wang, 'Low power 10 Gbit/s VCSEL driver for optical interconnect', Electronics Letters, vol. 39, no. 24, pp. 1743–1744, Nov. 2003

[13] L. Wenyuan, C. Sun, Q. Zhao, W. Xiong, 'Design of 12-channel 120Gb/s laser diode driver array in 0.18µm CMOS technology', in 2014 International Conference on Advanced Technologies for Communications (ATC 2014), pp. 27–31, Oct. 2014.

[14] S. Zhuo, et al., 'A 200 MHz 14 W Pulsed Optical Illuminator with Laser Driver ASIC and On-Chip DLL-Based Time Interpolator for Indirect Time-of-Flight Applications', IEEE Transactions on Circuits and Systems II: Express Briefs, vol. 70, no. 2, pp. 396–400, Feb. 2023.

[15] A. Yamagishi, H. Nosaka, M. Muraguchi, T. Tsukahara, 'A phase-interpolation direct digital synthesizer with an adaptive integrator', IEEE Trans. Microwave Theory Technology, vol. 48, no. 6, pp. 905–909, Jun. 2000.

[16] S. Arora, P. T. Balsara, D. Bhatia, 'Digital Pulsewidth Modulation (DPWM) Using Direct Digital Synthesis', IEEE J. Emerg. Sel. Topics Power Electron., vol. 10, no. 4, pp. 4231–4244, Aug. 2022

[17] B. Madhavan, A. F. J. Levi, 'Low-power 2.5 Gbit/s VCSEL driver in 0.5 µm CMOS technology', Electronics Letters, vol. 34, no. 2, pp. 178–179, Jan. 1998.

979-8-3503-7832-0/24 $31.00 © 2024 IEEE

Gold Nanorods as Plasmonic-based Optical Biosensor for Creatinine Detection

Basyirah Zulkifli
Institute of Microengineering and Nanoelectronics, Universiti Kebangsaan Malaysia
Bangi, Selangor, Malaysia
p110935@siswa.ukm.edu.my

Ahmad Rifqi Md Zain
Institute of Microengineering and Nanoelectronics, Universiti Kebangsaan Malaysia
Bangi, Selangor, Malaysia
rifqi@ukm.edu.my

Tengku Hasnan Tengku Abdul Aziz
Institute of Microengineering and Nanoelectronics, Universiti Kebangsaan Malaysia
Bangi, Selangor, Malaysia
hasnan@ukm.edu.my

Nur Hidayah Azeman
Institute of Microengineering and Nanoelectronics, Universiti Kebangsaan Malaysia
Bangi, Selangor, Malaysia
nhidayah.az@ukm.edu.my

Mohd Suzeren Md Jamil
Faculty of Science and Technology, Universiti Kebangsaan Malaysia
Bangi, Selangor, Malaysia
suzeren@ukm.edu.my

Abstract— This work aims to develop a label-free and high-sensitivity biosensing platform for creatinine detection to enable early diagnosis of chronic kidney disease. We investigated the optical response of gold nanorods' localized surface plasmon resonance (LSPR) towards low-concentration creatinine biomarker. The unique optical properties of metallic nanorods with two plasmonic modes assert nanorods as a more robust platform than metallic nanospheres with sensing sensitivity improved at an increased aspect ratio. Utilizing the electrostatic adsorption from piranha substrate treatment, an adequate interparticle separation and uniform distribution of immobilized gold nanorods is obtained, allowing plasmon localization to be preserved for analyte sensing. The electrostatic interaction between the carbonyl group in creatinine molecules and the cationic headgroup capping the gold nanorods serves the adsorbate-induced sensing principle. The longitudinal LSPR mode shift yielded a linear relationship with increasing creatinine concentration, with sensitivity as high as 9.064 μm dL g^{-1} is recorded. The limit of detection (LOD) and the limit of quantification (LOQ) are determined to be 0.645 mg/dl and 0.712 mg/dl, respectively.

Keywords— *gold nanorods, localized surface plasmon resonance, plasmonic sensing, creatinine detection.*

I. INTRODUCTION

Noble-metal nanoparticles are known for their astonishing size-dependent optical properties leading to a wealth of extensive studies. The unique subwavelength structure has made it a straightforward candidate to interact with incident light without requiring a phase-matching medium, exhibiting strong extinction in the UV-visible spectra. When light is incident on the nanoparticles, resonance occurs as the photon frequency matches the plasmon frequency i.e. the collective oscillation of free electrons in the nanoparticle. This phenomenon is known as localized surface plasmon resonance (LSPR). The peak of the LSPR wavelength is intrinsically dependent on the nanoparticles' size, shape, and material, while also extrinsically determined by the dielectric constant in its local environment [1]. This dependency on the surrounding environment has put forth plasmonic nanoparticles as an auspicious tool in sensing applications.

This work aims to exploit gold nanorods (AuNRs) as a biosensing platform for low-concentration creatinine detection. Creatinine, the breakdown product from muscle activity ranks as the second most favorable biomolecule for clinical monitoring analysis due to its multiple associations with kidney diseases, muscular diseases, renal kidney dysfunction, and thyroid malfunction [2]. A severe renal kidney dysfunction, chronic kidney disease (CKD) is a serious public health issue that has been recognized to increase worldwide with severity in a broad range of diseases with high morbidity and mortality [3]–[5]. Most people with CKD are asymptomatic until the later stages which makes early diagnosis challenging [6]. Here, accurate biomarker detection for early diagnosis is critical to divert disease complications and slow down kidney dysfunctionality [7], [8]. Currently, the gold standard for creatinine detection is the calorimetric Jaffe method, a laboratory-bound protocol which is time-consuming and requires expert handling of the toxic picric acid used in the test [9]–[11]. The readings are also easily interfered with by temperature, pH, and other variables.

Detection of creatinine as the main biomarker at the point-of-care level has surged in demand to provide real-time and sensitive analysis to enhance monitoring performance. The plasmonic nanoparticle biosensing technique has attained substantial interest as an alternative due to its simple and direct sensing principle [12], [13]. This work presents a study on gold nanorods (AuNRs) as a biosensing platform for creatinine detection, encompassing label-free and real-time sensing mechanisms. The high sensitivity of the LSPR mode, namely the longitudinal-LSPR mode of the nanorod to the surrounding dielectric constant changes by creatinine concentration serves as a facile optical sensor for the biomarker. Low creatinine concentrations are used, replicating the normal value in human serum at ≤ 1.5 mg/dl [14].

II. MATERIALS AND METHOD

A. Materials and Reagents

Tetrachloroaurate trihydrate (HAuCl$_4$· 3H$_2$O, ≥ 99.9%), Sodium borohydride (NaBH$_4$, 99%), Silver nitrate (AgNO$_3$, ≥ 99.0%), Hydrochloric acid (HCl, 36.5% − 38%) and Creatinine anhydrous, (CRE, ≥ 98%) were all purchased from Sigma-Aldrich. Hexadecyltrimethylammonium bromide (CTAB, > 97%) was purchased from Systerm and Ascorbic acid (Vitamin C) was purchased from the real C. Deionized water (18 MΩ) was used in all experiments.

B. Gold Nanorods (AuNRs) Synthesis and Immobilization

AuNRs are synthesized by employing the silver-ion-assisted seed-mediated method following the procedure developed by Nikoobakht and El-Sayed [15]. Two solutions, the seed solution and the growth solution are prepared in this scheme. Briefly, in the seed preparation, gold salt HAuCl$_4$ (50 µl, 0.05 M) was mixed with hexadecyltrimethylammonium bromide (CTAB) solution (10 ml, 0.1 M). Then, ice-cold sodium borohydride NaBH$_4$ (0.6 ml, 0.01 M) was added into the solution while stirring at 1200 rpm for 2 minutes, resulting in light brownish colour. The seed solution is stored for at least 2 hours to ensure complete decomposition of the borohydride ions. For the preparation of growth solution, HAuCl$_4$ (0.2 ml, 0.05 M) was mixed with silver nitrate AgNO$_3$ (0.24 ml, 0.01 M) followed by the addition of CTAB (20 ml, 0.1 M). After a gentle stir, freshly prepared ascorbic acid (0.16 ml, 0.1 M) was added into the solution, resulting in a colourless solution. The pH is adjusted by introducing hydrochloric acid HCl (0.4 ml, 1.0 M) with a gentle mix. Finally, the seed solution (24 µl) was added. The solution is gently stirred for 30 second and left at room temperature for at least 18 hours. Upon the addition of seed solution, gold nanorods begins to grow in size and length. The gold nanorod are then purified through centrifugation at 6000 rpm twice and washed with deionized water to discard the surfactant and nanoparticle impurities.

For AuNRs immobilization, glass substrate was treated with piranha solution to maximize hydroxyl group [16]. The substrate was immersed in freshly prepared piranha solution (H$_2$SO$_4$/ H$_2$O$_2$; 4:1 v/v) for 40 minutes. The substrate was then washed thoroughly with deionized water and dried with N$_2$ gas. Finally, the treated substrate was immersed in colloidal AuNRs for 48 hours to allow AuNRs deposition.

C. Morphological Characterization

The as-prepared immobilized AuNRs is characterized using field emission scanning electron microscope (FESEM) to study its morphology, with machine model ZEISS SUPRA 55VP operated at an acceleration voltage of 3 kV.

D. Optical Measurement Response Setup

The sensor system setup consists of a bifurcated optical fiber with one end connecting to LS-1 tungsten halogen lamp Ocean Optics unit supplying the broadband VIS-NIR light, and the other end connected to HR2000 Series Ocean Optic Spectrometer. The immobilized gold nanorods substrate was laid flat at the bottom of creatinine solution. The probe is fixed to a position of approximately 0.5 mm perpendicular to the sample surface and the tip of the probe was immersed into creatinine solution. The spectral responses are recorded at varying creatinine concentrations at 0.2, 0.4, 0.6, 0.8, 1.0 and 1.2 mg/dl. The sensing response was measured upon activating the light source, which transmitted light towards the gold nanorods substrate, followed by the scattering of light that was eventually collected by the probe. Prior to sensing measurements, the baseline spectrum was made using a blank glass substrate to eliminate spectral feature of the substrate.

III. RESULTS AND DISCUSSION

A. Gold Nanorods Characterization

Fig. 1(a) shows the FESEM image of AuNRs immobilized on the glass substrate, displaying capsule-shaped nanorods. By measuring the size distribution of 100 nanorods, the average length and diameter are determined to be 71.95 nm and 20.21 nm respectively, giving its aspect ratio 3.56. The image exhibit uniform distribution of AuNRs without apparent aggregation among the nanorods, which is a common setback in the widely used spin coating method of nanoparticle immobilization [17]. This uniformity can be attributed to the immobilization method viz. dip coating of the piranha-treated substrate that deploys electrostatic interaction between the hydroxyl OH$^-$ group on the surface of the substrate, and the positively-charge AuNR capping, the CTA$^+$, from CTAB solution added during synthesis. Piranha solution is a strong oxidizing agent which hydroxylate the substrate surface, conveniently interact with positively charged nanoparticles such as our nanorods. The even distribution and adequate interparticle-separation of nanoparticles are especially important to preserve the localization of the plasmon of individual nanorods, inducing pronounced plasmon peaks which is an essential feature in plasmonic-based sensing applications.

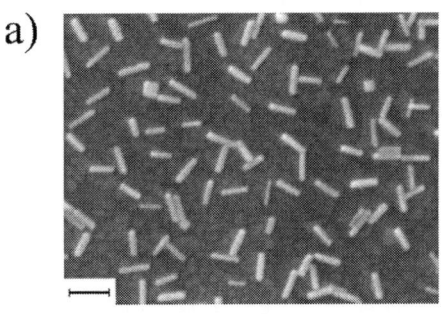

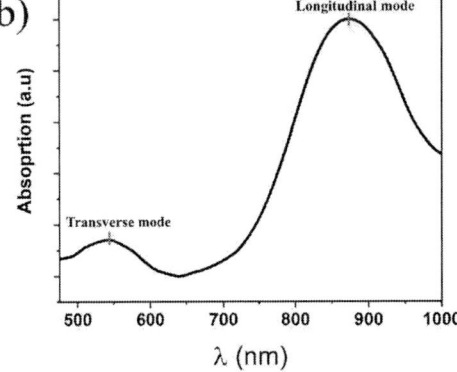

Fig. 1. (a) FESEM image of gold nanorods immobilized on a glass substrate with adequate interparticle spacing. The scale bar represents 100 nm. (b) The L-LSPR mode and T-LSPR mode of gold nanorods.

979-8-3503-7832-0/24 $31.00 © 2024 IEEE

Metallic nanorod is uniquely distinguished by its nanosphere counterparts, in which it exhibits two plasmonic modes, the transverse plasmon mode (T-LSPR) and longitudinal plasmon mode (L-LSPR). The T-LSPR and L-SPR are encoded in its anisotropic shape and correspond to the plasmon oscillation along the long axis and short axis, respectively. Fig. 1(b) shows the absorption spectrum of the immobilized AuNRs with T-LSPR peak wavelength (λ) at 545 nm, oscillating at a higher electromagnetic energy than the stronger L-LSPR peak wavelength at 873 nm. The spectral feature of the nanorods displays pronounced and almost symmetrical peaks in both T-LSPR and L-LSPR, which corresponds to the interparticle separation of the immobilized nanorods.

B. Plasmonic Sensing Response

Plasmonic-based sensing employs adsorbate-induced changes in the refractive index of the local nanorods' environment, resulting in plasmon mode shift, a central mechanism in plasmonic sensing. Here, creatinine serves as the analyte altering the refractive index, with the concentration of creatinine used ranging from 0.2 mg/dl to 1.2 mg/dl, recreating the normal level of creatinine in human serum secreted by the kidneys. The sensing response shows the L-LSPR mode shifting to longer wavelengths at increasing creatinine concentrations (Fig. 2(a)). The trend was analyzed by interpolating the L-LSPR peak shift at increasing creatinine concentration, yielding a linear relationship (Fig. (b)). Linear regression analysis generates a sensing sensitivity of 9.064 μm dL g^{-1}, with a good correlation coefficient of $R^2 = 0.989$. Following the ICH guidelines, the limit of detection (LOD)

and limit of quantification (LOQ) of the sensing response are determined to be 0.235 mg/dl and 0.712 mg/ml respectively. The linear shifting of the plasmonic peaks confirms the sensitivity of AuNRs towards creatinine molecules at local refractive index change. T-LSPR however, shows no correlation between the changes in the local refractive index and the presence of creatinine. This may be due to T-LSPR being lower in sensitivity than L-LSPR, as the aspect ratio also dictates the degree of sensitivity. Encoded in the longitudinal modes, the increase in the nanoparticle's aspect ratio increases its sensitivity by up to 10 times for an aspect ratio of 5.0 [18].

The adsorption of creatinine molecules on nanorod's surface may be attributed to the electrostatic interaction between the δ^- carbonyl group present in creatinine molecules with the cationic headgroup capping the AuNRs, as illustrated in Fig. 3. The yielded AuNRs are capped with a cationic CTA$^+$ surfactant molecules which consist of the hydrocarbon tail and quaternary ammonium headgroup facing outward, effectively making the AuNRs positively charged [19], [20]. These positive-charged nanorods are likely to adsorb via ion pairing with the electron-rich atom O at the partial negative δ^- group, the carbonyl group, in the creatinine molecule.

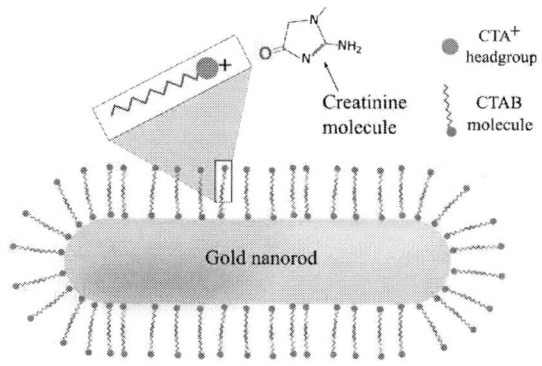

Fig. 3. Schematic illustration of gold nanorod capped with a bilayer of CTAB molecules, with its cationic headgroup interacting via ion pairing with the electron-rich O atom of the carbonyl group in the creatinine molecule.

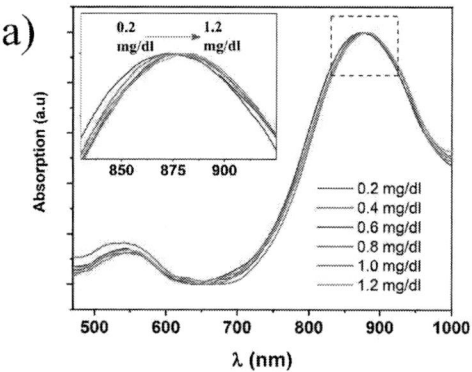

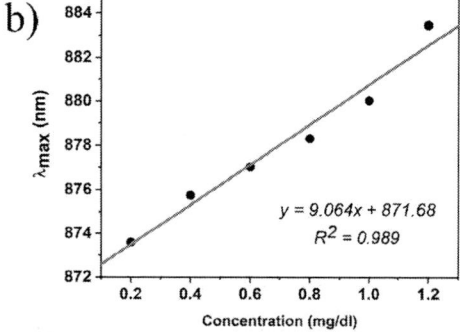

Fig. 2. (a) Gold nanorods L-LSPR absorption spectra at different creatinine concentration (mg/dl). Inset graph shows the spectra shift to longer wavelengths at increasing concentration. (b) Plot depicting linear trend between creatinine concentration and L-LSPR absorption peak

IV. CONCLUSION

CTAB-capped gold nanorod (AuNR) has shown to be an effective plasmonic nanoparticle for creatinine detection, owing to the electrostatic adsorption mechanism between the polar carbonyl group in creatinine molecule and CTA$^+$ ion covering the nanorods. By adopting the piranha treatment of substrate followed by substrate immersion in colloidal AuNRs, we have successfully immobilized AuNRs without aggregation, preserving the plasmon localization of individual nanorods for analyte detection. The longitudinal localized surface plasmon resonance (L-LSPR) of AuNRs responds linearly to the presence of creatinine at increasing concentration, with sensitivity as high as 9.064 μm dL g^{-1} is recorded. A low limit of detection (LOD) of 0.645 mg/dl and the limit of quantification (LOQ) of 0.712 mg/dl are determined. This finding serves as a building brick for the development of non-invasive optical sensing in early diagnosis and treatment [21]–[23].

ACKNOWLEDGMENT

This research was funded by the Malaysia Ministry of Higher Education (MOHE), Malaysia, through the Fundamental Research Grant Scheme (FRGS), grant number FRGS/1/2022/STG05/UKM/02/4 and the Universiti Kebangsaan Malaysia (UKM), through the Strategic Research Fund, grant number KRA-2023-001.

REFERENCES

[1] X. Huang, S. Neretina, and M. A. El-Sayed, "Gold nanorods: From synthesis and properties to biological and biomedical applications," *Adv. Mater.*, vol. 21, no. 48, pp. 4880–4910, 2009, doi: 10.1002/adma.200802789.

[2] M. Joffe, C. Y. Hsu, H. I. Feldman, M. Weir, J. R. Landis, and L. L. Hamm, "Variability of creatinine measurements in clinical laboratories: Results from the CRIC study," *Am. J. Nephrol.*, vol. 31, no. 5, pp. 426–434, 2010, doi: 10.1159/000296250.

[3] J. A. Vassalotti, R. Centor, B. J. Turner, R. C. Greer, M. Choi, and T. D. Sequist, "Practical Approach to Detection and Management of Chronic Kidney Disease for the Primary Care Clinician," *Am. J. Med.*, vol. 129, no. 2, pp. 153-162.e7, 2016, doi: 10.1016/j.amjmed.2015.08.025.

[4] D. N. Koye, D. J. Magliano, R. G. Nelson, and M. E. Pavkov, "The Global Epidemiology of Diabetes and Kidney Disease," *Adv. Chronic Kidney Dis.*, vol. 25, no. 2, pp. 121–132, 2018, doi: 10.1053/j.ackd.2017.10.011.

[5] B. L. Neuen, S. J. Chadban, A. R. Demaio, D. W. Johnson, and V. Perkovic, "Chronic kidney disease and the global NCDs agenda," *BMJ Glob. Heal.*, vol. 2, no. 2, pp. 7–10, 2017, doi: 10.1136/bmjgh-2017-000380.

[6] D. Oliveira *et al.*, "Chronic kidney disease Related papers Chronic kidney disease."

[7] R. J. Anderson, G. D. Bahn, N. V. Emanuele, J. B. Marks, and W. C. Duckworth, "Blood pressure and pulse pressure effects on renal outcomes in the Veterans Affairs Diabetes Trial (VADT)," *Diabetes Care*, vol. 37, no. 10, pp. 2782–2788, 2014, doi: 10.2337/dc14-0284.

[8] J. T. Wright *et al.*, "Effect of blood pressure lowering and antihypertensive drug class on progression of hypertensive kidney disease: Results from the AASK trial," *J. Am. Med. Assoc.*, vol. 288, no. 19, pp. 2421–2431, 2002, doi: 10.1001/jama.288.19.2421.

[9] T. Guinovart *et al.*, "Recognition and Sensing of Creatinine," *Angew. Chemie*, vol. 128, no. 7, pp. 2481–2486, 2016, doi: 10.1002/ange.201510136.

[10] D. A. Walsh and E. Dempsey, "Comparison of electrochemical, electrophoretic and spectrophotometric methods for creatinine determination in biological fluids," *Anal. Chim. Acta*, vol. 459, no. 2, pp. 187–198, 2002, doi: 10.1016/S0003-2670(02)00110-1.

[11] Y. He, X. Zhang, and H. Yu, "Gold nanoparticles-based colorimetric and visual creatinine assay," *Microchim. Acta*, vol. 182, no. 11–12, pp. 2037–2043, 2015, doi: 10.1007/s00604-015-1546-0.

[12] H. Guner *et al.*, "A smartphone based surface plasmon resonance imaging (SPRi) platform for on-site biodetection," *Sensors Actuators, B Chem.*, vol. 239, pp. 571–577, 2017, doi: 10.1016/j.snb.2016.08.061.

[13] M. Soler and L. M. Lechuga, "Boosting Cancer Immunotherapies with Optical Biosensor Nanotechnologies," *Eur. Med. J.*, no. December 2019, pp. 124–132, 2017.

[14] J. C. Verhave, P. Fesler, J. Ribstein, G. Cailar, and A. Mimran, "Measurements of Renal Function," vol. 46, no. 2, pp. 233–241, 2005, doi: 10.1053/j.ajkd.2005.05.011.

[15] B. Nikoobakht and M. A. El-Sayed, "Preparation and growth mechanism of gold nanorods (NRs) using seed-mediated growth method," *Chem. Mater.*, vol. 15, no. 10, pp. 1957–1962, May 2003, doi: 10.1021/cm0207321.

[16] A. De Barros, F. M. Shimizu, C. S. De Oliveira, F. A. Sigoli, D. P. Dos Santos, and I. O. Mazali, "Dynamic Behavior of Surface-Enhanced Raman Spectra for Rhodamine 6G Interacting with Gold Nanorods: Implication for Analyses under Wet versus Dry Conditions," *ACS Appl. Nano Mater.*, vol. 3, no. 8, pp. 8138–8147, 2020, doi: 10.1021/acsanm.0c01530.

[17] S. Nengsih, A. A. Umar, M. M. Salleh, and M. Oyama, "Detection of formaldehyde in water: A shape-effect on the plasmonic sensing properties of the gold nanoparticles," *Sensors (Switzerland)*, vol. 12, no. 8, 2012, doi: 10.3390/s120810309.

[18] A. D. Mcfarland and R. P. Van Duyne, "Single Silver Nanoparticles as Real-Time Optical Sensors with Zeptomole Sensitivity," 2003.

[19] M. K. Kadirov, A. I. Litvinov, and I. R. Nizameev, "Adsorption and Premicellar Aggregation of CTAB Molecules and Fabrication of Nanosized Platinum Lattice on the Glass Surface," 2014.

[20] A. Gurses, M. Yalcin, M. Sozbilir, and C. Dogar, "The investigation of adsorption thermodynamics and mechanism of a cationic surfactant , CTAB , onto powdered active carbon," vol. 81, pp. 57–66, 2003, doi: 10.1016/S0378-3820(03)00002-X.

[21] M. L. I. Uyang, R. A. S. Ingh, C. A. M. Arques, Z. Hang, and S. A. K. Umar, "2D material assisted SMF-MCF-MMF-SMF based LSPR sensor for creatinine detection," *Opt. Express*, vol. 29, no. 23, pp. 38150–38167, 2021.

[22] C. Pothipor, C. Lertvachirapaiboon, K. Shinbo, K. Kato, K. Ounnunkad, and A. Baba, "Detection of creatinine using silver nanoparticles on a poly(pyrrole) thin film-based surface plasmon resonance sensor," *Jpn. J. Appl. Phys.*, vol. 59, no. SC, 2020, doi: 10.7567/1347-4065/ab4a94.

[23] A. H.M. Almawgani *et al.*, "Creatinine Detection by Surface Plasmon Resonance Sensor Using Layers of Cerium Oxide and Graphene Over Conventional Kretschmann Configuration," *Plasmonics*, vol. 18, pp. 1743–1752, 2023.

Modeling of piezoelectric energy harvester: influence of electrode/piezoelectric width ratio on induced voltage

Ghulam Ali
School of Engineering and Built Environment
Griffith University
Nathan, Queensland, Australia
ghulam.ali@griffithuni.edu.au

Feng Xu
School of Engineering and Built Environment
Griffith University
Nathan, Queensland, Australia
feng.xu5@griffithuni.edu.au

Faisal Mohd-Yasin
School of Engineering and Built Environment
Griffith University
Nathan, Queensland, Australia
f.mohd-yasin@griffith.edu.au

Abstract— **Research in piezoelectric energy harvester (PEH) for powering micro-scale devices have been gaining momentum in the past decade. The modeling works are critical to accurately predict performances of PEH structures with different dimensions and active materials before committing to expensive fabrication processes. In this paper, a static modeling of unimorph cantilever PEH in transverse mode is performed. Equations for induced voltage are derived, taking onto consideration different width between electrode and piezoelectric layers (referred to as width ratio). To illustrate the results of this model, we plot the induced voltage versus input force and width ratio of PEH using five variants of Lead Zirconate Titanate (PZT) materials with parameters from commercial datasheets. Finally, the limitation of this model and possible use to improve the performances of PEH are elucidated.**

Keywords—Energy Harvester, PZT, Piezoelectric, Modeling, Transverse

I. INTRODUCTION

The research on the electrochemical batteries and energy storage have been at the forefront of science and engineering. This is due to their demand to power electronic devices such as wireless sensor networks, implantable and wearable devices, and mobile electronics. For low-power devices, energy harvesting could be an alternative to batteries[1]. One of the most ubiquitous energy sources is the vibration-based mechanical energy [2]. There are several transduction mechanisms to harvest and convert the vibration to electrical energy such as electromagnetic, electrostatic, piezoelectric, triboelectric, and magnetostrictive [3]. Among these approaches, piezoelectric energy harvester (PEH) is preferred for capturing mechanical energy from the ambient environments because its energy density is 3 to 5 times higher than electromagnetic and electrostatic devices. Furthermore, PEH is reliable, durable, and is able to withstand large amounts of strain. Most importantly, PEH can be fabricated at micro-scale dimension, which is important for integrations with circuits and sensors [4].

The simplest PEH structure is made of three-layered cantilever as shown in Fig. 1, which is commonly known as unimorph PEH. There have been numerous modelling works being proposed for this structure. Kenji Uchino of Pennsylvania State University is considered as one of the modelling pioneers and derives the static model to calculate the induced voltage [5]. Several follow-up models are mentioned herein. The first is the circuit equivalent model in 2004 [6]. Then, the dynamic model was published in 2009 [7]. One group studied the influence of length ratio of the non-piezoelectric/piezoelectric layers on the induced voltage [8], while another paper documents the impact of thickness ratio of the electrodes/piezoelectric layers on the generated power [9]. We would like to contribute to the literature by modelling the influence of electrode/piezoelectric width ratio on the induced voltage for unimorph cantilever PEH. We believe that the results of this model would be beneficial for those who wants to optimize the geometries of their structure and to reduce the fabrication costs of PEH. The organization of the rest of this paper is as follows. The mathematical modelling of the proposed PEH cantilever is given in the next section. The third section provides the simulation results of the proposed model from Matlab, which employs parameters from the commercial datasheets. After that, we elucidate the limitation of this model as well as its potential application in the fourth section. The fifth section concludes the paper with few suggestions for the future works.

II. MATHEMATICAL MODELING

Fig. 1 shows an unimorph cantilever piezoelectric energy harvester. The piezoelectric layer is assumed to be poled in the vertical direction to operate in the transverse mode. The piezoelectric layer is sandwiched between the upper and lower substrate (electrode) layers. The thicknesses of the bottom, middle, and top layers are given as t_s, t_p, and t_e respectively. The widths of the bottom, middle, and top layers are taken as W_1, W_2, and W_3 respectively, whereas the same length L is assumed for all layers. The distance of the neutral axis t_n from the zero reference point is obtained by the following relationship[10].

$$t_n = \frac{\sum_{i=1}^{n} t_i A_i}{\sum_{i=1}^{n} A_i} \tag{1}$$

Where A_i is the area of each layer and t_i is the distance from the reference point to the center of each layer. The eq (1) is modified as per the geometry of the Fig. 1. The resultant equation is given as follows.

$$t_n = \frac{-t_s^2 W_1 + t_p^2 W_2 + t_e^2 W_3 + 2t_p t_e W_3}{2(t_s W_1 + t_p W_2 + t_e W_3)} \tag{2}$$

979-8-3503-7832-0/24 $31.00 © 2024 IEEE

We further assume that the upper and lower layers have the same widths that is $W_1=W_3=R_wW$, where W is taken as the width of the middle piezoelectric layer ($W_2=W$) and R_w is the substrate / piezoelectric width ratio. Hence, eq (2) is modified as follows:

$$t_n = \frac{t_p^2 + R_W(t_e^2 - t_s^2 + 2t_pt_e)}{2\{t_p + R_W(t_s + t_e)\}} \qquad (3)$$

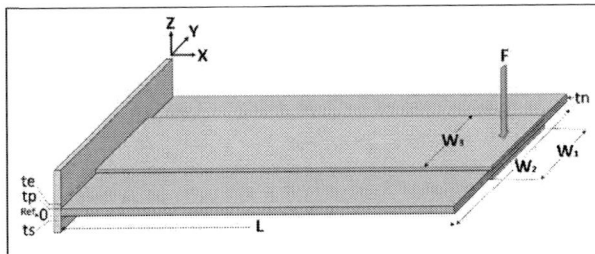

Fig. 1. Model of the unimorph cantilever-based piezoelectric energy harvester in transverse mode. The width ratio R_w is defined as the ratio between the width of upper or lower substrate to the width of the piezoelectric layer.

When the perpendicular concentrated force F at the cantilever tip is applied as shown in Fig. 1, the lateral stress σ in the piezoelectric layer is given as follows:

$$\sigma = \varepsilon E_p = \left\{\frac{(z - t_n)\, F\,(L - x)}{WD}\right\} E_p \qquad (4)$$

Where ε represents lateral strain at position (x, z) and E_p is the Young modulus of the piezoelectric material. The D represents the bending modulus per unit width and is given as follows[10]:

$$D = \sum_{i=1}^{n} E_i \int (h - t_n)^2\, dh \qquad (5)$$

Where E_i is the Young modulus of the respective layers. Eq (5) can be expanded as per geometry as follows:

$$D = \frac{1}{3}\left(E_st_s^3 + E_pt_p^3 + E_e(t_e^3 - t_p^3)\right) + t_n\left(E_st_s^2 - E_pt_p^2 + E_e(t_e^2 - t_p^2)\right) + t_n^2\left(E_st_s + E_pt_p + E_e(t_e - t_p)\right) \qquad (6)$$

Where E_s and E_e are the Young moduli of the lower and upper substrate layers.

The induced voltage at any given x is as follows:

$$V_{in}(x) = \int_0^{t_p} E_{in}(x, z)\, dz \qquad (7)$$

Where $E_{in}(x, z)$ is the induced electric field in the thickness direction in the piezoelectric layer and is given as follows [8]:

$$E_{in}(x, z) = g_{31}\sigma \qquad \text{for } 0 < x < L \qquad (8)$$

Where g_{31} is given as follows [7]:

$$g_{31} = \frac{d_{31}}{\varepsilon_r \varepsilon_o} \qquad (9)$$

Where d_{31} is the piezoelectric strain coefficient, ε_r is the piezoelectric dielectric constant and ε_o is the permittivity of

the free space. By using eq (4), eq (8), and eq (9) in eq (7), we obtain $V_{in}(x)$ as follows:

$$V_{in}(x) = \frac{E_p\, g_{31}\, F(L - x)}{WD}\left(\frac{t_p^2}{2} - t_pt_n\right)$$
$$\text{for } 0 < x < L \qquad (10)$$

The measured induced voltage between the two surfaces of the piezo layer is obtained as follows:

$$V_{in,avg} = \frac{1}{L}\int_0^L V_{in}(x)\, dx \qquad (11)$$

By using eq (10) and computing the integral in eq (11), we get:

$$\frac{V_{in,avg}}{F} = \frac{L}{2}\, g_{31}\, \frac{E_p}{WD}\left(\frac{t_p^2}{2} - t_pt_n\right) \qquad (12)$$

III. SIMULATION RESULTS

It is clear from eq (12) that the induced voltage is dependent on t_n, which is given in eq (3). Accordingly, t_n is highly dependent of R_w. To see how large is the influence of R_w to $V_{in,avg}$ in a practical sense, we run the MATLAB simulation using the piezoelectric materials with values from commercial datasheet. We employ copper alloy (UNS C80100) as the upper and lower substrate layers. The elastic modulus of the substrate layers is taken as 117 Gpa [10]. For the piezoelectric layer, we used four different PZT materials, PZT-4, PZT-5A, PZT-5H, and PZT-8 separately [11]. The material parameters required for the simulation of the PZT layer are summarized in Table I. Finally, the dimensions of the energy harvester such as length, width, and layer thicknesses closely follow the values taken from the PhD thesis at the University of Southampton [12], as given in Table II.

TABLE I. PZT PARAMETERS [11]

Material	Parameters		
	g_{31} (x 10^{-3} Vm/N)	ε_r	E_p (x 10^{10} N/m^2)
PZT-4	-10.6	1300	7.8
PZT-5A	-11.4	1700	6.6
PZT-5H	-8.5	3400	9.3
PZT-8	-10.9	1000	9.9

TABLE II. DIMENSIONS OF THE ENERGY HARVESTER [12]

Parameter	Value
L	20 mm
W	10 mm
t_e	20 um
t_p	80 um
t_s	15 um

The average induced voltage $V_{in,avg}$ in eq (12) is simulated in Matlab and the results are shown in Fig. 2. The curves are obtained by using the surface plot of Matlab. The average induced voltages are shown on z-axis, while the applied Force

F and the width ratio Rw are shown in x and y axis, respectively. The Fig. 2(a), 2(b), 2(c) and 2(d) show the simulated results for PZT-4, PZT- 5A, PZT- 5H and PZT-8, respectively. As expected from the mathematical modelling, the average induced voltage increases with the width ratio and applied force in the simulation results.

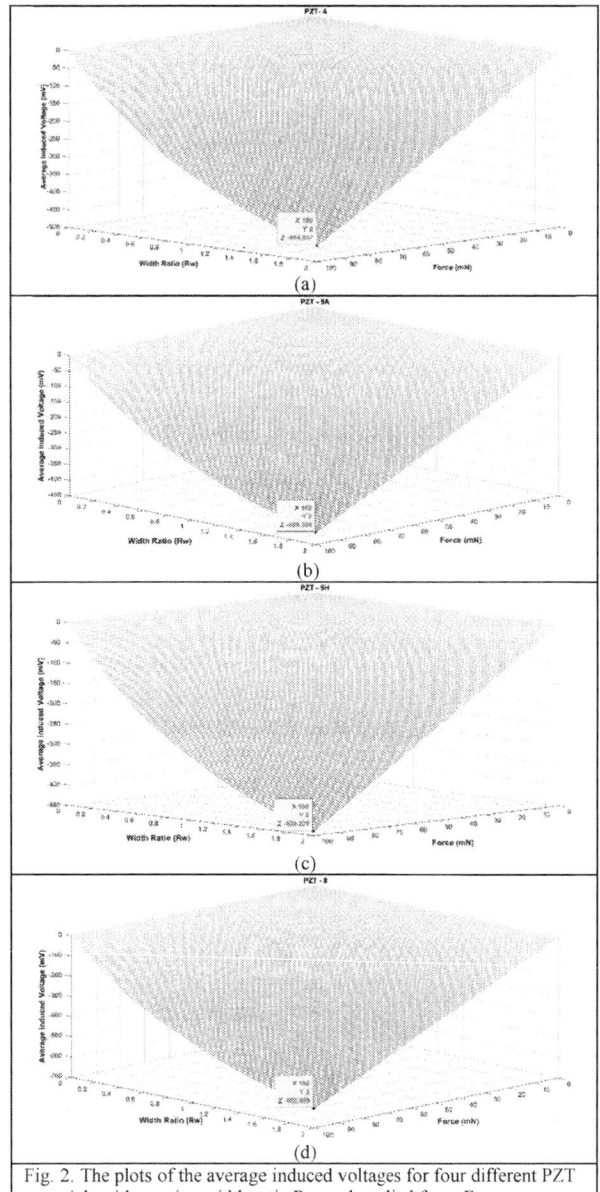

(a)

(b)

(c)

(d)

Fig. 2. The plots of the average induced voltages for four different PZT materials with varying width ratio Rw and applied force F.

For clarity, we define the width ratio (Rw) as follows. In the first case, Rw = 1 indicates that the substrate and the piezoelectric layers have the same width. This is the commonly used unimorph cantilever PEH structure, as the fabrication process is quite straightforward. In the second case, Rw > 1 means the substrate width is wider than the piezoelectric layer. Finally, the third case is to have Rw < 1, which is defined as piezoelectric layer having bigger width than the substrate layer. This case has been reported in previous works, in the case when the top/bottom electrode is consisted of an Ohmic contact [13] to save fabrication processes and simplify testing procedures.

Looking closely at Fig. 2, there are three major observations. First, the trends of the plots show that the average induced voltage increases linearly with the width ratio. This is perhaps the most important result from the derivation of eq. (12) of this model. The second observation has been well-reported by previous models [5]-[9] as well as in this paper. That is, the average induced voltage is directly proportional to the applied input force F. The third observation is related to the piezoelectric material. We observe from Fig. 2 that the minimum and maximum induced voltage is obtained when PZT-5A and PZT-8 are used as the piezoelectric layer, respectively. The main factor for the differing induced voltage between both materials is due to the value of the bending modulus per unit width, D. A closer inspection of eq (6), eq (9) and Table I indicates that the parameters ε_r and E_p are primarily responsible for these differences.

IV. PRACTICAL CONSIDERATIONS

Any mathematical model should be viewed with a pinch of salt as it does not always represent reality. In our case, eq (12) states that the induced voltage is linearly proportional to the value of width ratio Rw. There is a limit to this relationship, as the value of the induced voltage also depends on other parameters such as the length and thickness of the PEH structure. This is main reason for us to employ the dimensions from Table II [12] in simulating the results. Another option to find the limit of Rw is by using Finite Element Analysis [3][14] to generate the induced voltage of the PEH structure.

What is the possible use of this model? In general, since the fabrication of any micro-scale PEH is very expensive, the simulation works to optimize the dimension and choice of active materials of the PEH structure are very critical. This is the primary motivation for the previous works to come up with the static [5] and dynamic models [7], circuit modeling of PEH [6] for optimized coupling with the power management circuit [15], optimization of the length ratio of the non-piezoelectric/piezoelectric layers on the induced voltage [8], and optimizing the thickness ratio of the electrodes/piezoelectric layers on the generated power [9]. For this work, we believe that the optimization of the width ratio between non-piezoelectric/piezoelectric layers could lead to more efficient and cost-effective designs. This is especially true when the exotic materials [16]-[17] are being employed as the PE layer.

V. CONCLUSION

In this work, we present the static modeling of an unimorph cantilever-based PEH in transverse mode. The mathematical relation for obtaining average induced voltage is derived by considering the influence of the width ratio R_w. Our model indicates that the average induced voltage increases with the width ratio and applied input force.

Next, the Matlab plots for average induced voltage are obtained with varying width ratios and applied force. For practical analysis, different PZT materials namely PZT-4,

PZT-5A, PZT-5H, and PZT-8 are used for comparison using the data from the commercial datasheet [11]. The results show that the PZT-8 performs better in terms of induced voltage, while the PZT-5A shows the minimum voltage amplitude in comparison of the other PZT materials.

This work provides clear evidence the influence of R_w on the induced voltage. That is, R_w should not be assumed to be equal to unity to achieve optimized output voltage as the usual practice. For future works, the influence of R_w should also be modeled for other PEH cantilever structures such as the unimorph PEH operating in shear mode [11], multi-morph PEH [18], PEH with multiple cantilevers [19], and freestanding PEH [20]. Furthermore, a more advanced model should be proposed by taking into consideration the combined influences of R_w, as well as the length ratio [8] and thickness ratio [9] of the piezoelectric to non-piezoelectric layer [9], respectively.

ACKNOWLEDGMENT

We would like to express our gratitude to the School of Engineering and Built Environment as well as Queensland Micro and Nanotechnology Centre of Griffith University for providing the supports to perform this research and for awarding PhD scholarships to Ghulam Ali. This work was performed in part at the Queensland node of the Australian National Fabrication Facility, a company established under the National Collaborative Research Infrastructure Strategy to provide nano- and micro-fabrication facilities for Australia's researchers.

REFERENCES

[1] A. Iqbal, G. Walker, A. Iacopi and F. Mohd-Yasin, "Controlled sputtering of AlN (002) and (101) cystal orientations on epitaxial 3C-SiC-on-Si (100) substrate," Journal of Crystal Growth, vol. 440, pp. 76-80, 2016.

[2] A. Iqbal and F. Mohd-Yasin, "Comparison of seven cantilever designs for piezoelectric energy harvester based on Mo/AlN/3C-SiC," In 2015 IEEE Regional Symposium on Micro and Nanoelectronics (RSM2015), 2015, IEEE, pp. 1-4.

[3] J.M.R. Kudimi, F. Mohd-Yasin and S. Dimitrijev, "SiC-based piezoelectric energy harvester for extreme environment," Procedia Engineering, vol. 47, pp. 1165-1172, 2012.

[4] S. Saadon and O. Sidek, "A review of vibration-based MEMS piezoelectric energy harvesters," Energy Conversion and Management, vol. 52, no. 1, pp. 500-504, 2011.

[5] K. Uchino, Piezoelectric actuators and ultrasonic motors. Springer Science & Business Media, 1996.

[6] S. Roundy and P. K. Wright, "A piezoelectric vibration based generator for wireless electronics," Smart Materials and Structures, vol. 13, no. 5, p. 1131, 2004.

[7] Y. Yang and L. Tang, "Equivalent circuit modeling of piezoelectric energy harvesters," Journal of Intelligent Material Systems and Structures, vol. 20, no. 18, pp. 2223-2235, 2009.

[8] X. Gao, W.H. Shih and W.Y. Shih, "Induced voltage of piezoelectric unimorph cantilevers of different nonpiezoelectric/piezoelectric length ratios," Smart Materials and Structures, vol. 18, no. 12, p. 125018, 2009.

[9] Q. Wang, W. Dai, S. Li, J.A.S. Oh and T. Wu, "Modelling and analysis of a piezoelectric unimorph cantilever for energy harvesting application," Materials Technology, vol. 35, no. 9-10, pp. 675-681, 2020.

[10] M.H. Malakooti and H.A. Sodano, "Piezoelectric energy harvesting through shear mode operation," Smart Materials and Structures, vol. 24, no. 5, p. 055005, 2015.

[11] Data Sheet, 19-July-2024. [Online]. Available: https://www.bostonpiezooptics.com/assets/pdf/Ceramic_Materials.pdf

[12] S. L. Kok, "Design, fabrication and characterisation of free-standing thick-film piezoelectric cantilevers for energy harvesting," PhD thesis, University of Southampton, 2010.

[13] A. Iqbal, G. Walker, L. Hold, A. Fernandes, P. Tanner, A. Iacopi and F. Mohd-Yasin, "The sputtering of AlN films on top of on- and off-axis 3C-SiC (111)/Si (111) substrates at various substrate temperatures," Journal of Materials Science: Materials in Electronics, vol. 29, pp. 2434-2446, 2018.

[14] P. Karade, J.M.R. Kudimi and F. Mohd-Yasin, "Micro-scale piezoelectric energy harvester employing 3C-SiC and Si diaphragms," In 2023 IEEE 21st Student Conference on Research and Development (SCOReD), 2023, IEEE, pp. 93-96.

[15] K.M. Mui, M.K. Khaw and F. Mohd-Yasin, "Power management IC for a dual-input-triple-output energy harvester," Micromachines, vol. 11, no. 10, p. 937, 2020.

[16] T. Mori and S. Priya, "Materials for energy harvesting: At the forefront of a new wave," MRS Bulletin, vol. 43, no. 3, pp. 176-180, 2018.

[17] N. Sezer and M. Koç, "A comprehensive review on the state-of-the-art of piezoelectric energy harvesting," *Nano Energy*, vol. 80, p. 105567, 2021.

[18] Y. Zhang, T.F. Lu and W. Gao, "Equivalent homogeneous model of D31-mode longitudinal piezoelectric transducers," Journal of Intelligent Material Systems and Structures, vol. 28, no. 19, pp. 2651-2658, 2017.

[19] W. Jia, F. Bo, F. Jiwen, Z. Jiuchun and L. Chong, "A piezoelectric energy harvester based on multi-cantilevers and magnetic force," Energy Reports, vol. 8, pp. 11638-11645, 2022.

[20] S.W. Kok, N.M. White and N.R. Harris, "Fabrication and characterization of free-standing thick-film piezoelectric cantilevers for energy harvesting," Measurement Science and Technology, vol. 20, no.12, p. 124010, 2009.

Investigation of TID effects on electrical characteristics of GaN MIS-HEMT with LPCVD-grown SiN passivation

Chih-Yi Yang
International College of
Semiconductor Technology
National Yang Ming Chiao
Tung University
Hsinchu, Taiwan
abx50131@nycu.edu.tw

Chin-Han Chung
International College of
Semiconductor Technology
National Yang Ming Chiao
Tung University
Hsinchu, Taiwan
king@nycu.edu.tw

You-Chen Weng
Industry Academia Innovation
School
National Yang Ming Chiao
Tung University
Hsinchu, Taiwan
ycweng@nycu.edu.tw

Jui-Sheng Wu
Industry Academia Innovation
School
National Yang Ming Chiao
Tung University
Hsinchu, Taiwan
juisheng0226@gmail.com

Tsung-Ying Yang
International College of
Semiconductor Technology
National Yang Ming Chiao
Tung UniversityHsinchu,
Taiwan
al1504g4.st09@nycu.edu.tw

Edward-Yi Chang
International College of
Semiconductor Technology
National Yang Ming Chiao
Tung University
Hsinchu, Taiwan
edc@nycu.edu.tw

Abstract—This study examines the impact of total ionizing dose (TID) effects on GaN on Si metal-insulator-semiconductor high electron mobility transistors (MIS-HEMTs) for power applications, with a focus on the impact of the SiN passivation layer grown by low-pressure chemical vapor deposition (LPCVD). Devices featuring the LPCVD SiN passivation were subjected to various Co^{60} γ-ray irradiation doses (100 krad, 400 krad, 1 Mrad and 5 Mrad) under a grounded bias condition. The devices' characteristics were evaluated, including the drain leakage current ($I_{D, leakage}$), the threshold voltage (V_{TH}) shift, the on-resistance (R_{ON}), and the dynamic on-resistance (dyR_{ON}) before and after irradiation. This study provides insights into the performance and reliability of GaN-based power devices with LPCVD-grown passivation under ionizing radiation.

Keywords—GaN on Si, MIS-HEMTs, power device, LPCVD SiN passivation, total ionizing dose, dynamic R_{ON}, V_{TH} shifting.

I. INTRODUCTION

Gallium nitride (GaN) high electron mobility transistors (HEMTs) are being widely researched due to their potential to deliver superior performance for high power and high frequency applications. The exceptional properties of GaN-based HEMTs, such as the two-dimensional electron gas (2DEG) and large bandgap (~3.4 eV), enable enhanced carrier mobility and higher breakdown voltage (BV), leading to higher efficiency and reduced power consumption compared to traditional silicon-based devices. Additionally, GaN HEMTs have a higher tolerance to ionizing radiation, which is a critical factor for important applications such as

This work was financially supported in part (project number: NSTC 112-2622-8-A49-020, NSTC 112-2622-8-A49-013-SB, NSTC 112-2218-E-A49-018, and NSTC 113-2634-F-A49-008.) by the Co-creation Platform of the Industry-Academia Innovation School, NYCU, under the framework of the National Key Fields Industry-University Cooperation and Skilled Personnel Training Act, from the Ministry of Education (MOE), the National Development Fund (NDF), and industry partners in Taiwan..

space, nuclear power facilities, and accelerator facilities[1-3]. These features imply that GaN HEMTs can operate at low power loss, high switching speed, and high BV even in radiation environments.

The reliability of conventional GaN HEMTs has posed many challenges, such as the current collapse and high gate leakage current phenomenon that leads to degraded device performance. To address these issues, implementing the metal-insulator-semiconductor (MIS) structure, forming the MIS-HEMT, has been proposed as a potential solution, which has undergone extensive study in recent years [4-5]. The basic structure of a GaN MIS-HEMT, as depicted in Fig. 1, differs from that of a conventional Ga   N HEMT due to the combination of an insulating dielectric layer beneath the gate metal. Several reports have suggested that the GaN MIS-HEMT exhibits superior reliability and performance compared to conventional GaN HEMTs [6].

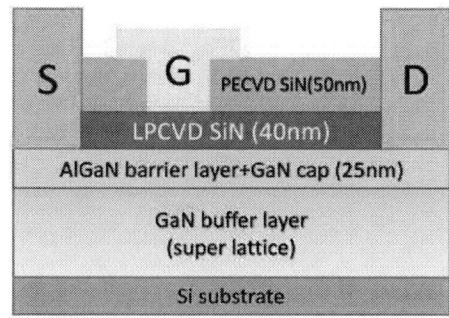

Fig.1 Schematic of a GaN MIS-HEMT with a LPCVD-grown SiN passivation.

The insulator materials used in the MIS-HEMTs highly influence their performance and reliability. SiN is one of the most commonly used materials and can be deposited using two different methods: plasma-enhanced chemical vapor deposition (PECVD) and low-pressure chemical vapor deposition (LPCVD). Compared to devices implementing the PECVD-grown SiN, those with LPCVD-grown SiN can achieve better electrical performance, such as a higher output current, a lower leakage current, and a more stable dynamic on-resistance (dyR_{ON}). The density of the SiN insulator deposited by LPCVD is much higher due to the higher deposition temperature (800-900 °C) compared to PECVD (< 400°C) [7], resulting in better quality of the SiN gate insulator and the GaN/SiN interface. The fewer traps located in the GaN/SiN interface, the better the interface quality, resulting in the suppression of dyR_{ON} degradation.

Despite the advantages, the combination of the gate dielectric layer in the MIS structure may potentially compromise the radiation hardness of the device through a mechanism similar to that observed in irradiated metal-oxide-semiconductor (MOS) structures in Si-based devices. [8-9]. When radiation creates electron-hole pairs (EHPs) within a device, the electrons tend to move out of the region faster than holes due to their higher mobility, which results in a higher likelihood of holes being captured by traps. This can cause trapped holes to accumulate within the device under continuous irradiation, leading to a degradation in device performance. This effect is known as the total ionizing dose (TID) effect. [10]. A comprehensive view of TID effects in GaN MIS-HEMTs was established by our previous work which focused on devices with a PECVD-grown SiN insulator. Thus, further investigation was required for devices with a SiN insulator grown by LPCVD.

In this work, electrical characterization was conducted for GaN MIS-HEMTs with the LPCVD-grown SiN gate insulator irradiated with Co^{60} gamma (γ) ray to study the impact of TID effects.

II. Experimental Details

In this investigation, GaN MIS-HEMT devices were meticulously engineered on Si substrates employing the precise technique of metal organic chemical vapor deposition (MOCVD). The epitaxial structure, intricately designed to exacting specifications, comprised a 4-μm super-lattice GaN buffer layer, a 22-nm $Al_{0.25}Ga_{0.75}N$ barrier layer, and a 3-nm GaN cap layer. Subsequent to this fabrication, ohmic metal stacks (Ti/Al/Ni/Au) were deposited on the wafers utilizing an E-gun evaporator. A crucial step ensued with the rapid thermal annealing (RTA) process at 800°C for 60 seconds, establishing ohmic contacts at the drain and source regions.

Further refinement was achieved through etching the samples with inductively coupled plasma reactive-ion etching (ICP-RIE) using Cl_2, defining mesa isolation. Passivation was then provided by depositing a SiN film, followed by the formation of the T-shaped gate metal. Delving into the investigation of TID impacts on denser SiN insulators, a 40-nm SiN layer deposited via LPCVD served as the gate insulator, forming the MIS structure.

The gate metal, comprising a Ti/Ni/Au stack, was deposited using an E-gun evaporator. Additionally, a 200-nm SiN layer was deposited as the passivation layer for the field plate process, wherein a Ti/Ni/Au stack served as the electrode metal. The setting of device dimensions included a 20 μm distance between the source and the drain (L_{SD}), a 15 μm distance between the gate and the drain (L_{GD}), a 2 μm gate length (L_G), and a substantial 120 mm gate width (W_G), allowing for the achievement of a remarkable maximum output current of 45 A with LPCVD SiN as the gate insulator. The overall process flow is illustrated in Figure 2.

Subsequent assessments involved I-V measurements conducted using the Keysight Technologies B1505A Power Device Analyzer in conjunction with the N1265A Ultra High Current Expander. dyR_{ON} measurements were performed utilizing the Keysight N1267A High Voltage/High Current Fast Switch Module, with Kelvin resistance measurement employed for enhanced precision, given the devices' inherent resistance of approximately 150 mΩ. The setup of dyR_{ON} is the same as previous work[11], which is illustrated in Fig. 3.

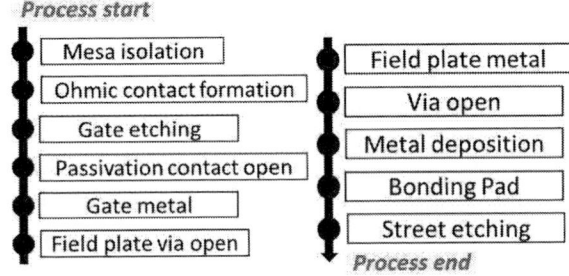

Fig.2 overall process flow of GaN MIS-HEMT.

For the meticulous implementation of dyR_{ON} measurement, a high drain voltage stress was applied to the device-under-test (DUT) under the off-state for 10 ms, employing various values of drain voltage stress (100, 200, 300, and 400 V). At the gate terminal, a voltage stress of -13.5±0.5 V (defined by V_{TH}-5V) was applied to keep the device at the off-state. Upon device activation, extraction of R_{ON} values with $V_D = 3$ V and $V_G = 0$ V facilitated the determination of the ratio of the initial R_{ON} ($R_{initial}$) to the stressed R_{ON} ($R_{stressed}$).

The experimental setup for irradiation, as illustrated in Fig. 4, involved the exposure of the devices to Co^{60} γ-ray irradiation at a dose of 100 krad, 400kard, 1 Mrad, and 5 Mrad at the Building of Isotope of Nuclear Science & Technology Development Center at National Tsing Hua University (NTHU), Taiwan. This approach ensured precise evaluation of the TID effects, providing invaluable insights into the devices' reliability and performance under extreme conditions.

979-8-3503-7832-0/24 $31.00 © 2024 IEEE

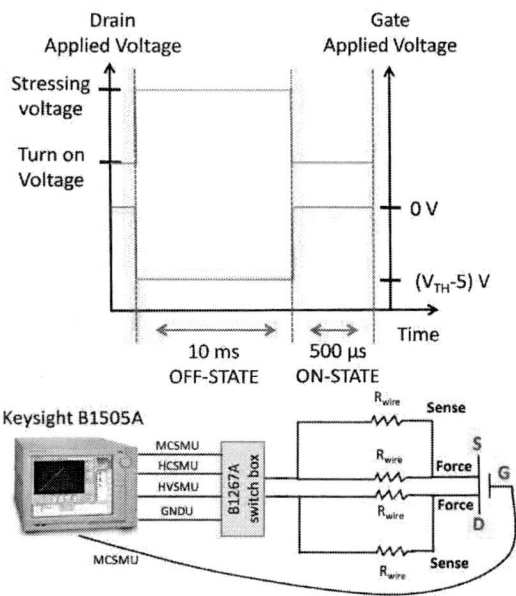

Fig.3 The setup of dyR$_{ON}$ measurement.

Fig.4 The experimental environment for irradiation.

III. RESULT AND DISCUSSION

The static electrical characteristics of the tested devices before and after Co^{60} irradiation are illustrated in Fig. 5. There are no obvious deviations of R$_{on}$ and leakage current (I$_{D, leakage}$) after the devices were irradiated by either a low dose (100 k and 400 k) or a high dose (1 M and 5 M); however, in Fig. 5(c), the threshold voltage (V$_{TH}$) showed a noticeable negative correlation shifts while devices suffered from higher irradiated dose. The results correspond to TID effect on GaN, which was studied comprehensively in the previous work [11]. While the device was exposed to Co^{60} gamma radiation, the EHPs would generate within dielectric layer, interface of SiN/GaN and epitaxial layer. In this case, most of the EHPs were generated and trapped into dielectric layer. The evidence is that R$_{ON}$ and I$_{D, leakage}$ didn't have any apparent changes while the V$_{TH}$ exhibited a shift. Once electrons/holes are trapped into either interface of SiN/GaN or the epitaxial layer, the R$_{ON}$ and the I$_{D, leakage}$ will be

influenced. Moreover, carrier concentration and mobility of the wafer before and after different irradiation dose. were evaluated in Fig. 6, which proves the epitaxial layer wasn't damaged by radiation.

For the dynamic performance, no obvious variations of dyR$_{ON}$ were observed after a low radiation dose (100 k and 400 krad). However, the higher exposed radiation dose increases, the dyR$_{ON}$ degraded more severely. This result shows more evidence of electron accumulation generated by radiation exposure, which causes worse current collapse phenomenon.

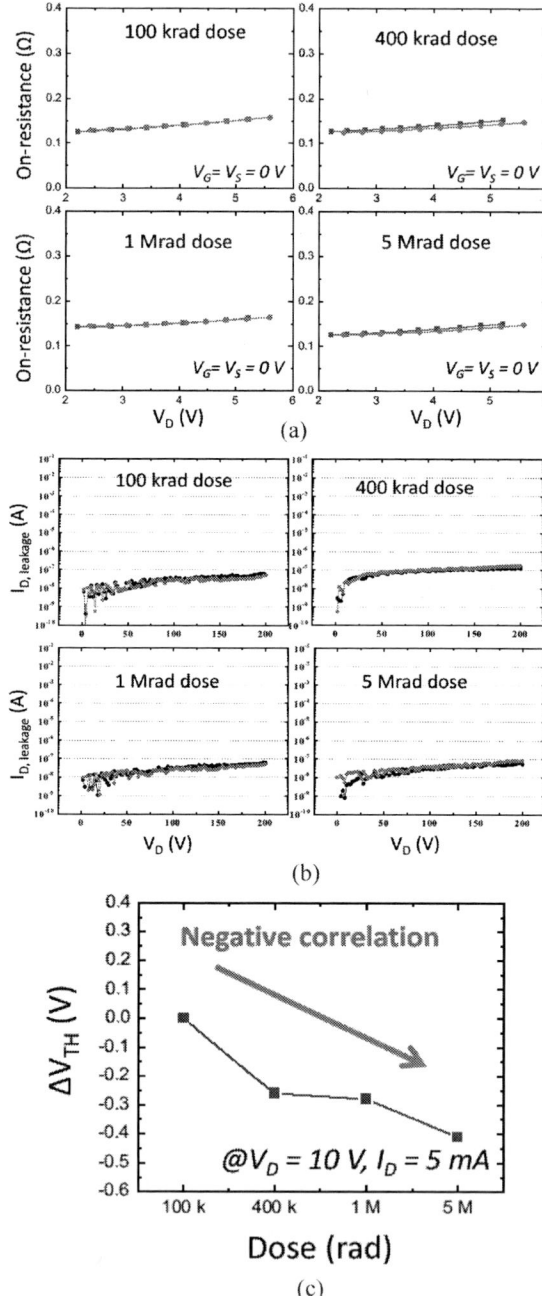

Fig.5 (a) R$_{ON}$, (b) I$_{D, leakage}$, and (c) V$_{TH}$ shift of devices before and after different irradiation dose.

979-8-3503-7832-0/24 $31.00 © 2024 IEEE

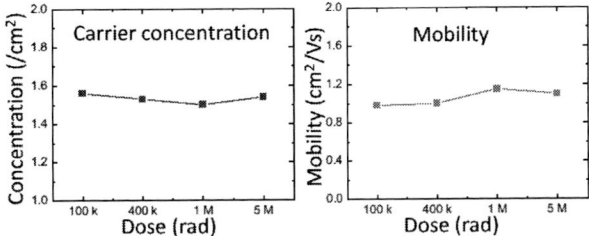

Fig.6 Carrier concentration and mobility of the wafer before and after different irradiation dose.

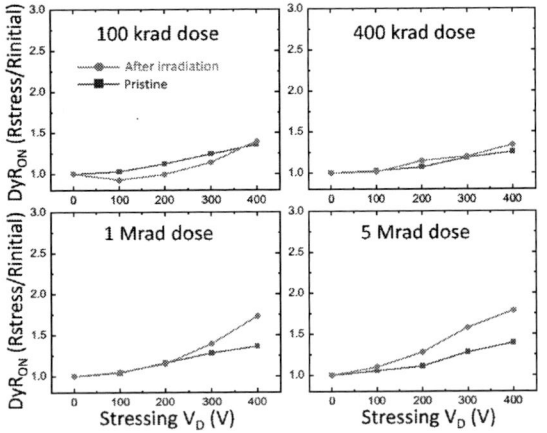

Fig.7 dyR$_{ON}$ of devices before and after different irradiation dose.

IV. CONCLUSION

In this work, the TID effects in GaN MIS-HEMTs with LPCVD-grown SiN passivation were fabricated and investigated. Up to 5 Mrad of Co^{60} γ ray was used to irradiate the devices. It was found that minimum changes were induced to the devices' R$_{ON}$ and I$_{D, leakage}$, even at a high dose. The carrier concentration and mobility also exhibited little variation. The V$_{TH}$ shifted more negatively as the dose increased, and the dyR$_{on}$ also degraded with increasing dose. The results provide insight into the TID response of the device implementing the LPCVD-grown SiN passivation.

REFERENCES

[1] K. J. Chen, O. Häberlen, A. Lidow, C. L. Tsai, T. Ueda, Y. Uemoto, and Y. Wu, "GaN-on-Si power technology: Devices and applications," *IEEE Trans. Electron Devices*, vol. 64, no. 3, pp. 779–795, Mar. 2017. doi: 10.1109/TED.2017.2657579

[2] Edward A. Jones, Fei Fred Wang and Daniel Costinett., "Review of Commercial GaN Power Devices and GaN-Based Converter Design Challenges". *IEEE Journal of Emerg. and Selec. Top. In Power Elect.*, vol. 4, no. 3, pp. 707-719, set. 2016. doi: 10.1109/JESTPE.2016.2582685

[3] A. Stocco, S. Gerardin, D. Bisi, S. Dalcanale, F. Rampazzo, M. Meneghini, G. Meneghesso, J. Grünenpütt, B. Lambert, H. Blanck, E. Zanoni, "Proton induced trapping effect on space compatible GaN HEMTs". *ELSEVIER Journal of Microelectronics Reliability*, vol 54, issues 9-10, pp. 2213-2216, Sep.-Oct. 2014. doi:10.1016/j.microrel.2014.07.120

[4] Marleen Van Hove, Sanae Boulay, Sandeep R. Bahl, Steve Stoffels, Xuanwu Kang, Dirk Wellekens, Karen Geens, Annelies Delabie, and Stefaan Decoutere, "CMOS Process-Compatible High-Power Low-Leakage AlGaN/GaN MISHEMT on Silicon" *IEEE Electron Device Letters*, vol. 33, no. 5, May 2012. doi: 10.1109/LED.2012.2188016

[5] Zhikai Tang, Qimeng Jiang, Yunyou Lu, Sen Huang, Shu Yang, Xi Tang, and Kevin J. Chen, "600-V Normally Off SiNx /AlGaN/GaN MIS-HEMT With Large Gate Swing and Low Current Collapse" *IEEE Electron Device Letters*, vol. 34, no. 11, Nov. 2013. doi: 10.1109/LED.2013.2279846

[6] Enrico Zanoni, Matteo Meneghini, Gaudenzio Meneghesso, Davide Bisi, Isabella Rossetto, Antonio Stocco Department of Information Engineering, University of Padova, "Reliability and failure physics of GaN HEMT, MIS-HEMT and p-gate HEMTs for power switching applications" *IEEE 3rd Workshop on Wide Bandgap Power Devices and Applications (WiPDA)*, 2-4 Nov. 2015 doi: 10.1109/WiPDA.2015.7369305

[7] Bing K. Yen, Richard L. White and Robert J. Waltman, Qing Dai, Dolores C. Miller, Andrew J. Kellock, Bruno Marchon, Paul H. Kasai, Michael F. Toney, Brian R. York, Hong Deng, Qi-Fan Xiao, Vedantham Raman, "Microstructure and properties of ultrathin amorphous silicon nitride protective coating" *J. Vac. Sci. Technol. A* 21, 1895–1904 (2003) doi: 10.1116/1.1615974

[8] Maruf A. Bhuiyan, Hong Zhou, Sung-Jae Chang, Xiabing Lou, Xian Gong, Rong Jiang, Huiqi Gong, En Xia Zhang, Chul-Ho Won, Jong-Won Lim, Jung-Hee Lee, Roy G. Gordon, Robert A. Reed, Daniel M. Fleetwood, Peide Ye. "Total-Ionizing-Dose Responses of GaN-Based HEMTs With Different Channel Thicknesses and MOSHEMTs With Epitaxial MgCaO as Gate Dielectric" *IEEE Transactions on Nuclear Science*, vol. 65, issue 1, pp. 46-52, Jan. 2018 doi: 10.1109/TNS.2017.2774928

[9] Kyung-ah Son, Anna Liao, Gerald Lung, Manuel Gallegos, Toshiro Hatake, Richard D. Harris, Leif Z. Scheick, William D. Smythe, "GaN-based High Temperature and Radiation-Hard Electronics for Harsh Environments" *Nanoscience and Nanotechnology Letters* pp. 89-95, June 2010 doi: 10.1117/12.852711

[10] Xiao Sun, Omair I. Saadat, Jin Chen, En Xia Zhang, Sharon Cui, Tomas Palacios, Daniel M. Fleetwood, and T. P. Ma, "Total-Ionizing-Dose Radiation Effects in AlGaN/GaN HEMTs and MOS-HEMTs" *IEEE Transactions on Nuclear Science*, vol. 60, no. 6, Dec. 2013 doi: 10.1109/TNS.2013.2278314

[11] C. Y. Yang et al., "A Comprehensive Study of Total Ionizing Dose Effect on the Electrical Performance of the GaN MIS-HEMT," in IEEE Transactions on Device and Materials Reliability, vol. 22, no. 2, pp. 276-281, June 2022, doi: 10.1109/TDMR.2022.3173000

979-8-3503-7832-0/24 $31.00 © 2024 IEEE

The Effects of Al₂O₃ Interlayer on the Ferroelectric Behavior of $Hf_{0.5}Zr_{0.5}O_2$ Thin Film for E-mode Ferroelectric Charge Trap Gate GaN HEMT

Jui-Sheng Wu
Industry Academia Innovation School
National Yang Ming Chiao Tung University
Hsinchu, Taiwan
juisheng0226@nycu.edu.tw

Tsung-Ying Yang
International College of Semiconductor Technology
National Yang Ming Chiao Tung University
Hsinchu, Taiwan
al1504g4.st09@nycu.edu.tw

You-Chen Weng
Industry Academia Innovation School
National Yang Ming Chiao Tung University
Hsinchu, Taiwan
ycweng@nycu.edu.tw

Chih-Yi Yang
International College of Semiconductor Technology
National Yang Ming Chiao Tung University
Hsinchu, Taiwan
abx50131@nycu.edu.tw

Yu-Tung Du
International College of Semiconductor Technology
National Yang Ming Chiao Tung University
Hsinchu, Taiwan
ado0975348979@gmail.com

Yi-Hsiang Wei
International College of Semiconductor Technology
National Yang Ming Chiao Tung University
Hsinchu, Taiwan
buffywei@gmail.com

Edward-Yi Chang
International College of Semiconductor Technology
National Yang Ming Chiao Tung University
Hsinchu, Taiwan
edc@nycu.edu.tw

Abstract—**This research primarily investigates the impact of adding an aluminum oxide (Al₂O₃) interlayer between titanium nitride (TiN) capping layers on the ferroelectric properties of hafnium zirconium oxide ($Hf_{0.5}Zr_{0.5}O_2$) thin films. Through ferroelectric hysteresis measurements, PUND measurements, XRD analysis, and TEM analysis, the influence of the Al₂O₃ interlayer on the ferroelectric properties and crystalline characteristics of $Hf_{0.5}Zr_{0.5}O_2$ is observed and examined.**

Keywords—ferroelectric, hafnium-zirconium oxide, ferroelectric charge trap gate stack, gallium nitride

I. INTRODUCTION

In recent years, hafnium-zirconium oxide ($Hf_xZr_{1-x}O_2$) has gained attention for its ferroelectric properties, offering distinct advantages over traditional ferroelectric materials. Hafnium-zirconium oxide exhibits remarkable thinness, good remnant polarization, a wide bandgap, and could be deposited by atomic layer deposition (ALD) systems [1][2]. The compatibility of $Hf_xZr_{1-x}O_2$ thin film with semiconductor fabrication processes makes it an attractive candidate for memory materials, including ferroelectric random-access memory (FeRAM) and ferroelectric gate field-effect transistors (FeFET).

Furthermore, the ferroelectric-charge-trapping gate stack has attracted a great deal of attention due to the concept of realizing E-mode operation for GaN-based HEMTs. Recently, a high V_{th} normally-OFF GaN HEMT with high drain current density was achieved by the implementation of a hybrid ferroelectric charge trap gate stack (FEG-HEMT), with a laminated $Hf_{0.5}Zr_{0.5}O_2$ ferroelectric layer stacked on top of a charge trapping layer [3][4]. The FEG-HEMT exhibits a higher positive V_{th} shift after charge injection, higher V_{th} stability, and reliable gate oxide with higher breakdown.

To induce ferroelectricity in hafnium-zirconium oxide thin films, the prevailing approach involves high-temperature annealing with titanium nitride (TiN) capping layers to induce a phase transition, resulting in an orthorhombic structure with ferroelectric properties [5]. However, experimental observations have revealed that during high-temperature processing, oxygen atoms tend to migrate, leading to the formation of titanium oxynitride (TiOxNy). The presence of TiOxNy has been associated with ferroelectric fatigue, compromising the material's reliability [6].

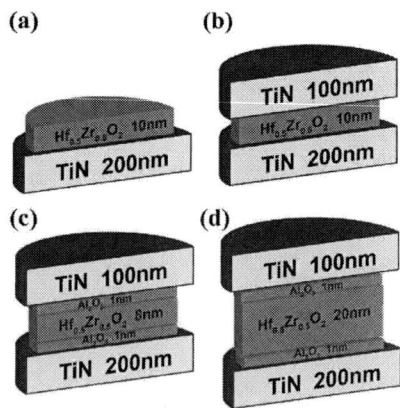

Fig. 1. Schematic structure of the four fabricated MFM (metal/ferroelectric/ metal), namely (a) $Hf_{0.5}Zr_{0.5}O_2$ (*10 nm*)/TiN, (b) TiN/$Hf_{0.5}Zr_{0.5}O_2$ (*10 nm*)/TiN, (c) TiN/Al₂O₃ (*1 nm*)/$Hf_{0.5}Zr_{0.5}O_2$ (*8 nm*)/Al₂O₃ (*1 nm*)/TiN, and (d) TiN/Al₂O₃ (*1 nm*)/$Hf_{0.5}Zr_{0.5}O_2$ (*20 nm*)/Al₂O₃ (*1 nm*)/TiN.

This study focuses on investigating the impact of adding additional aluminum oxide (Al_2O_3) interlayers on the ferroelectric properties of hafnium-zirconium oxide ($Hf_{0.5}Zr_{0.5}O_2$) thin films. Extensive characterization techniques were employed, including ferroelectric hysteresis measurements, pulse and negative voltage (PUND) measurements, X-ray diffraction (XRD) analysis, and transmission electron microscopy (TEM) analysis, to assess the influence of the aluminum oxide interlayer on the ferroelectric behavior and crystalline characteristics of hafnium-zirconium oxide.

II. EXPERIMENTAL DETAILS

In this work, four samples were fabricated and analyzed to demonstrate the thin film ferroelectric characteristics differences. The four samples of MFM (metal/ferroelectric/metal) fabricated was namely, (a) $Hf_{0.5}Zr_{0.5}O_2$ (*10 nm*)/TiN, (b) TiN/$Hf_{0.5}Zr_{0.5}O_2$ (*10 nm*)/TiN, (c) TiN/Al_2O_3 (*1 nm*)/$Hf_{0.5}Zr_{0.5}O_2$ (*8 nm*)/Al_2O_3 (*1 nm*)/TiN, and (d) TiN/Al_2O_3 (*1 nm*)/$Hf_{0.5}Zr_{0.5}O_2$ (*20 nm*)/Al_2O_3 (*1 nm*)/TiN. Figure 1 shows the schematic structure of the four fabricated MFM. All of the MFM structures was grown on Si substrate and the top TiN layer (samples b, c, d) and bottom TiN layer thickness is 100 nm and 200 nm, respectively. The TiN capping layers were grown by sputtering, while the ferroelectric thin film and Al_2O_3 interlayer were grown by thermal ALD using H_2O as the oxidant and tetrakis(dimethylamino)hafnium(IV) (Hf(NMe$_2$)$_4$, TDMAHf). Tetrakis(dimethylamido) zirconium (Zr(NMe$_2$)$_4$,

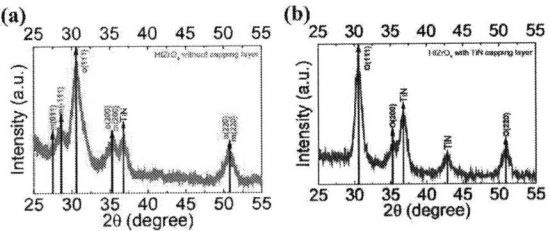

Fig. 2. XRD patterns of (a) $Hf_{0.5}Zr_{0.5}O_2$ (*10 nm*)/TiN, (b) TiN/$Hf_{0.5}Zr_{0.5}O_2$ (*10 nm*)/TiN.

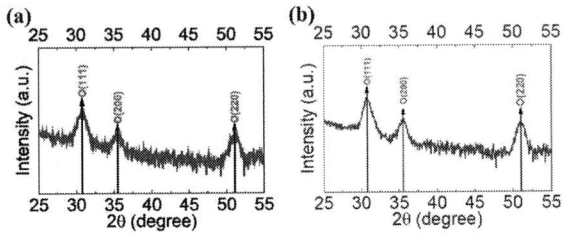

Fig. 3. XRD patterns of sample c: TiN/Al_2O_3 /$Hf_{0.5}Zr_{0.5}O_2$ (*8 nm*)/Al_2O_3/TiN and sample d: TiN/Al_2O_3 /$Hf_{0.5}Zr_{0.5}O_2$ (*20 nm*)/Al_2O_3/TiN.

TDMAZr), and trimethylaluminum (AlMe$_3$, TMA) as the precursors. Finally, all of the samples underwent a post deposition annealing of 400°C for 60 seconds. Sample (a) is set as a control group for a more complete analysis of the

samples. Samples (b) and (c) differ from the ferroelectric thin film stack, which sample (c) exists two Al_2O_3 interlayers. Although sample (c) has additional Al_2O_3 interlayers between the top and bottom TiN capping layers, the thickness of the oxide film is still the same as sample (b). The same oxide thickness of samples (b) and (c) were designed for a more suitable comparison, whereas when this ferroelectric stack will be future integrated into the ferroelectric charge trap gate stack E-mode GaN HEMT (FEG-HEMT). Furthermore, sample (d) was fabricated to show the improvements and the ferroelectric property differences when increasing the $Hf_{0.5}Zr_{0.5}O_2$ thin film thickness from 8 nm to 20 nm.

III. RESULTS AND DISCUSSION

Figure 2(a) shows the XRD analysis of sample (a), which is depositing $Hf_{0.5}Zr_{0.5}O_2$ without the top TiN capping layer. Through the XRD analysis, prominent diffraction peaks corresponding to m(-111) and o(111) are readily observable. Additionally, two minor diffraction peaks are observed, one arising from the mixture of o(200) and m(200), and the other from the mixture of o(220) and m(220). This suggests that the $Hf_{0.5}Zr_{0.5}O_2$ thin film lacking TiN coating exhibits a structure characterized by a partial monoclinic and partial orthorhombic nature.

In contrast, Figure 2(b) illustrates sample (b) of $Hf_{0.5}Zr_{0.5}O_2$ with the top TiN capping layer. XRD analysis reveals diffraction peaks corresponding to o(111), o(200), and o(220). While these three peaks indicative of the o-phase are observed, it is not conclusive evidence that $Hf_{0.5}Zr_{0.5}O_2$ possesses an orthorhombic structure. The similarity between the tetragonal and orthorhombic structures necessitates caution in interpreting these diffraction peaks, which could potentially represent a mixture of o-phase and t-phase. However, it is confirmed that the top TiN-coated sample effectively inhibit the nucleation of the monoclinic phase, thereby suppressing the monoclinic structure, as asserted in the literature due to the ability of TiN coating to mitigate volume expansion in thin films and exert shear stress on the monoclinic structure. Further confirmation of the orthorhombic structure would require P-E measurements. If $Hf_{0.5}Zr_{0.5}O_2$ exhibits no ferroelectricity, it can be inferred to be of the t-phase, whereas if it demonstrates ferroelectric properties, it can be inferred to possess an o-phase structure.

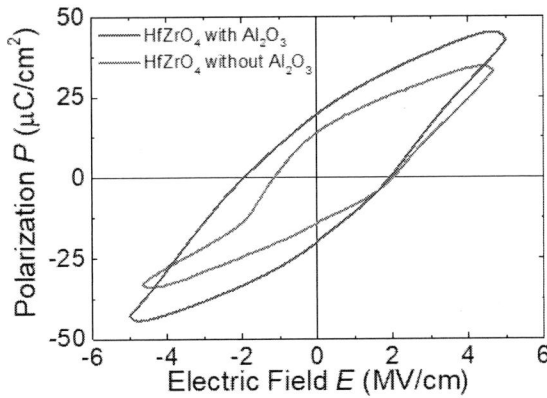

Fig. 4. Polarization *vs.* electric field measurements (P-E) of $Hf_{0.5}Zr_{0.5}O_2$ thin film without Al_2O_3 interlayer (sample b) and $Hf_{0.5}Zr_{0.5}O_2$ thin film with Al_2O_3 interlayer (sample c).

Both Figure 3(a) and Figure 3(b) exhibit distinct diffraction peaks corresponding to the three primary o-phase planes ((111), (200), and (220)), while lacking diffraction peaks indicative of the m-phase. Consequently, irrespective of the thickness of $Hf_{0.5}Zr_{0.5}O_2$, ranging from 8 nm to as thick as 20 nm, the utilization of Al_2O_3 as an interlayer effectively suppresses the growth of the monoclinic structure in $Hf_{0.5}Zr_{0.5}O_2$.

The underlying factor influencing the ferroelectric (FE) performance resides in the presence of the orthorhombic phase (space group $Pca2_1$, denoted as the o-phase), which consistently coexists with various polymorphs of HfO_2, namely monoclinic (space group $P2_1/c$, referred to as the m-phase), tetragonal (space group $P4_2/nmc$, designated as the t-phase), or even cubic (space group $Fm3m$, termed as the c-phase). Such coexistence presents a suboptimal characteristic of this ferroelectric material system. Notably, the absence of the most thermodynamically stable m-phase, typical in bulk material, suggests a propensity towards the formation of metastable phases, including the ferroelectric o-phase. In the context of $Hf_{0.5}Zr_{0.5}O_2$ (HZO), the prevalence of the m-phase varies considerably depending on its specific composition and annealing conditions, thereby influencing the remnant polarization (P_r) value detrimentally.

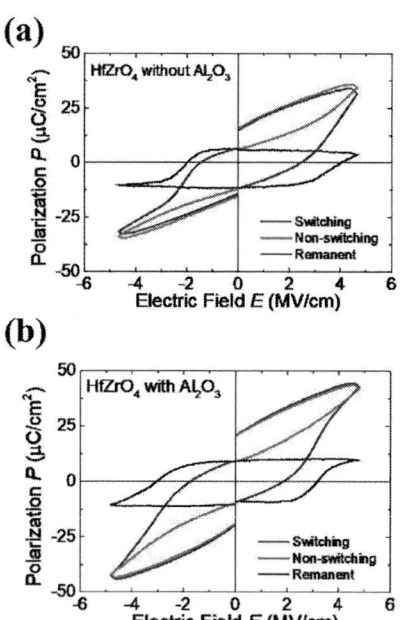

Fig. 5. Corrected P–E loops obtained using the PUND method of (a) $Hf_{0.5}Zr_{0.5}O_2$ thin film without Al_2O_3 interlayer (sample b) and (b) $Hf_{0.5}Zr_{0.5}O_2$ thin film with Al_2O_3 interlayer (sample c).

In the P-E measurement graph (Figure 4), it is observed that both $Hf_{0.5}Zr_{0.5}O_2$ specimens with TiN interlayer ($Hf_{0.5}Zr_{0.5}O_2$ without Al_2O_3) and those with Al_2O_3 interlayer ($Hf_{0.5}Zr_{0.5}O_2$ with Al_2O_3) exhibit ferroelectric properties. This indicates that Al_2O_3, similar to TiN, can prevent the formation of the m-phase in $Hf_{0.5}Zr_{0.5}O_2$, suppress grain growth, and induce a phase transition to the ferroelectric oIII-phase. Furthermore, in the specimens with Al_2O_3 interlayer, not only is there no significant deterioration in ferroelectric properties observed, but there also appears to be an increase in remnant

polarization (P_r), with the remnant polarization of the Al_2O_3 interlayer sample approximately 19 ($\mu C/cm^2$) compared to approximately 13 ($\mu C/cm^2$) for the TiN capping layer sample. Concerning the coercive field (E_c), it is noted that the graph of the TiN interlayer specimens exhibits a relatively asymmetric shape, with $-E_c$ being approximately only half of $+E_c$. In

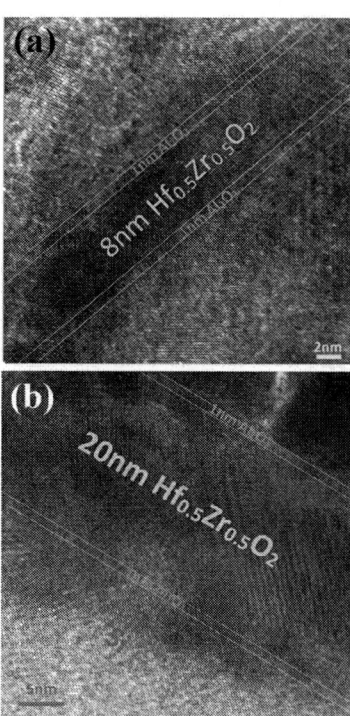

Fig. 6. TEM image of $Hf_{0.5}Zr_{0.5}O_2$ thin film (a) 8 nm (b) 20 nm with Al_2O_3 interlayer (sample c and d).

contrast, the E_c of the Al_2O_3 interlayer specimens is approximately 2 (MV/cm).

In order to mitigate potential influences stemming from leakage currents and parasitic charges, the positive-up negative-down (PUND) methodology was employed. Comparison between Figures 5(a) and 5(b) reveals that even with the insertion of an Al_2O_3 interlayer, there is minimal change in the remnant polarization (defined as the difference between the intersection of the remnant curve and the polarization axis, denoted as $2P_r$) of $Hf_{0.5}Zr_{0.5}O_2$. The sample with the Al_2O_3 interlayer exhibit a remnant polarization of approximately 9 ($\mu C/cm^2$), while those with the TiN interlayer show a similar remnant polarization, also around 9 ($\mu C/cm^2$). This indicates that specimens with the Al_2O_3 interlayer possess ferroelectric properties comparable to those with the TiN interlayer. Discrepancies between the remnant polarization measured here and in figure 4 are attributed to the use of PUND (Positive-Up Negative-Down) measurements, which exclude the influence of leakage current and dielectric displacement current, thus isolating the ferroelectric displacement current for more precise results.

Furthermore, from the graphs, it can be observed that the slope of the remnant curve at the intersection point on the X-axis is larger for specimens with the Al_2O_3 interlayer, indicating a faster switching when the electric field changes direction. Subsequently, comparing the non-switching red

979-8-3503-7832-0/24 $31.00 © 2024 IEEE

curves, it is noted that both specimens exhibit similar widths. These non-switching red curves are derived from the measured leakage current and dielectric displacement current, indicating comparable leakage currents between $Hf_{0.5}Zr_{0.5}O_2$ specimens with Al_2O_3 and TiN interlayers. However, overall, both samples exhibit relatively high leakage currents, likely due to the polycrystalline structure and thinness of $Hf_{0.5}Zr_{0.5}O_2$, facilitating the flow of leakage current.

From Figures 6(a) and 6(b), distinctions between the Al_2O_3 interlayer of 8 nm $Hf_{0.5}Zr_{0.5}O_2$ and 20 nm $Hf_{0.5}Zr_{0.5}O_2$ are observable under TEM imaging. Darker regions within the TEM images denote the $Hf_{0.5}Zr_{0.5}O_2$ thin film layer. Given the higher atomic numbers of Hf (atomic number 72) and Zr (atomic number 40) compared to aluminum (atomic number 13), titanium (atomic number 22), and silicon (atomic number 14), the $Hf_{0.5}Zr_{0.5}O_2$ layer appears distinctly darker in both Figures 6(a) and 6(b). Notably, the $Hf_{0.5}Zr_{0.5}O_2$ layer in Figure 6(b) exhibits a visibly greater thickness.

In Figure 6(a), flanking the central dark $Hf_{0.5}Zr_{0.5}O_2$ region, lighter-colored thin lines are discernible, presumed to be Al_2O_3 based on structural inference. The surrounding area exhibits slightly lighter-colored TiN. Additionally, multiple distinct striations within the $Hf_{0.5}Zr_{0.5}O_2$ layer denote its polycrystalline nature. Notably, irregularities are observed along the boundaries of the $Hf_{0.5}Zr_{0.5}O_2$ layer, potentially stemming from the non-uniformity noticed during the initial deposition of the underlying TiN layer.

In Figure 6(b), longer and broader striations are evident within the $Hf_{0.5}Zr_{0.5}O_2$ layer, attributed to the increase in grain size with film thickness. Similar irregularities are observed along the boundaries of the $Hf_{0.5}Zr_{0.5}O_2$ layer, and the upper and lower layers of Al_2O_3 are not clearly distinguishable,

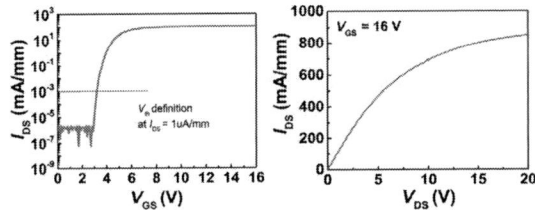

Fig. 7. Transfer characteristics and output characteristics of the E-mode FEG-HEMT applying the Al_2O_3/ $Hf_{0.5}Zr_{0.5}O_2$ oxide gate stack.

possibly due to mutual diffusion between Al_2O_3 and $Hf_{0.5}Zr_{0.5}O_2$ during annealing, compounded by the intrinsic thinness of Al_2O_3 (1 nm), rendering it difficult to discern.

Figure 7 shows the preliminary results of incorporating the Al_2O_3/ $Hf_{0.5}Zr_{0.5}O_2$ oxide stack into the gate of the GaN MIS-HEMT, creating the hybrid ferroelectric charge trap gate stack GaN HEMT (FEG-HEMT). The fabricated FEG-HEMT demonstrated an $V_{th} = 3.22$ V, $I_{DS, MAX}$ of 842 mA/mm at a gate voltage (V_{GS}) of 16 V and a R_{ON} of 9.1 Ω·mm.

IV. CONCLUSION

Experimental findings corroborate that even with the substitution of titanium nitride (TiN) by aluminum oxide (Al_2O_3) as the interlayer, hafnium zirconium oxide ($Hf_{0.5}Zr_{0.5}O_2$) retains its ferroelectric properties. Remnant polarization measurements conducted via P-E hysteresis loops indicate a remnant polarization of approximately 19 ($\mu C/cm^2$) for the sample with the Al_2O_3 interlayer, compared to approximately 13 ($\mu C/cm^2$) for those with the TiN interlayer. Subsequent PUND measurements enable precise deduction of the effects of leakage and dielectric displacement currents, sample with the Al_2O_3 interlayer is comparable to that of specimens with the TiN interlayer, both falling within a moderate range (approximately 9 $\mu C/cm^2$). Furthermore, even with $Hf_{0.5}Zr_{0.5}O_2$ thickness reaching 20 nm, it is still observed that aluminum oxide effectively suppresses the formation of the m-phase within the $Hf_{0.5}Zr_{0.5}O_2$ film. Additionally, TEM analysis reveals the polycrystalline nature of the grown $Hf_{0.5}Zr_{0.5}O_2$, consistent with XRD diffraction results. These investigations shows promising results for future integration into E-mode GaN FEG-HEMTs.

ACKNOWLEDGMENT

This work was financially supported in part (project number: NSTC 112-2622-8-A49-020, NSTC 112-2622-8-A49-013-SB, NSTC 112-2218-E-A49-018, NSTC 112-2622-8-A49-020 and NSTC 113-2634-F-A49-008.) by the Co-creation Platform of the Industry-Academia Innovation School, NYCU, under the framework of the National Key Fields Industry-University Cooperation and Skilled Personnel Training Act, from the Ministry of Education (MOE), the National Development Fund (NDF), and industry partners in Taiwan.

REFERENCES

[1] J. Mueller, E. Yurchuk, T. Schloesser, J. Paul, R. Hoffmann, S. Mueller, D. Martin, S. Slesazeck, P. Polakowski, J. Sundqvist, M. Czernohorsky, K. Seidel, P. Kuecher, R. Boschke, M. Trentzsch, K. Gebauer, U. Schroeder, and T. Mikolajick, "Ferroelectricity in HfO2 enables nonvolatile data storage in 28 nm HKMG," IEEE Symposium on VLSI Technology, 2012.

[2] A. Chernikova, M. Kozodaev, A. Markeev, D. Negrov, M. Spiridonov, S. Zarubin, O. Bak, P. Buragohain, H. Lu, E. Suvorova, A. Gruverman, and A. Zenkevich, "Ultrathin $Hf_{0.5}Zr_{0.5}O_2$ Ferroelectric Films on Si," ACS Appl. Mater. Interfaces, 2016, 8, 7232.

[3] C.-H. Wu, S.-C. Liu, C.-K. Huang, Y.-C. Chiu, P.-C. Han, P.-C. Chang, F. Lumbantoruan, C.-A. Lin, Y.-K. Lin, C.-Y. Chang, C. Hu, H. Iwai, and E. Y. Chang, "High V_{th} enhancement mode GaN power devices with high $I_{D,max}$ using hybrid ferroelectric charge trap gate stack," VLSI Symp. Tech. Dig., pp. T60-T61, June. 2017..

[4] J.-S. Wu, P.-H. Liao, S.-J. Chang, T.-Y. Yang, C.-Y. Teng, Y.-K. Liang, D. Panda, Q. H. Luc, and E. Y. Chang, "Superior Breakdown, Retention, and TDDB Lifetime for Ferroelectric Engineered Charge Trap Gate E-mode GaN MIS-HEMT," 2022 International Electron Devices Meeting (IEDM), San Francisco, CA, USA, 2022, pp. 847-850.

[5] Boscke, T. S.; Muller, J.; Brauhaus, D.; Schroder, U.; Bottger, U., "Ferroelectricity in hafnium oxide thin films," Appl. Phys. Lett. 2011, 99,

[6] Y. Goh and S. Jeon, "The effect of the bottom electrode on ferroelectric tunnel junctions based on CMOS-compatible HfO2," Nanotechnology 29(33), 335201 (2018).

Low Noise with High Linearity AlGaN/GaN HEMT Using Γ-Shaped Gate for Ka-Band Applications

Howie Tseng
International College of Semiconductor Technology
National Yang Ming Chiao Tung University
Hsinchu, Taiwan
howie.tseng.st11@nycu.edu.tw

Neng-Da Li
Department of Materials Science and Engineering
National Yang Ming Chiao Tung University
Hsinchu, Taiwan
lnd.en12@nycu.edu.tw

Yueh-Chin Lin
Department of Materials Science and Engineering
National Yang Ming Chiao Tung University
Hsinchu, Taiwan
nctulin@yahoo.com.tw

Mu-Yu Chen
Institute of Electronics
National Yang Ming Chiao Tung University
Hsinchu, Taiwan
kyle.ee11@nycu.edu.tw

Edward Yi Chang
Department of Materials Science and Engineering
National Yang Ming Chiao Tung University
Hsinchu, Taiwan
edc@nycu.edu.tw

Abstract—In this study, AlGaN/GaN high-electron mobility transistors (HEMT) with Γ-shaped gate structure was developed and analyzed for Ka-band application. Under the frequency of 28 GHz, the proposed device exhibits great minimum noise figure (NF_{min}) of 1.51 dB, which can be attributed to the enlarged cross-sectional area caused by Γ-shaped gate structure. Besides, a high third order intercept point (OIP3) value of 35.8 dBm is achieved, which is resulting from the field plate structure of the Γ-shaped gate. The results prove that the AlGaN/GaN HEMT with Γ-shaped gate structure is capable of having great microwave noise performance and high linearity for Ka-band applications.

Keywords—AlGaN/GaN HEMT, noise figure, linearity, Γ-shaped gate

I. INTRODUCTION

In recent years, communication technologies have undergone rapid development. To improve the quality and efficiency of a transceiver, low noise amplifier (LNA) is often implemented. Among all released frequency bandwidth, Ka-band (27-40 GHz) has various kinds of applications.

For a device to be applied in LNA circuits, the microwave noise performance must be improved. Besides, the demand for massive data transmission increases, in order to prevent the transmitted signal from distorting, a device with high linearity is required. As a result, developing a transistor with low noise and high linearity for Ka-band application has gradually become an important issue.

An AlGaN/GaN high-electron mobility transistors (HEMT) with Γ-shaped gate structure was implemented in this study. AlGaN/GaN HEMT was chosen for its large bandgap and high mobility. Besides, Γ-shaped gate has enlarged cross-sectional area of gate metal, which is believed to improve noise figure for its reduced gate resistance. Furthermore, Γ-shaped gate provides a field plate structure, which was proven as an efficient way to improve device linearity [1].

In this study, AlGaN/GaN HEMT with Γ-shaped gate structure was fabricated. To examine the device's microwave noise performance and linearity, the measurement of minimum noise figure (NF_{min}) and the two-tone linearity measurement under Ka-band frequency were conducted.

II. FABRICATION

Fig. 1 shows the cross-section schematic diagram of the proposed AlGaN/GaN HEMT. The device was grown on 4-inch SiC substrate. The fabrication started from ohmic contact region definition using I-line lithography, then the ohmic contact metal stack of Ti/Al/Ni/Au were deposited and annealed at 830°C to form ohmic contact. Mesa isolation was then conducted by B^+ ion implantation, for defining active region.

The gate formation process began thereafter. First layer of SiN_x was deposited to define gate stem height using plasma-enhanced chemical vapor deposition (PECVD). Afterward, we defined the gate structure by two lithography process. The first lithography process using stepper, and the follow-up SiN_x dry etching process using inductively coupled plasma (ICP), were applied to define the gate foot position. Afterward, to define the gate metal deposition area and to create the Γ-shaped structure, the second stepper lithography was conducted. The second lithography pattern was shifted slightly from the trench created in the first stepper lithography process, and the overlap between the patterns of the two lithography process managed to define the position for Γ-shaped gate. Schottky gate metals of Ni/Au were then deposited, followed by ICP dry etching on the SiN_x layer.

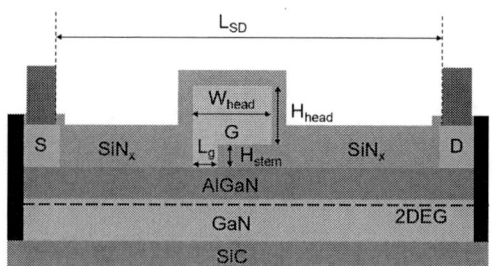

Fig. 1. Cross-section of AlGaN/GaN HEMT with Γ-shaped gate

After that, the second layer of SiN$_x$ was deposited also using PECVD to serve as passivation layer. Then, via opening process, which consisted of I-line lithography patterning and ICP dry etching, was conducted for future interconnect metal deposition. Finally, an Au metallization layer with thickness of 1.5 μm was deposited.

So far, the fabrication for the AlGaN/GaN HEMT with Γ-shaped gate finished. The device has gate length (L$_g$) of 150 nm, gate head width (W$_{head}$) of 500 nm, gate head height (H$_{head}$) of 350 nm, gate stem height (H$_{stem}$) of 150 nm, source-drain spacing (L$_{SD}$) of 2.5 μm. Besides, the fabricated devices have two kinds of gate finger peripheries: two fingers and four fingers, with each finger having gate width of 50 μm (namely, 2×50 μm and 4×50 μm).

III. RESULTS AND DISSCUSSION

A. DC Characteristics

Fig. 2(a) and (b) show the DC characteristics of 2×50 μm device with Γ-shaped gate structure. As shown in the figures, when gate-to-source voltage (V$_{GS}$) was 0 V and drain-to-source voltage (V$_{DS}$) was 20 V, the saturated drain-to-source current density (I$_{DSS}$) reached 893.0 mA/mm. Besides, the maximum transconductance (G$_{m,max}$) reached 320.7 mS/mm. The good G$_{m,max}$ and I$_{DSS}$ can be attributed to the scaled L$_g$ of 150 nm.

B. Microwave Noise Performance

Noise measurement was conducted within the frequency band of 24-45 GHz, which encapsules the Ka-band range. The noise characteristics of the proposed device was measured at V$_{DS}$ = 10 V, and the drain current density (I$_{DS}$) was biased at class AB. Fig. 3 (a) and (b) show the results of NF$_{min}$ and the associated gain versus frequency. Under the frequency of 28 GHz, for 2×50 μm and 4×50 μm devices, the NF$_{min}$ were 2.14 and 1.51 dB, respectively, and the associated gain were 5.23 and 6.11 dB, respectively. These results were comparable with other studies regarding AlGaN/GaN HEMT noise performance, as shown in Table I. The great noise characteristics can be attributed to the enlarged gate cross-sectional area caused by the Γ-shaped gate structure, which then led to the decrease in gate resistance. The results also verified the Fukui equation proposed by previous research [2], which stated that an optimized NF$_{min}$ can be acquired if R$_g$ decreases, which can be accomplished by applying Γ-shaped gate structure, as shown in this study.

C. Linearity

To determine the linearity of 2×50 μm device, the gate voltage swing (GVS, defined as 10% drop of G$_{m,max}$) was calculated, and the two-tone measurement at 28 GHz with load-pull system was conducted. After extracting from the DC characteristic, the proposed device exhibited GVS of 1.15 V. The good GVS implied that the device had good G$_m$ flatness, which is a crucial factor for good device linearity. For the two-tone measurement, the device was measured at V$_{DS}$ = 20 V, and I$_{DS}$ was biased at class A. As shown in Fig.4, the proposed device exhibited high third order intercept point (OIP3) of 35.8 dBm. The high OIP3 implied that the device was capable of maintaining main signal of the circuit under the interference of adjacent channels, which is also a crucial factor for good device linearity. The extracted

linearity results were comparable with other studies regarding AlGaN/GaN HEMT linearity, as shown in Table II. The above-mentioned results stated that the proposed device had good linearity, which could be attributed to the Γ-shaped gate structure. The Γ-shaped gate provided a small gate field plate structure, which is capable of improving device linearity by altering electric field distribution.

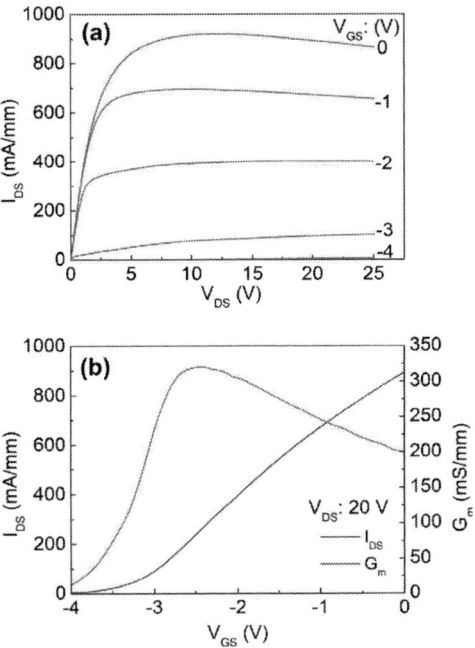

Fig. 2. (a) I$_{DS}$-V$_{DS}$ and (b) I$_{DS}$-V$_{GS}$ characteristics of AlGaN/GaN HEMT with Γ-shaped gate

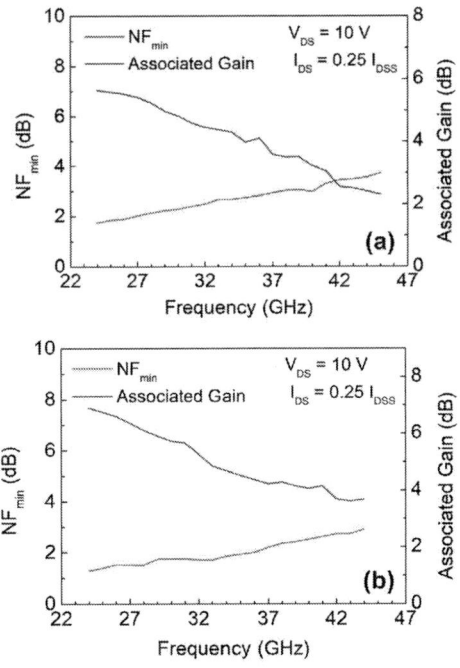

Fig. 3. Microwave Noise performance of (a) 2×50 μm and (b) 4×50 μm AlGaN/GaN HEMT with Γ-shaped gate

TABLE I. COMPARISON OF MICROWAVE NOISE PERFORMANCE OF THE FABRICATED DEVICE AND OTHER PUBLISHED DATA IN RECENT YEARS

Device	V_{DS} (V)	Freq. (GHz)	Device size (μm)	NF_{min} (dB)	Associated gain (dB)
This study	10	28	2×50	2.14	5.23
This study	10	28	4×50	1.51	6.11
[3]	10	28	2×50	2.2	6.20
[4]	6	26	2×75	1.7	8
[5]	6	30	2×75	2	7.5
[6]	6	18	2×50	1.88	8

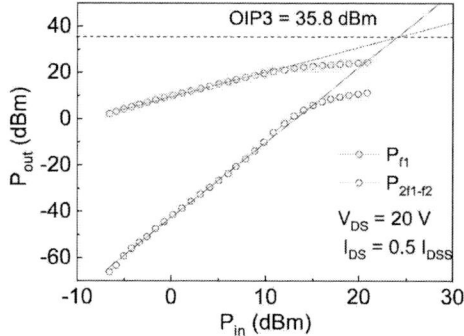

Fig. 4. Linearity of AlGaN/GaN HEMT with Γ-shaped gate

TABLE II. COMPARISON OF LINEARITY OF THE FABRICATED DEVICE AND OTHER PUBLISHED DATA IN RECENT YEARS

Device	V_{DS} (V)	Freq. (GHz)	Device size (μm)	OIP3 (dBm)
This study	20	28	2×50	35.8
[7]	10	6	2×50	26.6
[8]	10	30	2×37.5	34
[9]	7	30	2×25	32
[10]	28	10	2×75	27.7

IV. CONCLUSION

In this study, AlGaN/GaN HEMT with Γ-shaped gate structure was successfully fabricated. The device was measured with great NF_{min} of 1.51 dB and high OIP3 of 35.8 dBm at the frequency of 28 GHz, which can be attributed to the Γ-shaped gate for its enlarged cross-sectional area and its field plate structure. The results proved that the device with Γ-shaped gate structure has great microwave noise performance and high linearity for Ka-band applications.

ACKNOWLEDGMENT

This work was financially supported in part (project number: NSTC 112-2622-8-A49 -020, NSTC 112-2622-8-A49-013-SB, NSTC 112-2218-E-A49-018 , NSTC 112-2622-8-A49-020 and NSTC 113-2634-F-A49-008.) by the Co-creation Platform of the Industry-Academia Innovation School, NYCU, under the framework of the National Key Fields Industry-University Cooperation and Skilled Personnel Training Act, from the Ministry of Education (MOE), the National Development Fund (NDF), and industry partners in Taiwan.

REFERENCES

[1] C.-W. Hsu, Y.-C. Lin, C.-H. Yang, and E. Y. Chang, "Study of Low Noise with High Linearity AlGaN/GaN HEMTs by Optimizing Γ-Gate Structure for Ka-Band Applications," ECS Journal of Solid State Science and Technology, vol. 12, no. 7, p. 075005, 2023.

[2] H. Fukui, "Optimal noise figure of microwave GaAs MESFET's," IEEE Transactions on Electron Devices, vol. 26, no. 7, pp. 1032-1037, 1979.

[3] H. Tseng et al., "Improvement of AlGaN/GaN HEMT Noise Figure Using Thick Cu Metallization for Satellite Communication Applications," IEEE Journal of the Electron Devices Society, 2024.

[4] S. Piotrowicz et al., "12W/mm with 0.15 μm InAlN/GaN HEMTs on SiC technology for K and Ka-Bands applications," in 2014 IEEE MTT-S International Microwave Symposium (IMS2014), 2014: IEEE, pp. 1-3.

[5] S. D. Nsele et al., "Ka-band low noise amplifiers based on InAlN/GaN technologies," in 2015 International Conference on Noise and Fluctuations (ICNF), 2015: IEEE, pp. 1-4.

[6] W. Lu, V. Kumar, E. L. Piner, and I. Adesida, "DC, RF, and microwave noise performance of AlGaN-GaN field effect transistors dependence of aluminum concentration," IEEE Transactions on Electron Devices, vol. 50, no. 4, pp. 1069-1074, 2003.

[7] H.-C. Wang, H.-F. Su, Q.-H. Luc, C.-T. Lee, H.-T. Hsu, and E. Y. Chang, "Improved linearity in AlGaN/GaN HEMTs for millimeter-wave applications by using dual-gate fabrication," ECS Journal of Solid State Science and Technology, vol. 6, no. 11, p. S3106, 2017.

[8] M. Guidry et al., "Demonstration of 30 GHz OIP3/PDC> 10 dB by mm-wave N-polar deep recess MISHEMTs," in 2019 14th European Microwave Integrated Circuits Conference (EuMIC), 2019: IEEE, pp. 64-67.

[9] P. Shrestha et al., "High linearity and high gain performance of N-polar GaN MIS-HEMT at 30 GHz," IEEE Electron Device Letters, vol. 41, no. 5, pp. 681-684, 2020.

[10] Y.-n. Zhong and Y.-m. Hsin, "Linearity of algan/gan hemts with different gate-to-source length," in 2019 IEEE Workshop on Wide Bandgap Power Devices and Applications in Asia (WiPDA Asia), 2019: IEEE, pp. 1-3.

979-8-3503-7832-0/24 $31.00 © 2024 IEEE

Threshold Voltage Instability of GaN HEMTs with Thin Barrier AlGaN Technology

Tsung-Ying Yang
International College of
Semiconductor Technology
National Yang Ming Chiao
Tung University
Hsinchu, Taiwan
al1504g4.st09@nycu.edu.tw

Jui-Sheng Wu
Industry Academia Innovation
School
National Yang Ming Chiao
Tung University
Hsinchu, Taiwan
juisheng0226@gmail.com

You-Chen Weng
Industry Academia Innovation
School
National Yang Ming Chiao
Tung University
Hsinchu, Taiwan
ycweng@nycu.edu.tw

Chih-Yi Yang
International College of
Semiconductor Technology
National Yang Ming Chiao
Tung University
Hsinchu, Taiwan
abx50131@nycu.edu.tw

Edward-Yi Chang
International College of
Semiconductor Technology
National Yang Ming Chiao
Tung University
Hsinchu, Taiwan
edc@nycu.edu.tw

Abstract—In this work, the effect on threshold voltage (V_{th}) instability of AlGaN barriers using higher aluminum concentrations (29%) but only 5 nm will be discussed and compared to enhancement-mode (E-mode) devices using the gate recess process. The devices exhibit varying degrees of threshold voltage shift in the positive bias stress (PBS) test and the negative bias test (NBS) test. From the results, the FEG-HEMTs were able to have better V_{th} stability in the presence of PBS. In the case of NBS, the negative V_G influence will lose the electrons stored in the charge trap layer and cause the V_{th} to be shifted more.

Keywords—*GaN-HEMTs, ferroelectric charge trap gate stack, PBS, NBS, V_{th} instability.*

I. INTRODUCTION

Gallium Nitride (GaN) high-power devices are leading-edge technology designed for high-power and high-frequency applications. They feature a wide bandgap, allowing operation at high temperatures and electric fields without breakdown. GaN devices have high switching speeds due to their high carrier mobility, making them ideal for rapid switching applications like electric vehicle chargers and solar converters. With lower on-state resistance and reduced switching losses, GaN devices offer higher efficiency in high-power applications.

While GaN offers numerous advantages, we aim for it to operate in Enhancement-mode (E-mode) to save energy and ensure fault turn effect. Currently, there are several viable approaches: P-GaN, cascode, gate recess, and fluorine doping. Each has its own advantages and disadvantages. In this work, we employ thin barrier AlGaN instead of gate recess, along with a ferroelectric charge trap gate stack

This work was financially supported in part (project number: NSTC 112-2622-8-A49-020, NSTC 112-2622-8-A49-013-SB, NSTC 112-2218-E-A49-018, and NSTC 113-2634-F-A49-008.) by the Co-creation Platform of the Industry-Academia Innovation School, NYCU, under the framework of the National Key Fields Industry-University Cooperation and Skilled Personnel Training Act, from the Ministry of Education (MOE), the National Development Fund (NDF), and industry partners in Taiwan..

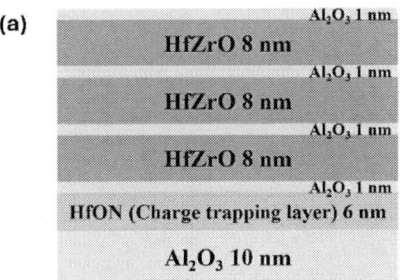

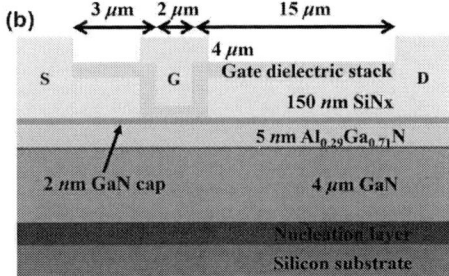

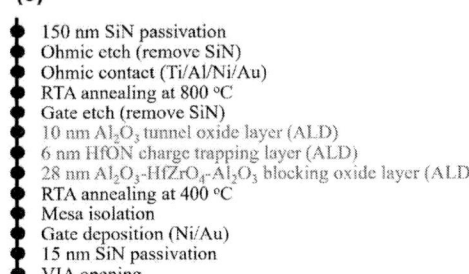

Fig. 1. (a) Ferroelectric charge trap gate stack structure. (b) Thin barrier FEG-HEMT device cross section. (c) Fabrication process flow.

structure. This allows us to achieve both high V_{th} and high current characteristics simultaneously [1] - [5].

In MIS-HEMTs, processes such as gate recess or the dielectric layer itself can generate excess traps, leading to

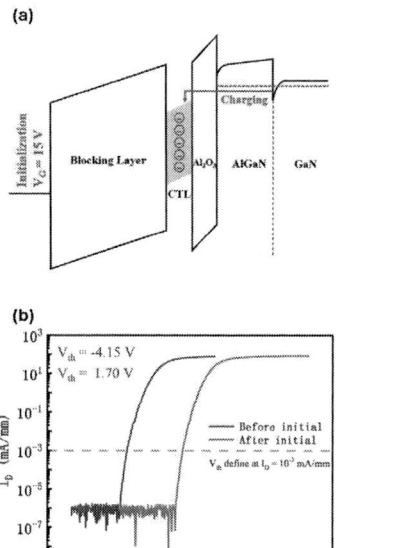

(a)

(b)

$V_{th} = -4.15$ V
$V_{th} = 1.70$ V

— Before initial
— After initial

V_{th} define at $I_D = 10^{-3}$ mA/mm

Fig. 2. (a) The principle of the initialization process. (b) V_{th} of components before and after initialization.

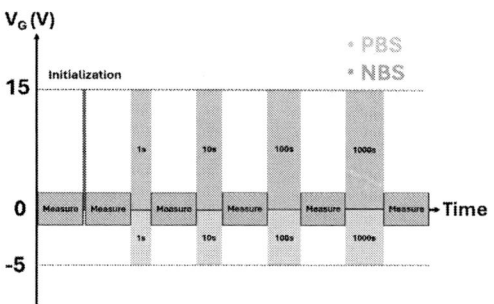

Fig. 3. Stress test conditions for PBS, NBS, and retention time.

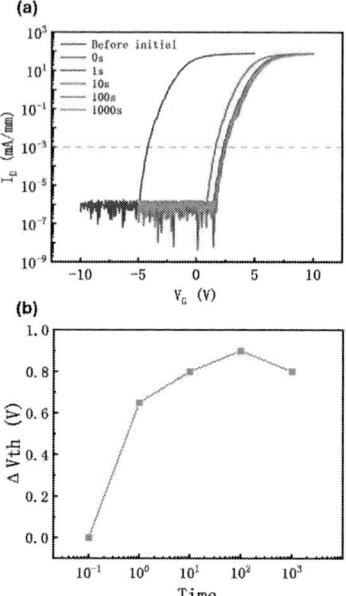

(a)

(b)

Fig. 4. PBS test after V_G equal to 15 V for 1, 10, 100, and 1000 seconds.

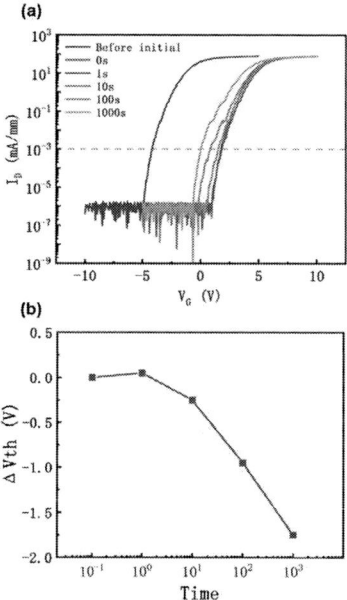

(a)

(b)

Fig. 5. NBS test after V_G equal to 15 V for 1, 10, 100, and 1000 seconds.

unstable V_{th}. Therefore, analyzing the stability of the device under different biases is essential. PBS and NBS are commonly used to assess the stability of V_{th} and investigate the quantity of traps [6]. In this study, they were also employed to evaluate thin barrier devices.

From the results, we can conclude that the device exhibits minimal impact under PBS, maintaining a stable V_{th} value. This helps to ensure stable operation when V_G is greater than 0 V. However, under NBS, the impact is more significant. If the operating environment involves V_G being less than 0 V, special design considerations or periodic initialization processes for the device may be necessary.

II. DEVICE FABRICATION

The device was fabricated using MOCVD to grow GaN epitaxial layers on Si(111) substrates. Following the epitaxial growth, a 150 nm SiN layer was deposited via LPCVD as passivation. Subsequently, ohmic etch was performed to remove the SiN beneath the source and drain regions, followed by deposition of Ti/Al/Ni/Au as ohmic metal, and annealing at 800 °C for 1 minute. A low-power SF$_6$ plasma was then used for gate etch to remove the SiN and define the gate region. The ferroelectric charge trap layer was grown via ALD, as shown in Fig. 1(a), followed by annealing at 400 °C for 10 minutes. Ni/Au was utilized as the gate metal.

Finally, a 15 nm SiN layer was deposited via PECVD as the top passivation layer, with VIA opened using ICP for subsequent measurements. The cross-section of the device is depicted in Fig. 1(b), with dimensions of 3 μm for source-to-drain distance, 2 μm for gate length, 15 μm for gate-to-drain distance, 4 μm for T-gate length, and 25 μm for gate width. The full process can be seen in Fig. 1(c).

III. RESULT AND DISCUSSION

Fig. 2(a) illustrates the initialization process of the GaN device. When a 15 V bias is applied to the gate for 1 ms while V_S and V_D are both 0 V, the 2DEG in the channel tunnels through the Al$_2$O$_3$ layer of the ferroelectric charge trap layer due to the 15 V voltage, storing electrons in the HfON charge trap layer. Additionally, within the high voltage of 15 V, the ferroelectric properties of HZO can also be activated. Once the initialization process is completed, the electrons stored in the HfON layer can deplete the 2DEG beneath the gate, resulting in the E-mode operation of the device. The ferroelectric properties of HZO contribute to the storage of electrons in HfON.

In Fig. 2(b), we observe the variation in V_{th} before and after the initialization process. Before the initialization process, the V_{th} of the device is -4.15 V, while after the initialization process, the V_{th} increases to 1.7 V, resulting in a V_{th} shift of 5.85 V. Here, V_{th} is defined when I_D is 10^{-3} mA/mm.

Next, we conducted V_{th} stability tests on the thin barrier devices, including PBS and NBS tests. The test conditions for these three tests are illustrated in Fig. 3. Each test begins with an initial measurement, followed by an initialization process, and then a first measurement as the baseline. Subsequently, stress tests of 1, 10, 100, and 1000 seconds were performed, with a V_{th} check conducted after each stress period.

In Fig. 4(a), after undergoing stress with V_G = 15 V, the V_{th} values observed after each stress period are as follows: 1.7 V, 2.35 V, 2.5 V, 2.6 V, and 2.5 V. In Fig. 4(b), it is evident that following the first second of stress, the device's V_{th} experiences the most significant increase, approximately 0.6 V. However, as time progresses, the V_{th} of the device remains relatively stable, with minimal variations observed after 10, 100, and 1000 seconds, all within the range of V_{th} equal to 2.5 V.

In Fig. 5(a), the V_{th} values observed after stress with V_G = -5 V are as follows: 1.8 V, 1.85 V, 1.55 V, 0.85 V, and 0.05 V. In Fig. 5(b), it is evident that the impact of NBS on V_{th} is much greater than that of PBS. This is because the principle behind the formation of E-mode by the FEG gate stack is similar to the progress-erase mechanism of flash memory. Therefore, under negative bias conditions, it is easier to push the charged 2DEG back into the GaN channel, resulting in a decrease in V_{th}.

IV. CONCLUSION

In this experiment, we used high aluminum concentration (29%) and thin (5nm) AlGaN thin barriers to fabricate E-mode FEG-HEMTs, aiming to replace the AlGaN gate recess process with thin barrier epitaxy. From the experimental results, we successfully replicated E-mode devices using FEG-HEMTs, achieving a V_{th} of 1.7V. Subsequently, we tested the stability of V_{th}, including PBS and NBS. The results showed that PBS exhibited more V_{th} shift in the initial 1 second, but after 10 seconds, it was almost unaffected by PBS, with V_{th} remaining at 2.5V. The NBS results showed more negative shifts, which we attribute to electrons in the charge trap layer being forced back into the channel under negative gate bias, causing significant V_{th} variation.

REFERENCES

[1] C.-H. Wu, S.-C. Liu, C.-K. Huang, Y.-C. Chiu, P.-C. Han, P.-C. Chang, F. Lumbantoruan, C.-A. Lin, Y.-K. Lin, C.-Y. Chang, C. Hu, H. Iwai, and E. Y. Chang, "High Vth Enhancement Mode GaN Power Devices with High ID, max Using Hybrid Ferroelectric Charge Trap Gate Stack," 2017 Symposium on VLSI Technology, 2017, pp. T60-T61, doi: 10.23919/VLSIT.2017.7998201.

[2] C.-H. Wu, P.-C. Han, S.-C. Liu, T.-E. Hsieh, F.J. Lumbantoruan, Y.-H. Ho, J.-Y. Chen, K.-S. Yang, H.-C. Wang, Y.-K. Lin, P.-C. Chang, Q.H. Luc, Y.-C. Lin, and E.Y. Chang, "High-Performance Normally-OFF GaN MIS-HEMTs Using Hybrid Ferroelectric Charge Trap Gate Stack (FEG-HEMT) for Power Device Applications," IEEE Electron Device Letters, vol. 39, no. 7, pp. 991-994, July 2018, doi: 10.1109/LED.2018.2825645.

[3] J.-S. Wu, P.-H. Liao, S.-J. Chang, T.-Y. Yang, C.-Y. Teng, Y.-K. Liang, D. Panda, Q. H. Luc, and E. Y. Chang, "Superior Breakdown, Retention, and TDDB Lifetime for Ferroelectric Engineered Charge Trap Gate E-mode GaN Mis-HEMT," 2022 International Electron Devices Meeting (IEDM), San Francisco, CA, USA, 2022, pp. 847-850, doi: 10.1109/IEDM45625.2022.10019471.

[4] H. -Y. Lee, C. -H. Lin, C. -C. Wei, J. -C. Yang, E. Y. Chang and C. -T. Lee, "AlGaN/GaN Enhancement-Mode MOSHEMTs Utilizing Hybrid Gate-Recessed Structure and Ferroelectric Charge Trapping/Storage Stacked LiNbO3/HfO2/Al2O3 Structure," in IEEE Transactions on Electron Devices, vol. 68, no. 8, pp. 3768-3774, Aug. 2021, doi: 10.1109/TED.2021.3090343.

[5] Y. Cai, Y. Zhang, Y. Liang, I. Z. Mitrovic, H. Wen, W. Liu and C.Zhao, "Low ON-State Resistance Normally-OFF AlGaN/GaN MIS-HEMTs With Partially Recessed Gate and ZrOx Charge Trapping Layer," in IEEE Transactions on Electron Devices, vol. 68, no. 9, pp. 4310-4316, Sept. 2021, doi: 10.1109/TED.2021.3100002.

[6] S. Zafa, Y. Kim, V. Narayanan, C. Cabral, V. Paruchuri, B. Doris, J. Stathis, A. Callegari, M. Chudzik, "A Comparative Study of NBTI and PBTI (Charge Trapping) in SiO2/HfO2 Stacks with FUSI, TiN, Re Gates," 2006 Symposium on VLSI Technology, 2006. Digest of Technical Papers., Honolulu, HI, USA, 2006, pp. 23-25, doi: 10.1109/VLSIT.2006.1705198.

1kV Vertical Breakdown Voltage AlGaN/GaN HEMTs on Si with AlN and AlGaN/AlN Superlattice Buffer Engineering

You-Chen Weng
Industry Academia Innovation
School
National Yang Ming Chiao
Tung University
Hsinchu, Taiwan
ycweng@nycu.edu.tw

Chin-Yi Yang
International College of
Semiconductor Technology
National Yang Ming Chiao
Tung University
Hsinchu, Taiwan
abx50131@nycu.edu.tw

Tsung-Ying Yang
International College of
Semiconductor Technology
National Yang Ming Chiao
Tung University
Hsinchu, Taiwan
al1504g4.st09@nycu.edu.tw

Chin-Han Chung
International College of
Semiconductor Technology
National Yang Ming Chiao
Tung University
Hsinchu, Taiwan
king@nycu.edu.tw

Tsung-Han Chiang
International College of
Semiconductor Technology
National Yang Ming Chiao
Tung University
Hsinchu, Taiwan
e19931010.c@nycu.edu.tw

Fu-Chin Tung
Mechanical and Mechatronics
Systems Research Labs
Industrial Technology Research
Institute
Hsinchu, Taiwan
fctung@itri.org.tw

Shih-Hsiang Lai
Mechanical and Mechatronics
Systems Research Labs
Industrial Technology Research
Institute
Hsinchu, Taiwan
shlai@itri.org.tw

Hung-Wei Yu
International College of
Semiconductor Technology
National Yang Ming Chiao
Tung University
Hsinchu, Taiwan
bachelor.mse96g@nycu.edu.tw

Edward Yi Chang
International College of
Semiconductor Technology
National Yang Ming Chiao
Tung University
Hsinchu, Taiwan
edc@nycu.edu.tw

Abstract—In this study, we investigated the device characteristics of AlGaN/GaN High-Electron-Mobility Transistors (HEMTs) fabricated on 6-inch silicon substrates using AlN and an $Al_{0.07}GaN/AlN$ superlattice. Transmission electron microscopy revealed a sharp interface between the AlN and the Si substrate, effectively reducing leakage from the Si. Additionally, it showed a distinct interface between the AlGaN and AlN layers within the superlattice structure. The AlGaN/GaN HEMTs exhibited a vertical breakdown voltage of approximately 1000 V and maintained ultra-low leakage up to 700 V, with a three-terminal off breakdown voltage exceeding 1500 V. These results highlight the significant potential of both HLH-AlN and the AlGaN/AlN superlattice on silicon substrates for high-power switching applications.

Keywords—GaN HEMT, AlGaN/AlN superlattice

I. INTRODUCTION

Nitride-based materials and their compound alloys have emerged as very promising options for high-power device applications. This growing interest is largely attributed to their wide bandgap, which enables efficient energy conversion, allowing devices to handle higher voltages and temperatures than traditional silicon-based semiconductors. This wide bandgap also results in GaN-based devices having higher breakdown voltages, which is crucial for high-power applications. Additionally, GaN materials exhibit high electron mobility, which facilitates faster operation and significantly reduces energy loss during device operation.

This means that electronic devices can perform more efficiently and at higher speeds. Moreover, GaN materials are known for their robust thermal stability, which allows them to maintain performance even in high-temperature environments. This thermal stability, combined with their exceptional electronic properties, makes GaN-based semiconductors particularly well-suited for operation in harsh and aggressive environmental conditions where reliability is crucial, such as in aerospace, military, and industrial applications. These attributes collectively position GaN-based semiconductors at the forefront of advancements in electronic and power systems technologies, offering a pathway to more efficient, reliable, and powerful electronic devices in the future.

For GaN-on-Silicon (Si) applications in high-power scenarios, one of the most significant challenges is addressing leakage current losses caused by issues with the Aluminum Nitride (AlN)/Si [1] interface and suboptimal buffer designs. At the AlN/Si interfaces form as a reverse channel, which we called the parasitic channels often form due to the diffusion of Ga atoms into the Si substrate. Various buffer structures have been explored to tackle this issue, including the step-graded (SG)[2], [3] AlGaN buffer, low- (Al)GaN buffer, low-temperature Al(Ga)N interlayer buffers, and (Al)GaN/AlN superlattice (SL) buffers [4], [5], [6], [7].

The SL buffer has recently become first choice for high-voltage operations. This is because it enhances the GaN film quality and facilitates the growth of thick epitaxial structures on silicon substrates. The SL buffer significantly improves key device performance metrics, including breakdown voltage and Current Collapse (CC).

II. EXPERIMENT

The AlGaN/GaN HEMT structures were grown using Veeco Propel™ metalorganic chemical vapor deposition (MOCVD) on 150 mm Si(111) wafers. Four distinct types of epitaxial layers were grown, as illustrated in Fig. 1.

Sample A features a sophisticated epitaxial structure beginning with a 200 nm AlN seed layer, followed by a 600 nm strained SG AlGaN buffer. This buffer includes layers of 150 nm $Al_{0.7}GaN$, 200 nm $Al_{0.5}GaN$, and 250 nm $Al_{0.25}GaN$, meticulously engineered from bottom to top. Continuing, there's a substantial 3000 nm GaN:C back-barrier layer that provides essential support. Finally, the structure culminates with a 400 nm GaN channel, designed to optimize performance metrics.

Sample B exhibits a comparable foundation, commencing with a 200 nm AlN seed layer and a similar 600 nm SG AlGaN buffer configuration. Notably, this sample integrates a 3,250 nm GaN/AlN superlattice (SL) buffer, comprising 27 nm GaN and 5 nm AlN per pair across 101 pairs, showcasing advanced epitaxial growth capabilities. A further 1,000 nm GaN:C back-barrier layer enhances structural integrity, complemented by a precise 300 nm GaN channel to facilitate efficient device operation.

Sample C introduces a distinctive approach with a 200 nm HLH-AlN seed layer, underscoring its specialized design philosophy. Following suit is a robust 600 nm SG AlGaN buffer arrangement, meticulously layered with 150 nm $Al_{0.7}GaN$, 200 nm $Al_{0.5}GaN$, and 250 nm $Al_{0.25}GaN$. This structure integrates a 3,250 nm GaN/AlN SL buffer, featuring 20 nm GaN and 5 nm AlN per pair over 101 pairs, emphasizing precision and reliability. A 1,000 nm GaN:C back-barier layer ensures robust performance, harmonizing seamlessly with a 300 nm GaN channel for optimized functionality.

Sample D continues the trend with a foundational 200 nm HLH-AlN seed layer, establishing a robust base for epitaxial growth. The 600 nm SG AlGaN buffer follows, meticulously composed of 150 nm $Al_{0.7}GaN$, 200 nm $Al_{0.5}GaN$, and 250 nm $Al_{0.25}GaN$ layers. Noteworthy in this structure is the integration of a 3250 nm $Al_{0.07}GaN$/AlN SL buffer, presenting 20 nm $Al_{0.07}GaN$ and 5 nm AlN per pair across 130 pairs, highlighting enhanced material properties. A 1000 nm GaN:C back-barier layer further fortifies structural integrity, complemented by a precise 300 nm GaN channel to ensure optimal device performance.

Across all samples, a consistent active layer design is employed, featuring a 22 nm $Al_{0.24}GaN$ barrier and a 1 nm GaN cap layer, underscoring uniformity in performance enhancement and material reliability.

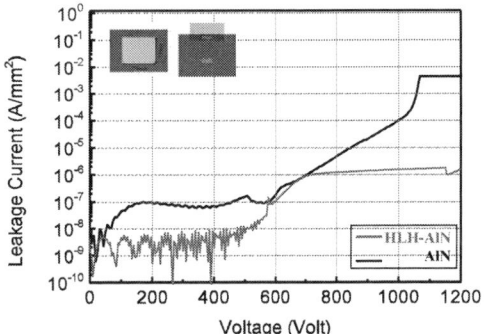

Fig. 1. The schematic of the different epitaxial device structures in this study

The fabrication process of AlGaN/GaN Devices involved two main steps: ohmic contact and ion implantation. Initially, a Ti/Al/Ni/Au ohmic metal stack was deposited using electron beam evaporation, followed by a lift-off process. Subsequently, the samples underwent annealing at 850 °C for 30 seconds in an N_2 atmosphere using rapid thermal annealing to achieve proper ohmic contacts.

Following this, the ion implantation isolation process was conducted. This step included the implantation of $B11^+$ ions at a dose of 3×10^{13} ions per cm^2 and an energy of 190 keV. This ion implantation step was crucial for defining the active region of the HEMTs, ensuring precise control over the device's operational characteristics.

III. RESULT AND DISCUSSION

Fig. 2. Forward vertical leakage current characteristics of AlGaN/GaN HEMT structure w/i and w/o HLH AlN buffer

Fig. 2 illustrates the forward vertical leakage current characteristics of various structures. The forward bias was swept from 0 V to 1,200 V at 25°C with a step size of 10 V, and the breakdown voltage (V_{bd}) was determined at a leakage current of 1×10^{-6} A/mm². Sample C demonstrates notably lower initial leakage current and a gradual increase in leakage current with voltage, suggesting that HLH-AlN holds substantial promise for high-voltage operation. These characteristics underscore its potential for enhancing device reliability and performance in demanding applications.

Fig. 3. GaN/AlN superlattice with different pair thickness (a) GaN/AlN= 27/5 nm/pair, (b) GaN/AlN= 20/5 nm/pair.

Fig. 3(a) displays the High-Resolution Transmission Electron Microscopy (HRTEM) image of the GaN/AlN superlattice (SL) buffer layer in Sample C. This layer consists of 27 nm GaN and 5 nm AlN layers, showing smooth and well-defined interfaces. These characteristics facilitate effective strain release and the efficient bending of Threading Dislocations (TDs), crucial for optimizing material quality in semiconductor devices.

Fig. 3(b) depicts a GaN/AlN superlattice structure with alternating layers of 20 nm GaN and 5 nm AlN. This illustration highlights the periodic arrangement of materials, essential for controlling crystal lattice strain and enhancing the structural integrity of semiconductor components.

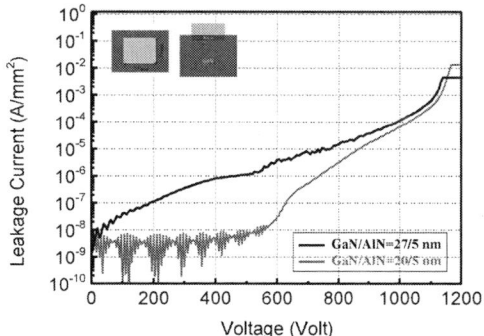

Fig. 4. Forward vertical leakage current characteristics of AlGaN/GaN HEMT with different superlattice pair thickness

From Fig. 4 and Fig. 5, it is evident that increasing the pair density of the GaN/AlN superlattice can effectively suppress leakage current. This suppression leads to enhanced film stability, as indicated by the increase in breakdown voltage (V_{bd}) from 520 V to 720 V. This observation underscores the critical role of superlattice design in optimizing the electrical properties and reliability of semiconductor films.

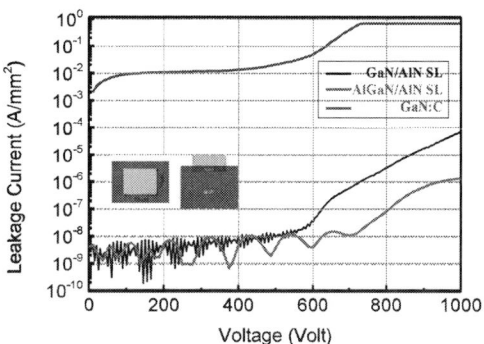

Fig. 5. Forward vertical leakage current characteristics of AlGaN/GaN HEMT structure different buffer layer

IV. CONCLUSION

AlGaN/GaN HEMTs on Si with HLH-AlN and a high pair density AlGaN/AlN superlattice were investigated for high-power applications. Compared to conventional AlGaN/GaN HEMTs with GaN:C and SG AlGaN buffer, AlGaN/GaN HEMTs with HLH-AlN and AlGaN/GaN buffer demonstrated a sharp interface at the superlattice buffer, high vertical breakdown voltage, and ultra-low leakage current at low operation voltage. These results indicate that AlGaN/GaN HEMTs with HLH-AlN and an AlGaN/AlN SL buffer are suitable for ultra-high-power applications.

ACKNOWLEDGMENT

This work was financially supported in part (project number: NSTC 112-2622-8-A49 -020, NSTC 112-2622-8-A49-013-SB, NSTC 113-2218-E-A49-020- , and NSTC 113-2634-F-A49-008.) by the Co-creation Platform of the Industry-Academia Innovation School, NYCU, under the framework of the National Key Fields Industry-University Cooperation and Skilled Personnel Training Act, from the Ministry of Education (MOE), the National Development Fund (NDF), and industry partners in Taiwan.

REFERENCES

[1] K. L. Lin *et al.*, "MOVPE high quality GaN film grown on Si (111) substrates using a multilayer AlN buffer," in *Physica Status Solidi (C) Current Topics in Solid State Physics*, 2008, pp. 1536–1538. doi: 10.1002/pssc.200778454.

[2] Y. L. Hsiao *et al.*, "Effect of graded Al xGa 1-xN layers on the properties of GaN grown on patterned Si substrates," *Jpn J Appl Phys*, vol. 51, no. 2 PART 1, Feb. 2012, doi: 10.1143/JJAP.51.025505.

[3] K.-L. Lin *et al.*, "Effects of AlxGa1−xN interlayer for GaN epilayer grown on Si substrate by metal-organic chemical-vapor deposition," *Journal of Vacuum Science & Technology B, Nanotechnology and Microelectronics: Materials, Processing, Measurement, and Phenomena*, vol. 28, no. 3, pp. 473–477, May 2010, doi: 10.1116/1.3385672.

[4] X. He *et al.*, "Step-Graded AlGaN vs superlattice: role of strain relief layer in dynamic on-resistance degradation," *Applied Physics Express*, vol. 15, no. 1, Jan. 2022, doi: 10.35848/1882-0786/ac3dc0.

[5] J. Su, E. A. Armour, B. Krishnan, S. M. Lee, and G. D. Papasouliotis, "Stress engineering with AlN/GaN superlattices for epitaxial GaN on 200 mm silicon substrates using a single wafer rotating disk MOCVD reactor," *J Mater Res*, vol. 30, no. 19, pp. 2846–2858, May 2015, doi: 10.1557/jmr.2015.194.

[6] A. Tajalli *et al.*, "High breakdown voltage and low buffer trapping in superlattice gan-on-silicon heterostructures for high voltage applications," *Materials*, vol. 13, no. 19, pp. 4271.1–4171.12, Oct. 2020, doi: 10.3390/MA13194271.

[7] J. Su, E. Armour, S. M. Lee, R. Arif, and G. D. Papasouliotis, "Uniform growth of III-nitrides on 200 mm silicon substrates using a single wafer rotating disk MOCVD reactor," *Physica Status Solidi (A) Applications and Materials Science*, vol. 213, no. 4, pp. 856–860, Apr. 2016, doi: 10.1002/pssa.201532708.

Effect of Silver Nanotriangle Orientation on Localised Surface Plasmon Resonance Sensing Performance

Muhammad Asif Ahmad Khushaini
Department of Applied Physics Faculty of Science & Technology Universiti Kebangsaan Malaysia 43600 Bangi, Selangor, Malaysia
asif@ukm.edu.my

Basyirah Zulkifli
Institute of Microengineering & Nanoelectronics (IMEN) Universiti Kebangsaan Malaysia 43600 Bangi, Selangor, Malaysia
P110935@ukm.edu.my

Nur Hidayah Azeman
Institute of Microengineering & Nanoelectronics (IMEN) Universiti Kebangsaan Malaysia 43600 Bangi, Selangor, Malaysia
nhidayah.az@ukm.edu.my

Tengku Hasnan Tengku Abdul Aziz
Institute of Microengineering & Nanoelectronics (IMEN) Universiti Kebangsaan Malaysia 43600 Bangi, Selangor, Malaysia
hasnanaziz@ukm.edu.my

Ahmad Ashrif A. Bakar
Faculty of Engineering and Built Environment, Universiti Kebangsaan Malaysia 43600 Bangi, Selangor, Malaysia
ashrif@ukm.edu.my

Ahmad Rifqi Md Zain
Institute of Microengineering & Nanoelectronics (IMEN) Universiti Kebangsaan Malaysia 43600 Bangi, Selangor, Malaysia
rifqi@ukm.edu.my

Abstract—This study focuses on the plasmonic characteristics of silver nanotriangles (AgNT) in localized surface plasmon resonance (LSPR) sensing applications. AgNTs are extremely valuable for LSPR sensing due to their strong plasmonic properties and high sensitivity. Among the various fabrication techniques for AgNTs, colloidal synthesis offers a cost-effective and relatively simple preparation method. However, the random distribution of AgNTs in colloids affects the reliability of sensors. Using the finite-difference time-domain (FDTD) simulation method, we examined the impact of the disordered distribution of AgNTs in colloids on their plasmonic properties. Our simulations demonstrate that the disordered ensemble of AgNTs in colloids results in uneven plasmonic properties. Additionally, we found that the non-monotonic response of LSPR sensors based on colloidal AgNTs can be attributed to the modal strong coupling phenomenon.

Keywords—localized surface plasmon, silver nanotriangles, LSPR sensor, FDTD simulation, strong coupling.

I. INTRODUCTION

Localized Surface Plasmon Resonance (LSPR) occurs when conduction electrons on the surface of metal nanoparticles oscillate in resonance with incident light. This results in a strong absorption and scattering of light at specific wavelengths. The resonance condition is highly sensitive to the surrounding environment, making LSPR a useful tool for sensing applications [1,2]. Silver nanoparticles are commonly used in LSPR due to their strong plasmonic properties. Compared to gold, silver (Ag) exhibits sharper and more intense plasmon resonances, making it more sensitive for detecting changes in the local environment [3]. The shape of the nanoparticle greatly influences the LSPR properties, and among various shapes, nanotriangles (NT) are particularly interesting.

Silver nanotriangles (AgNT) have unique optical properties due to their anisotropic shape. The sharp corners and flat surfaces of these triangles enhance the local electromagnetic field, leading to stronger LSPR signals. Numerous investigations have focused on the use of colloidal AgNT for sensor development [4,5]. However, using colloidal AgNT for sensor applications presents several challenges and issues, despite their promising properties. Achieving uniform and precisely controlled shape and size of silver nano triangles is challenging. Variations in these parameters can lead to inconsistent plasmonic properties and thus affect the reliability of the sensors. Colloidal silver nano triangles tend to aggregate over time, which can alter their plasmonic properties and reduce their effectiveness as sensors. Aggregation can be driven by van der Waals forces and other interparticle interactions.

In this work, the influence of AgNT orientation was examined. The investigation was performed using finite-difference-time-domain (FDTD) simulation by observing the resulting spectrum of the LSPR as a result of changing the orientation of the AgNTs. The sensitivity of resonance shifting as a result of the changes in the background refractive index of the simulation region was also investigated. In an actual LSPR experiment, the variation of refractive index mimics the use of different concentrations of the analyte. We also demonstrated that the non-monotonic behaviour of the LSPR-based sensor may also be explained by the modal strong coupling phenomenon.

II. METHODOLOGY

The simulation was conducted using Lumerical's FDTD software. This is done by simulating the system of incident plane wave impingement from the -x direction onto a single AgNT and the collection of AgNTs. The simulation was done by calculating the Poynting vector as a function of frequency, electric field, and magnetic field at each i point namely;

$$\langle S_i(\theta, \emptyset, \omega)\rangle = \frac{1}{2}\text{Re}(\boldsymbol{E_\theta H_\emptyset^*} - \boldsymbol{E_\emptyset H_\theta^*}) \quad (1)$$

where $\boldsymbol{E_\theta}, \boldsymbol{E_\emptyset}$ and $\boldsymbol{H_\emptyset^*}, \boldsymbol{H_\theta^*}$ are the electric field and the conjugate of magnetic field components at θ and $\emptyset$ spherical axes, respectively. The Poynting vector over the surface, $S_r(\theta, \emptyset, \omega)$ can be obtained by integrating over all i such as;

$$\langle S_r(\theta, \emptyset, \omega)\rangle$$
$$= \int_0^\pi \int_0^{2\pi} \langle S_i(\theta, \emptyset, \omega)\rangle a^2 \sin\theta \, d\emptyset d\theta . \quad (2)$$

The three-dimensional cartesian coordinate was used to build the surface of our simulation model as a cube with six faces namely $\pm x$, $\pm y$ and $\pm z$. Thus, the Poynting vector components are given by;

$$\langle S_{\pm x}(\theta, \emptyset, \omega)\rangle$$
$$\langle S_{\pm y}(\theta, \emptyset, \omega)\rangle \quad (3)$$
$$\langle S_{\pm z}(\theta, \emptyset, \omega)\rangle .$$

Equation (3) may then be integrated across the whole surface area to determine the optical power namely;

$$P_i = \int_{y_{min}}^{y_{max}} \int_{z_{min}}^{z_{max}} \langle S_{\pm x}(\theta, \emptyset, \omega)\rangle dydz$$
$$P_j = \int_{x_{min}}^{x_{max}} \int_{z_{min}}^{z_{max}} \langle S_{\pm y}(\theta, \emptyset, \omega)\rangle dxdz \quad (4)$$
$$P_k = \int_{x_{min}}^{x_{max}} \int_{y_{min}}^{y_{max}} \langle S_{\pm z}(\theta, \emptyset, \omega)\rangle dxdy$$

where $i = 1,2$, $j = 3,4$ and $k = 5,6$. Following that, the total optical power of scattered field is given by;

$$P_{tot} = P_i + P_j + P_k \quad (5)$$

while the total scattering cross-sections (SCS) can be calculated with;

$$C_{tot} = \frac{P_{tot}}{I_{inc}} \quad (6)$$

where I_{inc} is the intensity of the incident light.

III. RESULTS AND DISCUSSIONS

A. Effect of orientation of AgNT

The relative magnitude of SCS of the NPs is greatly influenced by its shape and size. Compared to the symmetric NP such the spherical NP, the present of sharpness or edges on the NP resulted in the red shift of the absorption spectra. This is because, multiple modes can present at the same time on the non-spherical NP which leads to less pronounced, red-shifted resonance peaks [6]. Consequentially, the resonance spectra of the non-spherical NP are susceptible to its orientation with respect to the incident light. It is demonstrated in Figure 1(a), varying the angle of AgNT resulted in changing the SCS spectra significantly.

The magnitude of spectral peaks are observed to have reduced by rotating the orientation single AgNT particle by 20° and 40° in respect to its initial orientation (inset of Figure 1(a)). However, further rotating the particle with 60° resulted in an increase of its spectral peak as indicated by the green curve in the figure. With 90° rotation, despite slight reduction in peak magnitude, it is still significantly higher than the peak obtained from the initial orientation.

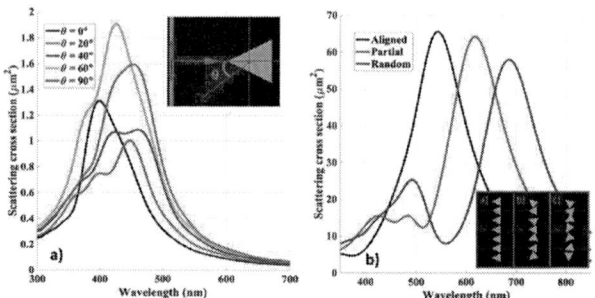

Fig. 1. (a) The change of CSC responses by rotating single AgNT particle, (b) The change of CSC responses by varying the degree of disorderly of an AgNT ensemble. Purple arrow in both of insets represent the incident light coming from -x direction (from left to right).

It is believed that this occurrence is originated from the enhancement of the field as the result of two edges of AgNT moving closer to the incident field. This can be seen in Figure 2(a) where the field enhancement is contributed strongly by the triangle edges. By rotating the triangle, these edges moving closer to the incident field hence contributing more to the scattered field (Figure 2(b)).

To exploit an ensemble of AgNTs for sensor application, it is important to investigate the optical response in relation to its randomness. We first arranged the collection of AgNT in three different orientations, namely, (inset of Figure 1(b)), a) ordered, where the AgNTs are aligned to one another; b) partial ordered, where several AgNTs are rotated randomly; and c) random, where all AgNTs in the ensemble are rotated randomly. As shown in Figure 1(b), there is significant red-shifting on the resonance wavelength as the ensemble becomes more disordered.

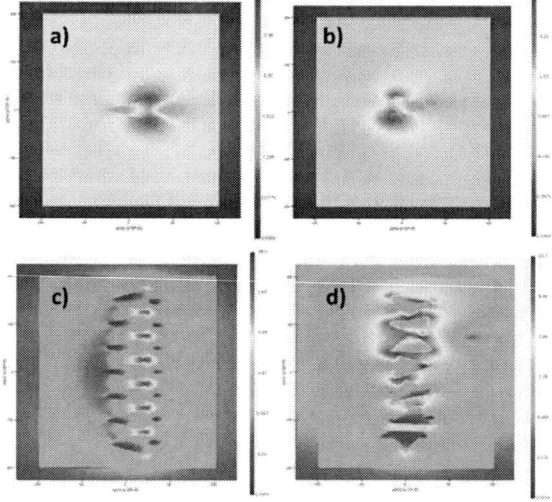

Fig. 2. Field distribution of (a) a single AgNT particle without the rotation, (b) a single AgNT particle with 60° rotation in respect to its initial orientation, (c) an ordered AgNT ensemble and, (d) a disordered AgNT ensemble. Incident light coming from left to right i.e. from -x direction.

Moreover, increasing the degree of disorder in the ensemble reduces the magnitude of SCS. This is due to the fact that more modes are incoherent with one another. Comparing the field distributions from the ordered and disordered ensembles—Figures 2(c) and 2(d), respectively—makes this more evident. It is also shown that the coherent response demonstrated in an ordered ensemble resulted in more localised enhancement as compared to the sparse enhancement demonstrated by a disordered ensemble.

B. Effect of background refractive index

The principle behind a plasmonic sensor is that varying the concentration of the analyte leads to a variation in the refractive index, which results in a shift of the plasmon resonance. To model the experimental procedure of varying concentrations of the analyte, the value of the background refractive index varies. Here, only the reflected field onto the left monitor (P_1) is considered. We model the system by first distributing randomly the AgNT particles into the simulation mesh. Following that, the incident field illuminates the ensemble from -x direction and the reflectance spectral is recorded to observe the LSPR minimum.

By referring to Figure 3(a), varying the background refractive index between n = 1.43 to n = 1.51 (referred to Range 1) resulted in red-shifting the LSPR resonance. Similar behavior can be obtained when the values of refractive index vary between n = 2.35 to n = 2.73 (referred to Range 2) as shown in Figure 3(b). However, the variation between each of the refractive index is larger in Figure 3(b) as compared to Figure 3(a). It is also important to note that using Range 1, more curves coincided to one another as compared to the spectra obtained from Range 2, where only the curves with n = 2.35 and n = 2.41 were seen to have overlapped to each other.

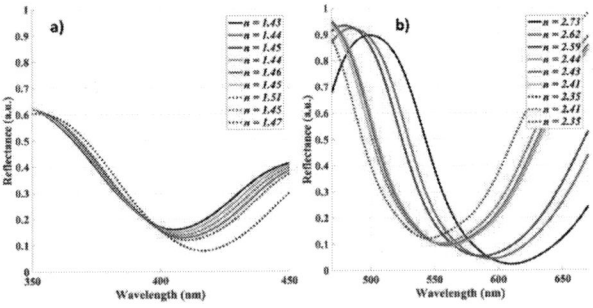

Fig. 3. The determination of refractive index of ligand-analyte system of (a) Range 1 and (b) Range 2. It is important to note that, some of the curves overlapped to each other as the result of using the same value of refractive index.

On the other hand, Figure 4 provide the comparison between the field distribution of both ranges at the resonance and at the wavelength smaller than the resonant. It is demonstrated that at the resonant, both Range 1 dan Range 2 experience high field enhancement at the magnitude of 9.21 and 10.8, respectively (Figure 4(a) and (c)). This is close to twice of field enhancement obtained from the systems that were illuminated with the off resonant incident light (Figure 4(b) and (d)). It is also shown in Figure 4(a) and (c), the enhancement is more localized as compare to sparse enhancement demonstrated by the field distribution of the off resonant systems. The reason mainly due to the dipole behavior of the particles diminishes as the incident wavelength becoming smaller than the size of the particles.

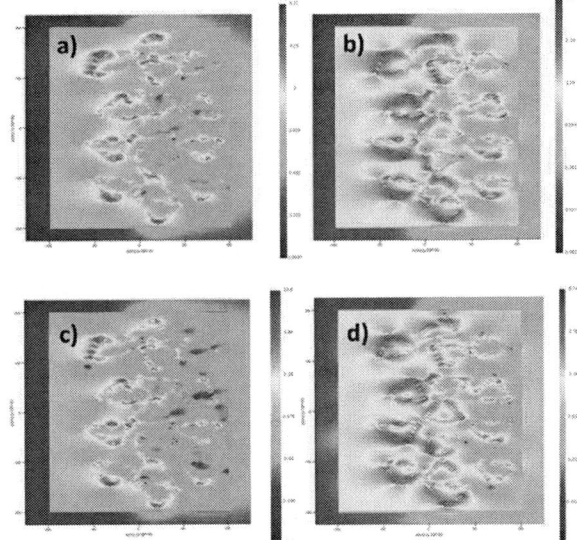

Fig. 4. Field distribution of (a) Range 1 (n = 1.43) at the resonant wavelength, (b) Range 1 (n = 1.43) at the wavelength smaller than the resonant wave, (c) Range 2 (n = 2.35) at the resonant wavelength, (b) Range 2 (n = 2.35) at the wavelength smaller than the resonant wave.

C. Modal strong coupling

Another possible explanation for the non-monotonic behaviour of the LSPR-based sensor is due to the presence of a strong coupling regime in the system. A strong coupling occurs when the rate of energy transfer between the plasmons and quantum emitters exceeds the total rate of losses in the system [7]. As a result, the energy of the system is no longer the same as that of the uncoupled system but a plasmon-emitter hybrid.

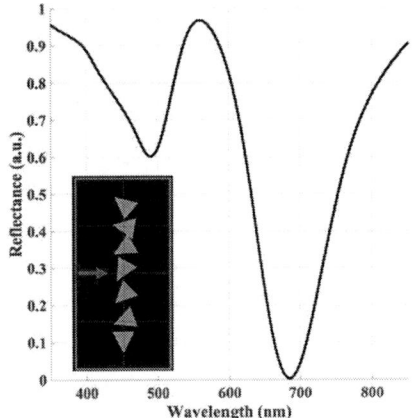

Fig. 5. The demonstration of the spectral splitting as the result of the strong coupling. The simulation was performed using an ensemble of AgNT particles arranged in random orientation.

One of the determining characteristics of the strong coupling is the observation of spectral splitting as the consequence of an avoided crossing between two eigenenergies of a newly formed plasmon-emitter hybrid [8]. By performing the simulation of incident light with a wavelength range of 350–850 nm and illuminating an ensemble of AgNT particles arranged in random orientation, this spectral splitting can be observed, as shown in Figure 5.

The supposedly monotonic behaviour of resonant wavelength shifting as the result of ligand-analyte interaction may have been obscured by the effect of the formation of a strong coupling regime in the system.

IV. CONCLUSION

Silver nanotriangles, AgNT are highly effective for LSPR-based sensing applications due to their strong plasmonic properties and high sensitivity. Advances in fabrication and functionalization techniques continue to enhance their performance and expand their application range in various fields, from medical diagnostics to environmental monitoring. When employed for sensing applications, colloidal AgNT, despite its simple production process, presents a variety of issues. As demonstrated in this work, disorderly ensemble of the AgNT in the colloid leads to uneven plasmonic properties, which could affect the reliability of the sensors. Furthermore, as the simulation results suggest, the inability of any AgNT-based sensor to exhibit a linear operating range can also be explained by the formation of the modal strong coupling phenomenon. this study provides valuable insights into the challenges associated with AgNT-based sensors, particularly regarding their uneven plasmonic properties and non-linear operating ranges due to the modal strong coupling phenomenon.

REFERENCES

[1] Chen, J.-S., Chen, P.-F., Lin, H. T.-H., & Huang, N.-T. (2020). A Localized surface plasmon resonance (Lspr) sensor integrated automated microfluidic system for multiplex inflammatory biomarker detection. *The Analyst*, *145*(23), 7654–7661.

[2] Yang, Y., Murray, J., Haverstick, J., Tripp, R. A., & Zhao, Y. (2022). Silver nanotriangle array based LSPR sensor for rapid coronavirus detection. *Sensors and Actuators B: Chemical*, *359*, 131604.

[3] Hang, Y., Wang, A., & Wu, N. (2024). Plasmonic silver and gold nanoparticles: Shape- and structure-modulated plasmonic functionality for point-of-caring sensing, bio-imaging and medical therapy. *Chemical Society Reviews*, *53*(6), 2932–2971.

[4] Neri, G., Fazio, E., Mineo, P. G., Scala, A., & Piperno, A. (2019). Sers sensing properties of new graphene/gold nanocomposite. *Nanomaterials*, *9*(9), 1236.

[5] Wu, C., Zhou, X., & Wei, J. (2015). Localized surface plasmon resonance of silver nanotriangles synthesized by a versatile solution reaction. *Nanoscale Research Letters*, *10*(1), 354.

[6] Lu, X., Rycenga, M., Skrabalak, S. E., Wiley, B., & Xia, Y. (2009). Chemical synthesis of novel plasmonic nanoparticles. *Annual Review of Physical Chemistry*, *60*(1), 167–192.

[7] Khushaini, M. A. A., Azeman, N. H., Abdul Aziz, T. H. T., Bakar, A. A. A., & Zain, A. R. M. (2024). Harnessing quantum plexcitons for enhanced sensitivity and selectivity of creatinine sensor. *Sensors and Actuators B: Chemical*, *412*, 135748.

[8] Pelton, M., Storm, S. D., & Leng, H. (2019). Strong coupling of emitters to single plasmonic nanoparticles: Exciton-induced transparency and Rabi splitting. *Nanoscale*, *11*(31), 14540–14552

Comparative performances of 60nm and 80nm gold nanoparticles based mode-locker for erbium doped fiber laser

Noor Zirwatul Ahlam Naharuddin
Faculty of Electrical and Electronics Engineering Technology
University Malaysia Pahang Al-Sultan Abdullah
Pahang, Malaysia
*zirwatul@umpsa.edu.my

Maisarah Mansor
School of Engineering and Technology
Sunway University
Selangor, Malaysia
maisarahma@sunway.edu.my

Nor Hadzfizah Mohd Radi
Faculty of Electrical and Electronics Engineering Technology
University Malaysia Pahang Al-Sultan Abdullah
Pahang, Malaysia
hadzfizah@umpsa.edu.my

Mohd Adzir Bin Mahdi
Fakulti Kejuruteraan
Universiti Putra Malaysia
Selangor, Malaysia
mam@upm.edu.my

Rosyati Hamid
Faculty of Electrical and Electronic Engineering Technology
Universiti Malaysia Pahang Al-Sultan Abdullah
Pahang, Malaysia
rosyati@umpsa.edu.my

Abstract— The study evaluates the influence of a nanocomposite-based gold nanoparticle (Au-NPs) saturable absorber (SA) with particle sizes of 60 and 80 nm on the generation of ultrashort pulses in an erbium-doped fiber laser. The saturable absorber was created by coating a nanocomposite of Au-NPs and polydimethylsiloxane (PDMS) onto the surface of a microfiber using a spin coater. The Au-NP-based SA was subsequently incorporated into a ring cavity of an erbium-doped fiber laser, and its performance was evaluated by comparing the SA properties and pulse quality for both 60nm and 80nm Au-NP-based SAs. The study discovered that Au-NPs larger than 60nm led to inadequate optical performance for mode-locking applications. This decrease was noticeable in the shift of the center wavelength of the optical spectrum from about 1559nm to 1532nm and an increase in pulse duration from 940fs using 60nm Au-NPs SA to 1229fs using 80nm Au-NPs SA. The findings suggest that the ideal size for Au-NPs, ensuring an efficient mode-locker, should not exceed 60nm. This previously unpublished study is crucial in determining the optimal gold nanoparticle size for producing ultrashort pulse output that fulfills the specific needs of a given system.

Keywords—Gold nanoparticles, saturable absorber, microfiber

I. INTRODUCTION

The hunt for novel materials with significant absorption properties, perfect for producing ultrafast pulse lasers, is still underway. This includes investigating metal nanomaterials, specifically nanostructured metals known for their surface plasmon resonance (SPR) effects. SPR is the phenomenon in which light causes electrons on the surface of metallic particles to vibrate[1]. Currently, metals like gold and silver nanoparticles have different uses in microbiology domains like pathogen and cellular detection[2][3]. The way light is restricted by the plasmonic structures of these metal nanoparticles is vital for building novel nonlinear optical systems. Due to their distinct absorption patterns in the SPR area, gold nanoparticles (Au-NPs) have been widely studied and applied, indicating intriguing promise for the production of ultrashort pulse lasers.

By modifying the size and structure of Au-NPs, supported by their remarkable surface-to-volume ratio, SPR may be adjusted [4], in which absorption wavelength moves to longer wavelengths with increasing particle size, enabling flexibility in developing optical properties for varied applications.

Recent studies have highlighted the potential of various kind of Au-NPs as an alternative matter for constructing mode-locked SA, focusing on shape such as nanorods [5], and types like rose-gold nanoparticles[6]. Our previous research explored the impact of optical pulse profiles for Au-NPs sizes ranging from 10nm to 60nm[7]. However, there is a gap in the literature concerning the effects of sizes beyond 60nm. To the best of our knowledge, no reports have investigated this aspect. This study delves into the comparative performance of Au-NPs for sizes of 60nm and 80nm-based saturable absorbers (SAs).

II. METHODOLOGY

A. Saturable Absorber Fabrication

In this research project, an adiabatic microfiber is fabricated using the Vytran GPX-3400 workstation, featuring a total length of 60.5 mm and a waist diameter of 10 μm. The nanocomposite utilized in the study is created by blending Au-NP powders, specifically selecting sizes of 60nm and 80nm from nanoComposix, with polydimethylsiloxane (PDMS). PDMS is a hydrophilic polymer with a refractive index value of approximately 1.4039 [7]. This lower refractive index ensures that the evanescent wave can effectively propagate and reflect back to the core area of the microfiber with minimal loss. The fabrication method begins with 5 mg of Au-NPs powder being sonicated in 10 mL of isopropanol (IPA) for 30 seconds. Then, 1 g of PDMS is mixed with 2 mL of the Au-NP colloid, stirred, and heated at 89 °C until the IPA has completely evaporated. Next, 0.1 g of curing agent is added, and the air bubbles are removed by placing the mixture in a vacuum chamber for a maximum of two hours. A spin coater set to 4000 rpm for 5 minutes is used to deposit the ready nanocomposites on the tapered region. The fabricated SAs are allowed to dry before being measured in the ring cavity of the Erbium-Doped Fiber Laser (EDFL), as shown in Fig. 1.Experimental Setup

B. Experimental Setup

Fig.1 illustrates the ring cavity configuration, featuring an 8.5-meter Lucent HP980 Erbium-Doped Fiber (EDF) as the gain medium, with an absorption coefficient ranging from 3.5 to 5.5 dB/m within the 1530 nm to 1550 nm wavelength range. The cavity is forward-pumped by a 980 nm laser diode via a 980/1550 nm wavelength division multiplexer (WDM). The Au-NP-SA is strategically positioned between an isolator and a polarization controller (PC) to optimize light-matter interaction while ensuring unidirectional light propagation. A 60/40 coupler is placed after the PC, redirecting 60% of the light back into the cavity and channeling the remaining 40% to analytical instruments such as an Optical Spectrum Analyzer (OSA) and an Optical Power Meter (OPM). Optical pulse and signal-to-noise ratio (SNR) measurements are conducted using an autocorrelator (A.P.E. PulseCheck) and radio frequency-spectrum analyzer, respectively. Additionally, a Tektronix TDS 3012C oscilloscope, equipped with a 5 GHz Thorlabs SIR5-FC InGaAs FC/PC-coupled photodetector, is employed for output pulse train measurement.Before the measurement is done, transmission loss for both SAs have been measured, the insertion loss for 60nm Au-NPs SA is 5.65 dB while 8.56 dB for 80nm Au-NPs SA.

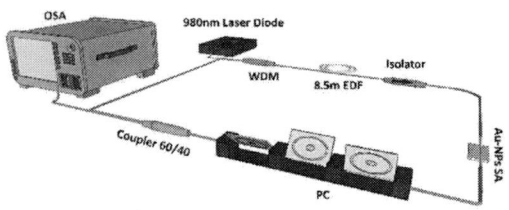

Fig. 1. Experimental setup of EDFL cavity

III. RESULT AND DISCUSSION

Fig. 2 illustrates the optical spectra of SA-10 obtained by an Optical Spectrum Analyzer (OSA) with a 0.02 nm bandwidth resolution. Three separate situations were analyzed: continuous wave (CW) laser threshold, pulse laser threshold, and maximum pump power. Fig. 2(a) shows the CW laser threshold reached at 60 nm SA with 11.6 mW, whereas the self-starting mode-locked power is 32 mW. The mode-locked pulse's central wavelength is 1559.1 nm. Compared to the threshold mode-locked condition, the maximum pump power of 250.2 mW shifted the working wavelength to 1559.3 nm. This minor shift is associated with the absorption properties of the Au-NPs, where the scattering properties of particles larger than 40 nm begin to dominate over their absorption qualities [8]. These spectra were observed to be free from any CW parasitic lasing, confirming stable pulse laser operation [9]. Kelly's sidebands appear symmetrical, implying that the fiber laser was operating in the soliton domain with anomalous dispersion [10][11].

Meanwhile, for 80 nm SA, the threshold for self-starting pulse operation is higher, with a significant increase from 11.6 mW (for 60 nm SA) to 161.0 mW, as depicted in Fig. 2(b). It can also be observed that the central wavelength shifted significantly to 1532 nm, with the formation of Kelly's sidebands appearing unsymmetrical, indicating an unstable generated soliton. This condition is related to the nonlinear effect introduced by the SA, which impacts the laser stability and causes a shift in the central wavelength.

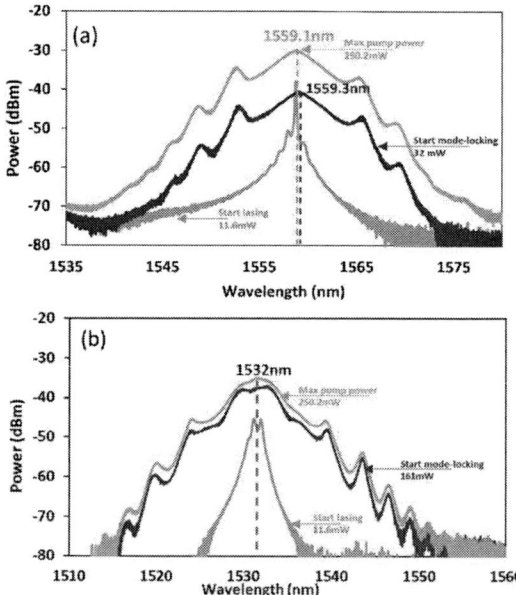

Fig. 2. Optical spectrum for lasing, threshold and maximum power mode-locking for (a) 60nm Au-NPs and (b) 80nm Au-NPs

Fig. 3 shows the autocorrelation trace of ultrashort mode-locked pulses for 60nm Au-NPs and 80nm Au-NPs SAs, whereas Fig. 3(a) and (b) indicate that the observed pulse is in agreement with the sech2 profile. The measured pulse duration of 940 fs is attained for 60nm Au-NPs SA while 1229fs for 80nm Au-NPs SA respectively. However, there are temporal pedestals detected at the edge of main pulse peak for SA-80 as depicted in Fig. 3(b). The developed pedestal on the autocorrelated traces, verified by the small wings at both sides, was due to the distortion related to the phase-mismatched between linear and nonlinear dispersions during the soliton propagation in the cavity, signifying instability of the generated pulse [12].

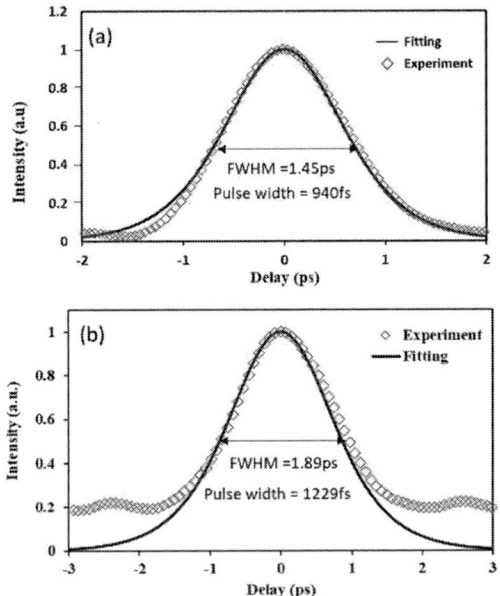

Fig. 3. Autocorrelation traces (a) 60nm Au-NPs SA and (b) 80 nm Au-NPs SA

Fig. 4 (a) and (b) presents the oscilloscope trace of mode-locked fiber laser for both SAs. The fundamental pulse train was attained by employment of ~33.7 m cavity length. Based on the finding, repetition rate of 6.36 MHz (157ns) was achieved for 60nm Au-NPs SA, while attained 6.45MHz (155ns) for 80nm Au-NPs SA.

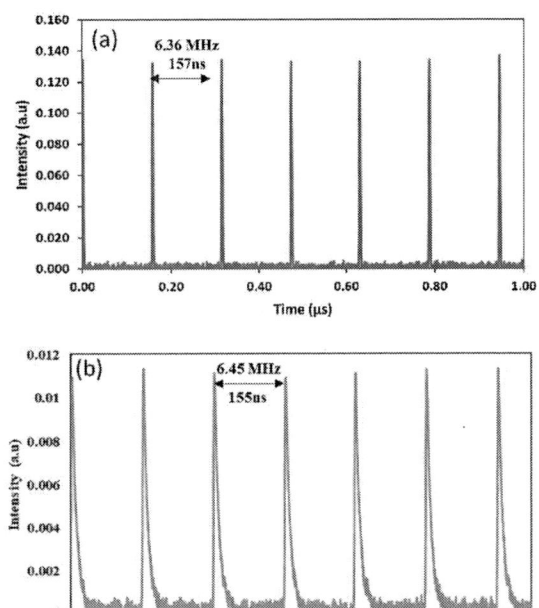

Fig. 4. Oscilloscope trace of pulse laser output for (a) 60nm Au-NPs and (b) 80nm Au-NPs SAs

The stability of the mode-locked pulses generated by the Au-NP SA was further validated by signal-to-noise ratio (SNR) measurement, as shown in Fig. 5(a) and Fig. 5(b). In the experiment, the RF spectrum analyzer was configured with a 100 MHz scanning span and a 30 kHz resolution bandwidth. As shown in Fig. 5(a), the first-order peak has an SNR value of 59.5 dB, indicating that the mode-locked fiber laser operated with excellent stability, as its SNR value is above 50 dB [13]. No Q-switching fluctuations were detected. From the measured trace, the fundamental cavity peak is located at 6.36 MHz, consistent with the fundamental cavity repetition rate observed from the oscilloscope trace. For the 80 nm Au-NPs SA, the obtained SNR value is 57 dB, and the cavity peak is located at 6.47 MHz. While the SNR value of the 80 nm saturable absorber is considered satisfactory, the dispersion effects caused by the nanocomposite layer of these absorbers viewed as more optimal for mode-locked pulses centred at 1559 nm[16][17]. The 60 nm-based nanocomposite saturable absorber achieves a good balance between nonlinear effects and dispersion, thereby promoting more stable pulse generation.

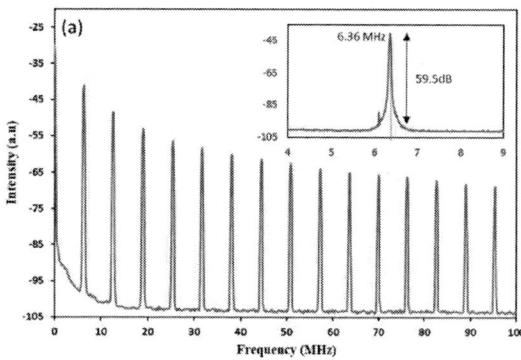

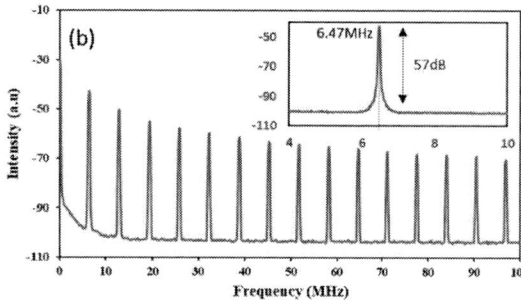

Fig. 5. RF spectrum trace of pulse laser output using (a) 60nm Au-NPs and (b) 80nm Au-NPs SAs

IV. CONCLUSION

In summary, the performance of Au-NPs SA was thoroughly examined based on two different nanoparticle sizes, namely 60nm and 80nm. The results indicate that the optimal size for fabricating a mode locker is 60nm, as beyond this size, the optical performance significantly degrades. This degradation is evident in the shift of the central wavelength of the optical spectrum from approximately 1559nm to 1532nm and an increase in pulse duration from 940fs (using 60nm Au-NPs SA) to 1229fs (using 80nm Au-NPs SA). Moreover, the quality of the pulses deteriorates. These findings emphasize that the ideal size for Au-NPs, ensuring an optimum mode-locker, should not exceed 60nm. As far as we are aware, this finding has not been reported before and is of considerable importance for customizing ultrashort pulse output according to the specific requirements of system performance.

ACKNOWLEDGMENT

This project was financially supported by the Fundamental Research Grant (Grant Number RDU220328) and the Tabung Persidangan Dalam Negara (TPDN) from Universiti Malaysia Pahang Al-Sultan Abdullah. Special thanks are extended to the Wireless and Photonic Networks Research Centre Of Excellence (WiPNET) at Universiti Putra Malaysia (UPM).

REFERENCES

[1] L. Catanzaro, V. Scardaci, M. Scuderi, M. Condorelli, L. D'Urso, and G. Compagnini, "Surface plasmon resonance of gold nanoparticle aggregates induced by halide ions," Mater. Chem. Phys., vol. 308, no. June, p. 128245, 2023, doi: 10.1016/j.matchemphys.2023.128245.

[2] M. Hegde, P. Pai, M. G. Shetty, and K. S. Babitha, "Gold nanoparticle based biosensors for rapid pathogen detection: A review," Environ. Nanotechnology, Monit. Manag., vol. 18, no. August, p. 100756, 2022, doi: 10.1016/j.enmm.2022.100756.

[3] P. Pourali, V. Dzmitruk, O. Benada, M. Svoboda, and V. Benson, "Conjugation of microbial-derived gold nanoparticles to different types of nucleic acids: evaluation of transfection efficiency," Sci. Rep., vol. 13, no. 1, pp. 1–14, 2023, doi: 10.1038/s41598-023-41567-7.

[4] Z. Sadiq, S. H. Safiabadi Tali, H. Hajimiri, M. Al-Kassawneh, and S. Jahanshahi-Anbuhi, "Gold Nanoparticles-Based Colorimetric Assays for Environmental Monitoring and Food Safety Evaluation," Crit. Rev. Anal. Chem., vol. 0, no. 0, pp. 1–36, 2022, doi: 10.1080/10408347.2022.2162331.

[5] Y. L. Hongyu Luo, Zhe Kang, Ying Gao, Hanlin Peng, Jianfeng Li, Guanshi Qin, "Large aspect ratio gold nanorods (LAR-GNRs) for mid-infrared pulse generation with a tunable wavelength near 3 μ m," Opt. Express, vol. 27, no. 4, pp. 4886–4896, 2019.

[6] S. W. H. I.S. Nabe, N.F. Zulkipli , A.H.A. Rosol , A.A. Rahman, "Rose gold nanoparticles film for generating Q-switched and mode-locked pulses," Results Opt., vol. 1, no. September, p. 100007, 2020, doi: 10.1016/j.rio.2020.100007.

[7] N. Z. A. Naharuddin, M.H. Abu Bakar, N. Tamchek, M.T. Alresheedi, A.F. Abas, C.S. Goh, N.H. Zainuddin, M.A. Mahdi, "Effect of gold-nanoparticle size on microfiber saturable absorber for mode-locked erbium-doped fiber lasers," Optik (Stuttg)., vol. 276, no. February, p. 170631, 2023, doi: 10.1016/j.ijleo.2023.170631.

[8] M. A. E.-S. Prashant K. Jain, Kyeong Seok Lee, Ivan H. El-Sayed, "Calculated Absorption and Scattering Properties of Gold Nanoparticles of Different Size, Shape, and Composition: Applications in Biological Imaging and Biomedicine," J. Phys. Chem. B, vol. 110, pp. 7238–7248, 2006.

[9] M. P. Yanqi Ge, Zhongjun Li, Han Zhang, Jaroslaw Sotor, "Fundamental and harmonic mode-locking at 21 μm with black phosphorus saturable absorber," Opt. Express, vol. 25, no. 15, p. 16916, 2017, doi: 10.1364/oe.25.016916.

[10] W.-C. X. Xu-De Wang, Ai-Ping Luo, Hao Liu, Nian Zhao, Meng Liu, Yan-Fang Zhu, Jian-Ping Xue, Zhi-Chao Luo, "Nanocomposites with gold nanorod/silica core-shell structure as saturable absorber for femtosecond pulse generation in a fiber laser," Opt. Express, vol. 23, no. 17, p. 22602, 2015, doi: 10.1364/oe.23.022602.

[11] S. Xu, A. Turnali, and M. Y. Sander, "Group-velocity-locked vector solitons and dissipative solitons in a single fiber laser with net-anomalous dispersion," Sci. Rep., vol. 12, no. 1, pp. 1–8, 2022, doi: 10.1038/s41598-022-10818-4.

[12] C. Pask and A. Vatarescu, "Spectral approach to pulse propagation in a dispersive nonlinear medium," J. Opt. Soc. Am. B, vol. 3, no. 7, p. 1018, 1986, doi: 10.1364/josab.3.001018.

[13] W.-C. X. Xu-De Wang, Zhi-Chao Luo, Hao Liu, Meng Liu, Ai-Ping Luo, "Microfiber-based gold nanorods as saturable absorber for femtosecond pulse generation in a fiber laser," Appl. Phys. Lett., vol. 105, no. 16, p. 161107, 2014, doi: 10.1063/1.4899133.

[14] Transl. J. Magn. Japan, vol. 2, pp. 740–741, August 1987 [Digests 9th Annual Conf. Magnetics Japan, p. 301, 1982.

[15] M. Young, The Technical Writer's Handbook. Mill Valley, CA: University Science, 1989.

[16] H. A. Haus, "Mode-locking of lasers," in IEEE Journal of Selected Topics in Quantum Electronics, vol. 6, no. 6, pp. 1173-1185, Nov.-Dec. 2000, doi: 10.1109/2944.902165

[17] Agrawal, G. P. (2012). Nonlinear Fiber Optics (5th ed.). Academic Press.

Enhancing Localized Surface Plasmon Resonance Response for Albumin Detection by Optimizing the Lateral Size of Hexagonal Gold Nanoparticles

Silva Nurfasha
Study Program of Physics
Universitas Pendidikan Indonesia
Bandung, Indonesia
silvanurfasha@upi.edu

Chandra Wulandari
Study Program of Physics
Institut Teknologi Bandung
Bandung, Indonesia
33322303@mahasiswa.itb.ac.id

Lilik Hasanah
Study Program of Physics
Universitas Pendidikan Indonesia
Bandung, Indonesia
lilikhasanah@upi.edu

Ahmad Aminuddin
Study Program of Physics
Universitas Pendidikan Indonesia
Bandung, Indonesia
aaminudin@upi.edu

Yanurita Dwi Hapsari
Department of Physics
Faculty of Science and Data Analytics
Institut Teknologi Sepuluh Nopember
Surabaya, Indonesia
yanuritadh@yahoo.com

Mohammad Arifin
Study Program of Physics
Universitas Pendidikan Indonesia
Bandung, Indonesia
mohammad_arifin@upi.edu

Roer Eka Pawinanto
Study Program of Industrial Automation and Robotics Engineering Education
Universitas Pendidikan Indonesia
Bandung, Indonesia
Roer_eka@upi.edu

Budi Mulyanti
Study Program of Electrical Engineering
Universitas Pendidikan Indonesia
Bandung, Indonesia
bmulyanti@upi.edu

Abstract— The kidney is a human organ with numerous vital functions, mainly for blood filtration. Damage to the kidneys can be detected from albumin levels in urine, higher levels of albumin in urine indicate the kidney's filtration function is impaired. The latest breakthrough in the medical field is the use of localized sensor plasma resonance (LSPR) in the field of biosensors to detect albumin protein in urine. Using Finite-Difference Time-Domain (FDTD) simulation, the lateral size of hexagonal gold nanoparticles (AuNPs) is varied to increase the optimal LSPR. AuNPs are well-known as the most promising material for plasmonic applications. In several studies, the hexagonal geometry was reported to increase the surface area of nanoparticles as well as enhance the LSPR signal. Varying the geometry of the hexagonal AuNPs will affect their sensitivity and optical properties. In this study, the lateral size of the hexagonal AuNPs varied from 10 nm to 60 nm, with a controlled thickness of 30 nm. The absorption graph blue shifts with the peak decreasing until the graph has almost no peak as the lateral size decreases. The lateral size of 40 nm at a thickness of 30 nm was found to have the highest shifting response towards the change of refractive index, indicating good sensitivity for sensors. Simulation of hexagonal AuNPs-based LSPR detection of albumin concentrations of 0.35 – 5.71 mM produces a sensitivity of 2.59 nm/mM, equivalent to 245.96 nm/RIU. The competitive performance and ease of detection procedures in real applications show the potential of the hexagonal AuNPs-based LSPR sensor to be further explored in developing alternative albumin detection techniques.

Keywords: LSPR, hexagonal, gold nanoparticles, albumin, FDTD

I. INTRODUCTION

Albumin is a protein produced in the liver as the most dominant plasma protein with a percentage of 50% in healthy individuals [1]. Albumin levels also can be an indicator of various body conditions, including damage of the kidneys [2]. Chronic kidney disease (CKD) are affecting 5–10% of people in the world. The patient with CKD experienced the bleeding on their kidney, causing damage to the walls of blood vessels [3]. Kidney disease can be detected by the urine albumin to creatine ratio (uACR) value, where the problematic kidney is at a uACR ratio $\geq$ 30 mg/g [4]. This method is the blood; when the kidneys drain the blood, this protein will be carried back with the blood. If a high level of albumin is found in the urine, damage to the kidneys is identified [2]. Measurement of albumin levels in urine is an important step in the early detection of kidney disease, which can help in better management and treatment. Therefore, the development of the detection technique of albumin become important to improve kidney disease treatment.

Localized Surface Plasmon Resonance (LSPR) is a unique optical property displayed by gold nanoparticles that refers to the collaborative oscillation of electrons in the conduction band of gold nanoparticles that resonate with a specific wavelength of incident light [5]. The sensitivity of LSPR is very high to changes in the surrounding environment of noble metal nanoparticles, which has been explored to prove that its application is very good for detecting proteins, DNA, and other chemical and biological substances [6, 7]. LSPR is effectively used for biosensors because it has high sensitivity [8]. LSPR sensors have a small sensing area, which can be beneficial in applications where space is limited [9].

Among of the most used metal nanoparticles (e.g. gold and silver nanoparticles), the gold nanoparticles (AuNPs) are commonly used especially in the biosensor application due to their highest sensitivity [10]. Numerous studies have reported the application of AuNPs in optical biosensing such as surface-enhanced Raman spectroscop (SERS) and LSPR, they compete to obtain the most significant performance in detecting various substances. The exploration of various nanoparticles morphplogy was found the hexagonal AuNPs to provide the enhancement of the sensitivity. Further development of hexagonal AuNPs is still conducted, especially to enhance their optical properties and sensitivity to the surrounding environment [11]. Compared to triangular AuNPs and nanobipyramids, the sensitivity to refractive index of hexagonal AuNPs is higher [12]. Some studies show that the optimization performance of hexagonal AuNPs could be obtained by controlling its the variables that determine the sensitivity of AuNPs are thickness and lateral size [11, 13].

In this study, the hexagonal AuNPs was used in the development of LSPR biosensor for albumin detection. Using the finite-difference time-domain (FDTD) methods, the hexagonal AuNPs were optimized by varying lateral size to obtain the sensing enhancement. The absorption spectrum of albumin was discovered using the FDTD approach [14]. The sensing performance was evaluated by perform the simulation in different refractive index of background. The change of refractive index was assumed as the change of analyte concentration. Finally, from this study we will obtain the optimum lateral size of hexagonal AuNPs for the detection of albumin.

II. METHODOLOGY

The FDTD method was used to construct hexagon AuNPs by adjusting the thickness and lateral size. In Figure 1, the parameters varied for AuNP are the thickness (h) and the lateral size (l). The lateral size of hexagonal AuNPs was varied by 10, 20, 30, 40, 50, and 60 nm with the thickness kept constant at 30 nm. Research conducted by Farooq, et al. (2022) showed that 30 nm is the optimal thickness for AuNPs [16].

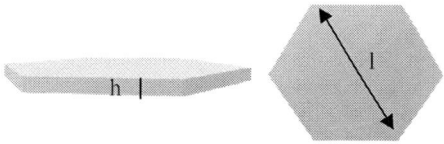

Fig. 1. Hexagonal design of gold nanoparticles (AuNPs)

In the FDTD simulation, the AuNPs hexagon was made, and then several monitors were added to calculate the scattering and absorption. The monitors used consist of frequency domain field and power, reflective index, and field time. The source is the total field scattering field (Figure 2) [17, 18]

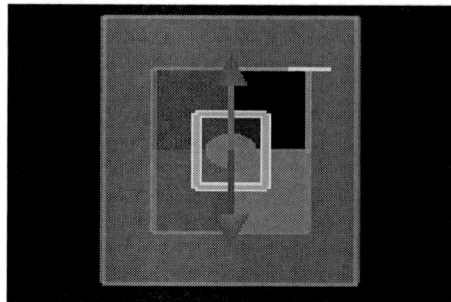

Fig. 2. Structural model of the hexagonal AuNPs in the FDTD simulation

The absorption graphs of each size of hexagonal AuNPs were taken at two different background index to see the peak shift of the absorption. The index applied in this study were 1.33 and 1.34. For each size, the peak shift was calculated. The largest peak shift indicates that the AuNPs have best performance due to the biosensor application which must detect the change in the surrounding medium. Referring to the application of LSPR AuNPs is used to detect albumin content in urine, this simulation used FDTD methods by making the background index value equal to the refractive index value of albumin as listed in Table 1 [19].

The refractive index of bovine serum albumin (BSA) is almost the same as that of human albumin, so the refractive index of BSA is used as the refractive index of the LSPR AuNPs hexagon background in FDTD simulations [20]. The sensitivity of the simulated LSPR sensor was measured from the slope in the plot concentration vs. absorption peak position.

TABLE I. Albumin concentrations and their refractive index values.

Concentration (mM)	Reflective Index
0.35	1.33
0.67	1.34
1.79	1.35
2.35	1.36
4.09	1.37
4.82	1.38
5.71	1.39

III. RESULT AND DISCUSSION

Using the FDTD methods, hexagonal AuNPs whose lateral size is varied from 10 nm to 60 nm with a constant thickness of 30 nm are simulated at a background index of

1.33 [21]. The absorption values obtained are shown in Figure 3.

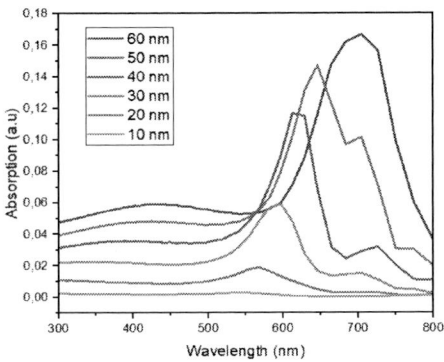

Fig. 3. Hexagon AuNPs absorption in the background of index 1.33

From the simulation results of hexagonal AuNPs with lateral sizes of 10 nm – 60 nm and a thickness of 30 nm at the same background index of 1.33, it can be seen that the peak of the curve shifts upwards as the lateral size gets larger [22]. The peak is higher at large sizes, and the peak slopes at small sizes. Rai, et al. (2016) have conducted research and shown that the size and shape of NPs play an important role in the collective oscillation of free conduction electrons when generating LSPR peaks [23]. However, from this trend, we still cannot conclude the best size that is suitable for detection applications. Therefore, in the next step of this study, we conducted the simulation for each variation to respond to the change in background index from 1.33 to 1.34. From this simulation, we will conclude the best hexagonal lateral size from the absorption peak shifting (ΔAbs).

Each LSPR AuNPs size that has been determined is simulated in FDTD in two different background indices, namely background indices 1.33 and 1.34. The absorption peak position and shifting was listed on the Table 2.

TABLE II. The absorption peak shifted from the refractive index of 1.33 to 1.34 for each lateral size.

Lateral size (nm)	Peak position (nm) n = 1.33	Peak position (nm) n = 1.34	Shifting ΔAbs
10	5.42	5.42	0
20	5.55	5.69	0.14
30	5.97	5.97	0
40	6.28	6.29	0.16
50	6.47	6.47	0
60	7.25	7.26	0.01

Each size variation was simulated at index 1.33 and index 1.34 and compared with each other. The peak shift is found to be the largest among other variations at a thickness of 30 nm with a lateral size of 40 nm. with a peak shifting of 0.16 nm. After obtaining the most optimal size value, the size is simulated at the refractive index value which equal to the listed albumin concentration as shown in Table 1. From this result, the hexagonal AuNPs with a lateral size of 40 nm will used to respond the

changes of refractive index. As shown in Figure 4, the absorption peak position resulting from the higher refractive index was blue-shifted. The shifting indicates the sensing capability of hexagonal AuNPs as LSPR materials.

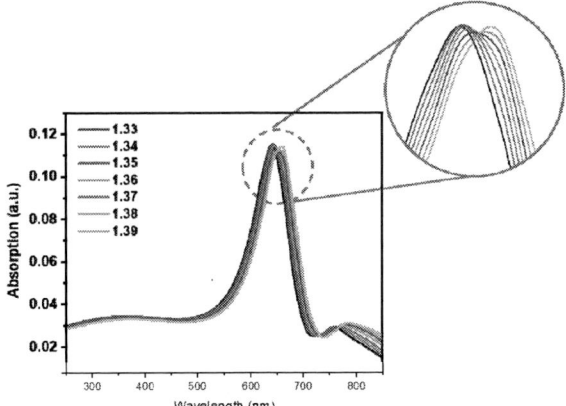

Fig. 4. LSPR absorption of hexagonal AuNPs at a lateral size of 40nm at the refractive index of albumin

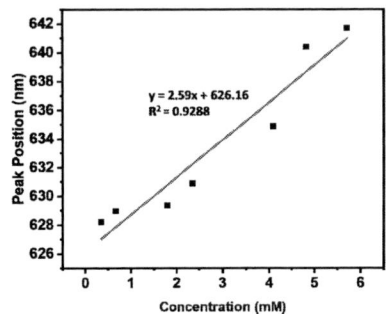

Fig. 5. Shows a plot of each absorption peak at a certain concentration.

By finding each peak at each index, the sensitivity value is determined as the slope of the concentration-to-peak position plot [23]. Figure 5 shows the plot of the absorption peak of 40 nm lateral-size hexagonal AuNPs against the concentration of albumin. The albumin concentration in the plot corresponds to Table 1. The slope of the graph, obtained from the results of each peak at indices 1.33, 1.34, 1.35, 1.36, 1.37, and 1.39 against the concentration of albumin, yields a sensitivity of 2.59 nm/mM, equivalent to 245.96 nm/RIU. This performance is notable when compared to the reported localized surface plasmon resonance (LSPR) sensitivities for albumin detection, which typically range from 100 to 200 nm/RIU, indicating a highly effective detection method.

IV. CONCLUSION

The hexagonal geometry with gold material was chosen for the formation of LSPRs that can optimize the absorbency and sensitivity of their use to detect the presence of albumin in urine. The size of LSPR AuNPs that has the best absorption is 30 nm with a lateral size of 40 nm. After being simulated at each refractive index of albumin at a concentration of 0.35 – 5.71 mM, the sensitivity of hexagonal LSPR AuNPs produced a sensitivity of 2.59 nm/mM, equivalent to 245.96 nm/RIU. indicating that LSPR with hexagon-shaped AuNPs material is quite good at detecting albumin in urine.

V. ACKNOWLEDGEMENT

This work is supported by grant from Directorate of Higher Education, Ministry of Education and Culture, Republic of Indonesia.

REFERENCES

[1] Bihari, S., Bannard-Smith, J., & Bellomo, R. (2020). Albumin as a drug: Its biological effects beyond volume expansion. *Critical Care and Resuscitation, 22*(3), 257–265. https://doi.org/10.1016/s1441-2772(23)00394-0

[2] Sheinenzon, A., Shehadeh, M., Michelis, R., Shaoul, E., & Ronen, O. (2021a). Serum albumin levels and inflammation. *International Journal of Biological Macromolecules, 184*, 857–862. https://doi.org/10.1016/j.ijbiomac.2021.06.140

[3] Vahdat, S., & Shahidi, S. (2020). D-dimer levels in chronic kidney illness: A comprehensive and Systematic Literature Review. *Proceedings of the National Academy of Sciences, India Section B: Biological Sciences, 90*(5), 911–928. https://doi.org/10.1007/s40011-020-01172-4

[4] U.S. Department of Health and Human Services. (n.d.). *Quick reference on UACR & GFR - NIDDK.* National Institute of Diabetes and Digestive and Kidney Diseases. https://www.niddk.nih.gov/health-information/professionals/advanced-search/quick-reference-uacr-gfr

[5] Palani, S., Kenison, J., Sabuncu, S., Huang, T., Çivitçi, F., Esener, S. C., & Nan, X. (2023). Multispectral Localized Surface plasmon resonance (MSLSPR) reveals and overcomes spectral and sensing heterogeneities of single gold nanoparticles. *ACS Nano, 17*(3), 2266–2278. https://doi.org/10.1021/acsnano.2c08702

[6] Ravindra, P. (2009). Protein-mediated synthesis of gold nanoparticles. *Materials Science and Engineering: B, 163*(2), 93–98. https://doi.org/10.1016/j.mseb.2009.05.013

[7] Shabaninezhad, M., & Ramakrishna, G. (2019). Theoretical investigation of size, shape, and aspect ratio effect on the LSPR sensitivity of hollow-gold nanoshells. *Journal of Chemical Physics, 150*(14). https://doi.org/10.1063/1.5090885

[8] Hammond, J., Bhalla, N., Rafiee, S., & Estrela, P. (2014). Localized surface plasmon resonance as a biosensing platform for developing countries. *Biosensors, 4*(2), 172–188. https://doi.org/10.3390/bios4020172

[9] Zhang, H., Zhou, X., Li, X., Gong, P., Zhang, Y., & Zhao, Y. (2023). Recent advancements of LSPR fiber-optic biosensing: Combination methods, structure, and prospects. *Biosensors, 13*(3), 405. https://doi.org/10.3390/bios13030405

[10] Yin, H., Guo, Y., Cui, X., Lu, W., Yang, Z., Yang, B., & Wang, J. (2018). Plasmonic and sensing properties of vertically oriented hexagonal gold nanoplates. *Nanoscale, 10*(31), 15058–15070. https://doi.org/10.1039/c8nr04463e

[11] Park, G., Min, K., Kwon, H., Yoon, S., Park, S. W., Kwon, J., Lee, S., Jo, J., Kim, M., & Kim, S. K. (2021). Strain-Induced modulation of localized surface plasmon resonance in ultrathin hexagonal gold nanoplates. *Advanced Materials, 33*(38). https://doi.org/10.1002/adma.202100653

[12] Yockell-Lelièvre, H., Lussier, F., & Masson, J. (2015). Influence of the particle shape and density of Self-Assembled gold nanoparticle sensors on LSPR and SERS. *Journal of Physical Chemistry C, 119*(51), 28577–28585. https://doi.org/10.1021/acs.jpcc.5b09570

[13] Park, G., Min, K. S., Kwon, H., Yoon, S., Park, S., Kwon, J., Lee, S., Jo, J., Kim, M., & Kim, S. K. (2021). Strain-induced modulation of localized surface plasmon resonance in ultrathin hexagonal gold nanoplates. *Advanced Materials, 33*(38). https://doi.org/10.1002/adma.202100653

[14] Sharma, S., & Gupta, S. (2021). Detection Of Albumin And Urea Concentration In Urine Using 2d Photonic Crystals. *Journal of Physics, 2007*(1), 012015. https://doi.org/10.1088/1742-6596/2007/1/012015

[15] Liu, H., Chen, C., Zhang, Y., Bai, B., & Tang, S. (2019). A high-sensitivity methane sensor with localized surface plasmon resonance behavior in an improved hexagonal gold nanoring array. *Sensors, 19*(21), 4803. https://doi.org/10.3390/s19214803

[16] Farooq, S., Wali, F., Zezell, D. M., de Araujo, R. E., & Rativa, D. (2022). Optimizing and quantifying gold nanospheres based on LSPR label-free biosensor for dengue diagnosis. *Polymers, 14*(8), 1592. https://doi.org/10.3390/polym14081592

[17] Cheng, L., Zhu, G., Liu, G., & Zhu, L. (2020). FDTD simulation of the optical properties for gold nanoparticles. *Materials Research Express, 7*(12), 125009. https://doi.org/10.1088/2053-1591/abd139

[18] Tira, C., Tîra, D. S., Simon, T., & Astilean, S. (2014). Finite-Difference Time-Domain (FDTD) design of gold nanoparticle chains with specific surface plasmon resonance. *Journal of Molecular Structure, 1072*, 137–143. https://doi.org/10.1016/j.molstruc.2014.04.086

[19] Wang, S., Sun, X., Ding, M., Peng, G., Qi, Y., Wang, Y., & Ren, J. (2018). The investigation of an LSPR refractive index sensor based on periodic gold nanorings array. *Journal of Physics D, 51*(4), 045101. https://doi.org/10.1088/1361-6463/aaa06a

[20] Shajari, D., Bahari, A., & Gill, P. (2018a). Fast and simple detection of bovine serum albumin concentration by studying its interaction with gold nanorods. *Colloids and Surfaces A: Physicochemical and Engineering Aspects, 543*, 118–125. https://doi.org/10.1016/j.colsurfa.2018.02.008

[21] Shin, D. O., Jeong, J.-R., Han, T. H., Koo, C. M., Park, H.-J., Lim, Y. T., & Kim, S. O. (2010). A plasmonic biosensor array by Block Copolymer Lithography. *Journal of Materials Chemistry, 20*(34), 7241. https://doi.org/10.1039/c0jm01319f

[22] Do, P. Q., Huong, V. T., Phuong, N. T., Nguyen, T.-H., Ta, H. K., Ju, H., Phan, T. B., Phung, V.-D., Trinh, K. T., & Tran, N. H. (2020a). The highly sensitive determination of serotonin by using gold nanoparticles (AU NPS) with a localized surface plasmon resonance (LSPR) absorption wavelength in the visible region. *RSC Advances, 10*(51), 30858–30869. https://doi.org/10.1039/d0ra05271j

[23] Rai, V. N. & Srivastava, A. K. Correlation between optical and morphological properties of nanostructured gold thin ☐lm. JSM Nanotech-nol. Nanomed. 4, 1039 (2016)

Bright-dark soliton pairs generation in an erbium-doped fiber laser utilizing gamma alumina saturable absorber

Norita Mohd Yusoff
Institute of Nanoscience and
Nanotechnology (ION2),
Universiti Putra Malaysia
43400, UPM Serdang, Selangor,
Malaysia
noritayusoff@upm.edu.my

Eng Khoon Ng
Department of Engineering,
University of Cambridge,
Cambridge CB3 0FA, United
Kingdom
ekn27@cam.ac.uk

Mohd Zul Hilmi Mayzan
Ceramic and Amorphous Group
(CerAm), Faculty of Applied
Sciences and Technology,
Pagoh Higher Education Hub,
Universiti Tun Hussein Onn
Malaysia, 84600 Panchor,
Johor, Malaysia
zulhilmi@uthm.edu.my

Mohd Adzir Mahdi
Wireless and Photonic Networks
Research Centre,
Faculty of Engineering,
Universiti Putra Malaysia,
43400 UPM Serdang,
Selangor, Malaysia
mam@upm.edu.my

Abstract—We report the experimental observation of bright-dark soliton pairs in an erbium-doped fiber laser utilizing gamma aluminum oxide saturable absorber (γ-Al_2O_3-SA) operating at 1.56 µm for the first time. The γ-Al_2O_3-SA was fabricated using simple adhesion of material on ferrule tip using index-matching gel. Dual-wavelength and narrow linewidths in optical domain with alternate intensity spike and dip that repeats for every 9.0 MHz repetition frequency manifested the formation of bright-dark soliton pairs. These findings contribute to the future exploration of metal oxide as saturable absorbing material for various pulse dynamics and fundamental laser physics in the topic of ultrafast photonics.

Keywords—Alumina, bright-dark soliton pairs, mode-locked, saturable absorber

I. INTRODUCTION

In recent years, the research directions in the field of ultrafast photonics have shifted towards a compact laser cavity that capable to produce more than one type pulse emissions. In particular, the efforts on exploring the dark pulses have attracted a great deal of attention, as they could co-exist in dissipative soliton [1] and traditional soliton [2] operating in normal and anomalous dispersion laser cavities, respectively. As opposed to the intensity peak on the constant background noise commonly termed as the "bright" pulses, the "dark" pulse characterized themselves as the intensity dip in the time domain. The uniqueness of dark pulse is its less susceptible to fiber loss and background noise, this such pulse type may finds demand in the application of high-order coherent communications [3].

In the framework of passive mode-locking technique, various materials such as graphene [2], carbon nanotubes [4], topological insulator [5], transition metal dichalcogenide [6], and black phosphorus [7] have been employed as saturable absorbing device for bright, dark, and co-exist of both pulses. However, this pulse phenomenon has not yet be reported in the category of metal oxides so far.

The exploration of aluminum oxide (Al_2O_3) as a saturable absorbing device has gained significant attention in recent years. Al_2O_3 exists in multiple crystal phases in nature, namely alpha (α), beta (β), and gamma (γ), to name a few [8]. Technically, the stable α-Al_2O_3 possesses superior mechanical hardness, high resistance to chemicals and temperature, and high binding energy due to its ceramic properties [9]. Associated with these features, this material has been incorporated into saturable absorber fabrication for the high energy and microsecond width of Q-switched pulses. Up to 16.9 nJ [10] and 150.3 nJ [11] pulse energies have been recorded at 1.0 and 1.56 µm wavelength bands, respectively. Besides that, the recent literature has demonstrated the feasibility of the metastable phase of γ-Al_2O_3 to generate ultrafast pulses via the mode-locking technique [9]. Furthermore, the nonlinear properties of this material have also been verified through z-scan measurement, whereby the nonlinear refractive index and nonlinear absorption coefficient are 3.67×10^{-8} cm^{2}W^{-1} and 6.6×10^{-4} cm^{2}W^{-1}, respectively [12]. Despite the high bandgap energy of 6.7 eV [13], the optical characterization reveals a stronger absorbing capability of γ-Al_2O_3 in the near-infrared region due to the multiple sub-band absorptions in defective structure [9].

While the investigation on the metastable state of γ-Al_2O_3 is in its early stages, we report the bright-dark soliton pairs in an erbium-doped fiber laser (EDFL) cavity by deliberately altering polarization controller states. The generated pulses repeat for every 9.0 MHz repetition frequency while maintaining the dual-wavelength spectrum centered at 1562.61 and 1563.47 nm. This work paves the way for exploring the metastable phase of aluminum oxide for future advancement of ultrafast fiber laser applications.

II. SATURABLE ABSORBER FABRICATION AND CHARACTERIZATION

In the experiment, gamma aluminum oxide (γ-Al_2O_3, 99%) with average size of 8-12 nm was purchased from ACS Material and used without further purification. The fabrication of SA device was done using a straightforward method incorporating index matching gel, as previously reported in [14]. The experiment was begun by transferring the gel was onto glass substrate and a scoop of γ-Al_2O_3 powders was added thereafter. After the mixing process of these substances, a clean fiber tip was tapped on the γ-Al_2O_3/gel composite. The thickness of the attached gel was gently reduced by using clean scotch tape, and consequently, it was mechanically aligned to another clean ferrule via an FC/PC connector. The transmission characteristic of the fabricated γ-Al_2O_3-SA device is presented in Fig. 1, having 67-69% transmission across the 1530-1600 nm wavelength region, which translated to the 1.7 dB loss at 1560 nm. The lower loss as compared to the previous works (3.5 dB [11] and 4.7 dB [9]) is mainly due to the small amount of nanopowders used and the inclusion of index-matching gel to minimize the scattering effect between two ferrules.

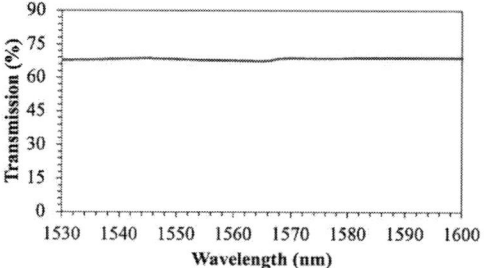

Fig. 1. Transmission of γ-Al$_2$O$_3$-SA with respect to wavelength.

III. LASER CAVITY CONFIGURATION

The all-fiber EDFL in ring scheme is schematically depicted in Fig. 2. The laser gain was provided by a spool of 5 m erbium-doped fiber having a peak absorption of 5.5 dB/m at 1530 nm. The erbium ions in the EDF was excited by a laser diode operating at 980 nm via 980/1550 nm wavelength division multiplexer (WDM). Meanwhile, the unidirectional light propagation at the forward direction was made possible by splicing the polarization independent isolator (PI-ISO). The saturable absorbing device, γ-Al$_2$O$_3$-SA, was inserted in between the PI-ISO and optical coupler (OC). A small portion of the intracavity light was tapped through 20% port of OC, whereas the remaining 80% of light was directed to the polarization controller (PC) and supplied to the signal port of WDM as optical feedback. The PC was utilized to optimize the laser cavity per round trip. With the total cavity length of 22.70 m, the net group velocity dispersion was calculated to be -0.26 ps^2.

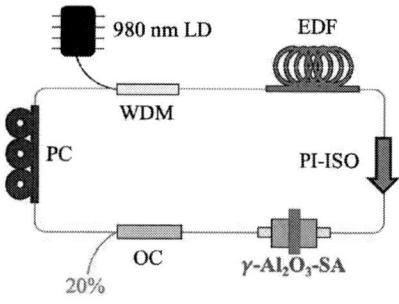

Fig. 2. EDFL configuration.

As all the passive optical components had polarization insensitive features, the effect of nonlinear polarization evolution was eliminated. All measurements were performed using Yokogawa optical spectrum analyzer (model AQ6370B) and Tektronix digital phosphor oscilloscope (model TDS3012C) coupled with Kyphotonics InGaAs photodetector (KY-PDM-2G-I).

IV. PULSE PERFORMANCES

In the experiment, the first continuous laser emission was observed at 38.0 mW, whereas the mode-locked operation was started to appear when the pump power was gradually increased to 80.0 mW. The formation of bright-dark soliton pairs was then investigated in the temporal domain at the pump power of slightly beyond threshold level (85.0 mW). Fig. 3 depicts the characteristics of bright-dark soliton pairs at different intracavity polarization states. In the PC state 1 shown in Fig. 3(a), strong interaction between bright and dark solitons were observed as they are mutually entangled to each other with minimal spacing of 9.25 ns. Further rotating the PC angles, the separation between these two solitons increased up to 63.00 ns in the PC state 5, implying the weakest interaction between them. However, the generated pulses are equidistant to each other, propagated at 111 ns round-trip time.

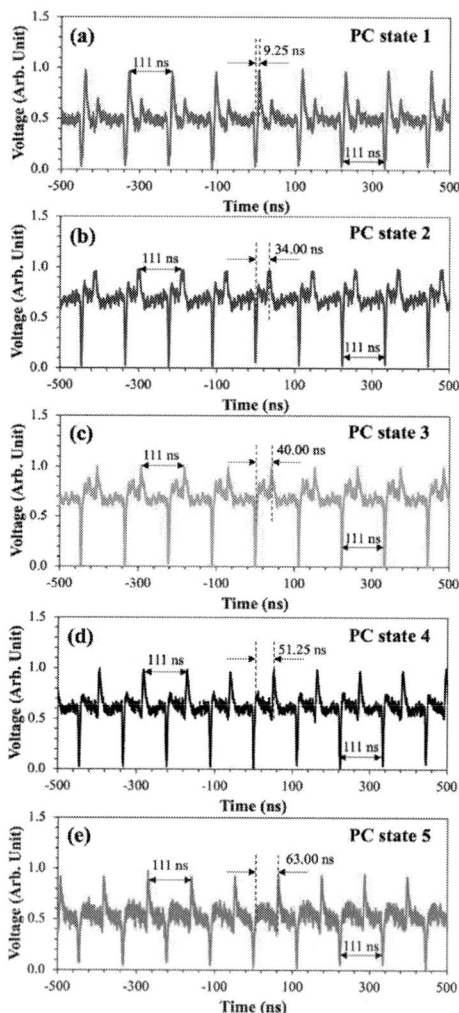

Fig. 3. Oscilloscope trace with respect to different PC angles.

Meanwhile, in the optical domain, the spectral outputs presented in Fig. 4 illustrates the narrow linewidth of dual-wavelength emissions with M-shape profile, having maximum spacing of 0.86 nm. This simultaneous interaction between bright and dark solitons gives rise to the incoherent coupling between them [3], subsequently induced the nonlinear effect of four-wave mixing [15]. From the results, the optical spectrum retained dual-peak characteristics centered at 1562.61 and 1563.47 nm at any polarization states, while the maximum wavelength drift are 0.02 and 0.05 nm, respectively.

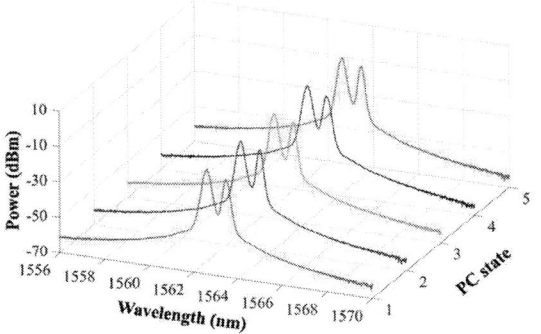

Fig. 4. Optical spectrum with respect to different PC angles.

V. CONCLUSIONS

In this research work, γ-Al$_2$O$_3$ successfully sandwiched between two ferrules with the aid of index matching gel as adhesive material. The deployment of this SA in ring cavity EDFL delivered bright-dark soliton pairs that repeat at 9.0 MHz repetition frequency. The effect of these solitons were investigated by adjusting the intracavity polarization states. This work provides a reference for the future exploration of various pulse emissions in a particular laser cavity and underlying physics within the topic of ultrafast photonics.

ACKNOWLEDGMENT

This work was supported by the Ministry of Higher Education Malaysia under the Geran Putra Inisiatif (GPI/2023/9757300).

REFERENCES

[1] L. Wang, "Coexistence and evolution of bright pulses and dark solitons in a fiber laser," Optics Communications, vol. 297, 129-132, 2013.

[2] Z. Li, M. Li, X. Hou, L. Du, L. Xiao, T. Wang, and W. Ma, "Generation of mode-locked states of conventional solitons and bright-dark solitons in graphene mode-locked fiber laser," Frontiers of Optoelectronics, vol. 16, 12, 2023.

[3] A. Fülöp, M. Mazur, A. Lorences-Riesgo, O. B. Helgason, P. Wang, Y. Xuan, D. E. Leaird, M. Qi, P. A. Andrekson, A. M. Weiner, and V. Torres-Company, "High-order coherent communications using mode-locked dark-pulse Kerr combs from microresonators," Nature Communications, vol. 9, 1598, 2018.

[4] H. H. Liu and K. K. Chow, "Dark pulse generation in fiber lasers incorporating carbon nanotubes," Optics Express, vol. 22, 24, 2014.

[5] B. Guo, Y. Yao, J. Tian, Y. F. Zhao, S. Liu, M. Li, and M. Quan, "Observation of bright-dark soliton pair in a fiber laser with topological insulator," IEEE Photonics Technology Letters, vol. 27, 7, 2015.

[6] R. Zhao, G. Li, B. Zhang, and J. He, "Multi-wavelength bright-dark pulse pair fiber laser on rhenium disulfide," Optics Express, vol. 26, 5, 2018.

[7] T. Wang, W. Zhang, J. Wang, J. Wu, T. Hou, P. Ma, R. Su, Y. Ma, J. Peng, L. Zhan, K. Zhang, and P. Zhou, "Bright/dark switchable mode-locked fiber laser based on black phosphorus," Optics and Laser Technology, vol. 123, 105948, 2020.

[8] K. Y. Paranjpe, "Alpha, beta and gamma alumina as a catalyst-A review," Pharma Innovation Journal, vol. 6, 11, 2017.

[9] N. F. Pikau, N. Mohd Yusoff, A. R. Sarmani, F. D. Muhammad, M. T. Alresheedi, E. K. Ng, and M. A. Mahdi, "Gamma alumina as a saturable absorbing material for C- and L-band ultrafast mode-locked fiber lasers," Results in Physics, vol. 61, 107759, 2024.

[10] S. K. M. Al-Hayali, S. Selleri, and A. H. Al-Janabi, "Dual-wavelength passively Q-switched ytterbium-doped fiber laser based on aluminum oxide nanoparticle saturable absorbers," Chinese Physics Letters, vol. 34, 11, 2017.

[11] N. Mohd Yusoff, J. L. Y. Chyi, E. K. Ng, A. R. M. Zain, M. T. Alresheedi, M. Z. H. Mayzan, and M. A. Mahdi, "Bulk α-alumina embedded tapered fiber as Q-switcher in an erbium-doped fiber laser," Physica Scripta, vol. 99, 065522, 2024.

[12] A. F. Alamouti, M. Nadafan, Z. Dehghani, M. H. M. Ara, and A. V. Noghreiyan, "Structural and optical coefficients investigation of γ-Al$_2$O$_3$ nanoparticles using Kramers-Kronig Relations and z-scan technique,", Journal of Asian Ceramic Societies, vol. 9, 1, 2021.

[13] J. Cañas, J. C. Piñero, F. Lloret, M. Gutierrez, T. Pham, J. Pernot, and D. Araujo, "Determination of alumina bandgap and dielectric functions of diamond MOS by STEM-VEELS, Applied Surface Science, vol. 461, 2018.

[14] H. Asghar, R. Ahmed, M. Sohail, Z. A. Umar, and M. A. Baig, "Q-switched pulse operation in erbium-doped fiber laser subject to CsS nanoparticles-based saturable absorber deposit directly on the fiber ferrule," Optical Materials, vol. 134, 113109, 2022.

[15] Q. Pang, X. Zhu, L. Shi, B. Xu, R. Weng, J. Wang, C. Zhou, M. Fan, W. Tang, and W. Xia, "Generation of bright-dark soliton pairs in mode-locked fiber laser based on WSe$_2$/MoSe$_2$ heterojunction," Infrared Physics and Technology, vol. 136, 105069, 2024.

979-8-3503-7832-0/24 $31.00 © 2024 IEEE

Physical Characterization of Immersion Method-Based Porous Paper Towards Sensing Application

Gan Shin Pyng
Faculty of Electrical Engineering
Universiti Teknologi Malaysia (UTM)
Johor, Malaysia
shinpyng-2018@graduate.utm.my

Mastura Shafinaz Zainal Abidin
Faculty of Electrical Engineering)
Universiti Teknologi Malaysia (UTM)
Johor, Malaysia
m-shafinaz@utm.my

Abstract—Humidity sensors are essential to many applications, from climate control systems to healthcare. Paper-based humidity sensors have recently gained significant interest due to their simplicity, cost-effectiveness, and unique paper substrate properties. Paper is made of cellulose fibers that are woven together in a complex way to form a porous structure. These pores allow liquids to permeate the paper and stay inside the structure, which determines the absorption capacity. This study investigates the characterization of porous paper that has been formed through the immersion technique. The normal printing paper samples were immersed in different solutions, such as hydrochloric acid (HCl) and sodium hydroxide (NaOH), to yield porous paper. The effect of solution concentration on paper porosity and morphology has been observed. Based on the present findings, the use of HCl has demonstrated superior outcomes compared to NaOH in modifying paper porosity properties. The higher concentrations of HCl have been associated with enhanced porosity in the modified paper samples. The changes in porosity properties were achieved due to the removal of carbonate (CO_3) filler from the paper cellulose structure. The highest percentage of porosity is 56.5%. These results offer important new understandings of optimizing chemical treatment techniques for customizing paper characteristics to fulfill certain application needs in humidity sensors.

Keywords—*porous paper, hydrochloric acid, sodium hydroxide, immersion*

I. INTRODUCTION

Paper is widely used for many different purposes, such as writing, printing, packaging, and filtering. The properties of paper, particularly its porous structure, directly impact the quality and functionality of paper products. Understanding and characterizing the formation of porous paper is crucial for optimizing paper manufacturing processes, enhancing product quality, and developing new materials with specific properties. In line with global carbon emissions control efforts, paper-based electronic products, like humidity sensors, have been prioritized for development nowadays. For such applications, porosity possibly improves performance by increasing the sensitivity, response time, flexibility, and stability of paper-based humidity sensors [1].

The porous structure of paper can be characterized by interconnected void spaces or pores, which influence its absorbency, strength, printability, and other key attributes. There are several methods to produce porous paper, such as chemical treatment [1],[2], mechanical treatment [3], additives [4], foaming technique [5], microbial treatment [6], and electrospinning [7].

In paper science and engineering, chemical treatment is an important method to study the porous structure of paper. Under this chemical treatment method, the immersion technique allows for precise control over the porosity of the paper by adjusting the experimental parameters such as concentration, duration of treatment, and type of chemical used [2]. This feature enables manufacturers to tailor the paper's porosity to meet specific application requirements. The immersion technique would provide uniform pore distribution throughout the paper and result in consistent porosity across the entire paper surface. This uniformity ensures reliable performance and quality in applications with critical pore size and distribution. There is one research study about modifying the porosity of the paper using the immersion technique, as reported by [1].

This study explores how the immersion's parameters, such as concentration and drying condition, would affect the characteristics of the porous paper, experimental observations, and analytical techniques. The optimization in the making process of porous paper using the immersion technique is targeted for humidity sensor applications.

II. METHODOLOGY

This study used the experimental setup in Fig. 1 to immerse the normal printing paper samples. Paper is primarily cellulose fibers, which contain Carbon, Oxygen, and Hydrogen as basic elements. It can also be present in compound form, such as carbonate. There were two types of immersing solutions that have been tested, acid and alkaline solutions. Under hydrochloric acid (HCl), it can selectively break down cellulose fibers, thus creating pores and increasing porosity [8]. For an alkaline-type immersion solution, sodium hydroxide (NaOH) is used due to its capability to swell cellulose fibers, leading to pore formation and enhancing porosity [9].

979-8-3503-7832-0/24 $31.00 © 2024 IEEE

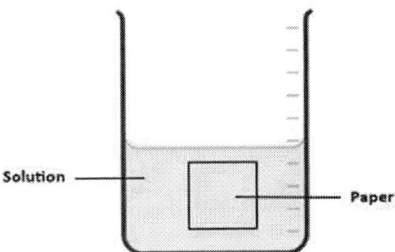

Fig. 1. Immersion setup

The normal printing paper samples (IK Yellow, 70 gsm) were immersed respectively in HCl and NaOH solution for 1 minute at room temperature before rinsing with deionized water and drying at room temperature for 1 day. Basically, the normal printing paper was cut into square-shaped samples in dimensions 5 cm in length. The effect of solution type and different concentrations were observed in this study. For each experiment condition, one sample has been prepared. The samples were weighed before and after the experiment to determine the porosity percentage. Porosity is typically calculated as the ratio of void volume to total volume, expressed as a percentage which can be calculated by applying the following equation by considering the density of cellulose as 1.5g/cm³ [1]:

$$ Porosity\ (\%) = (1 - \frac{Psample}{1.5\ g/cm^3}) \times 100 \qquad (1) $$

where $Psample$ (g/cm³) is the density of the sample. The $Psample$ can be determined from the mass of the sample after the experiment divided by the volume of the sample (length x width x thickness = 0.25 cm³).

Besides that, the analysis of porous paper samples also included morphology and elemental characterization using a Scanning Electron Microscope (SEM) (HITACHI TM3000 and JEOL JSM-IT300LV), Energy-dispersive X-ray spectroscopy (EDX), and porosity calculation.

III. RESULTS AND DISCUSSION

A. Comparison between Porous Paper Produced by Acidic and Alkaline Solution

The SEM-EDX results for the porous paper using HCl and NaOH solutions, compared to untreated normal printing paper, are shown in Fig. 2. From the top view of the samples, it is found that a significant difference between the 3 samples with the number of white spots appeared in between the cellulose fibers. These white spots were presumed to represent carbonate (CO_3) elements [1]. Compared to untreated paper samples, the HCl-porous paper sample had fewer white spots than NaOH porous paper sample. EDX spectrum further supports this observation.

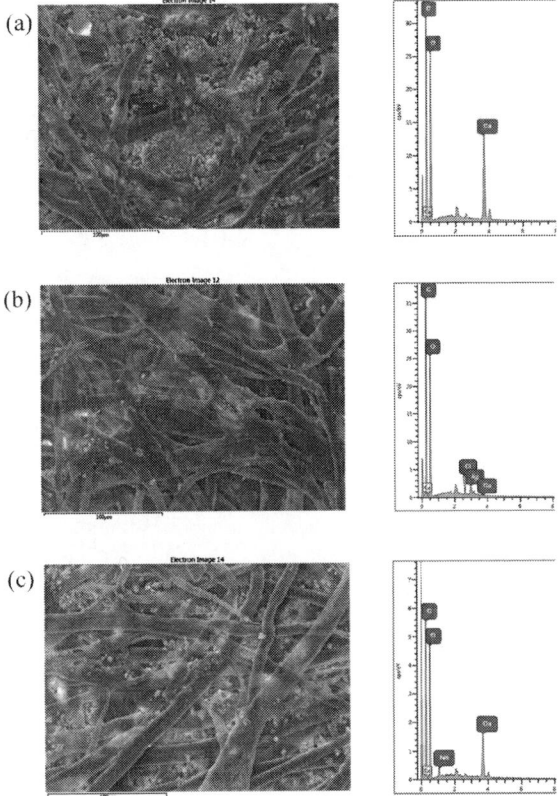

Fig. 2. SEM and EDX images of the samples (a) Untreated printing paper (b) HCl porous paper (c) NaOH porous paper

Based on EDX spectrum, the untreated printing paper sample shows higher amounts of oxygen, carbon, and calcium. Basically, the presence of calcium increases the quality of the printing paper and helps regulate its color, texture, and brightness [10]. However, after immersion in HCl acid, it showed increased carbon and oxygen and decreased calcium and chloride. Based on the chemical reaction, as shown in (2), it justified the cause of calcium element reduction from 6.7% to 0.2% due to the removal of CO_3 and the formation of calcium chloride, as mentioned in [1]. The chloride content was also significantly present at 0.1%. This shows that the immersion process with HCl acid altered the structure of the cellulose paper, resulting in increased porosity.

$$ CaCO_3 + 2HCl \rightarrow CaCl_2 + H_2O + CO_2 \qquad (2) $$

For NaOH porous paper sample, it exhibited increased oxygen, carbon, calcium, and sodium content. The surface reaction between paper samples with NaOH could be represented by (3); the analysis shows that the NaOH solution did not effectively reduce carbonate. Besides that, calcium remained relatively high, from 6.7% to 4.4%. This indicates that the NaOH solution might not have the same efficacy in removing carbonate as the HCl solution.

$$CaCO_3 + 2NaOH \rightarrow Na_2CO_3 + Ca(OH)_2 \qquad (3)$$

The paper immersed in HCl solution undergoes better modification by effectively removing CO3 elements than untreated paper and paper immersed in NaOH solution. HCl solution is a preferable choice for certain paper modification processes.

B. Effect of Different HCL Concentration Immersion

After comparing the porous paper's morphology produced by immersion in HCl and NaOH solutions, it was found that the HCl porous paper sample gave better results than NaOH porous paper sample. Therefore, further investigations of porous paper with HCl solutions immersion at different concentrations ranging from 0.1 mol to 0.5 mol were conducted. This is to reveal important insights into the relationship between acid concentration and porosity enhancement.

Based on the SEM-EDX results shown in Fig. 3, the sample immersed in 0.5 mol HCl has fewer white spots than 0.1 mol HCl. Based on these observations, it can be predicted that a higher concentration of HCl leads to more significant modifications to the paper structure in terms of porosity compared to lower concentrations. The increased removal of CO3, as indicated by the reduction in white spots, reflects the stronger effect of higher HCl concentration on altering the composition and structure of the paper.

Based on the EDX results, the calcium element in the untreated paper is measured at 13%, while the sample immersed in 0.1 mol HCl shows a calcium element of 11%, and the sample immersed in 0.5 mol HCl shows 7%. These results indicate that paper immersed in different concentrations of HCl can indeed remove varying percentages of calcium from the paper structure. Based on (2), it is evident that the concentration of HCl used for immersion directly impacts the percentage of carbonate removed from the paper. The Sample immersed in 0.5 mol HCl shows a better result in terms of CO3 removal compared to the sample immersed in 0.1 mol HCl. This aligns with the earlier prediction that higher concentrations of HCl lead to more significant modifications in the paper structure, including removing carbonate and other minerals.

The porosity calculation is based on (1), and Fig. 4 shows the porosity change with different concentrations of HCl solution. The figure shows that when the concentration of HCl increases, porosity increases. This means that higher concentrations of HCl are likely to induce more extensive chemical reactions with the paper fibers, leading to greater dissolution of components like calcium carbonate and enhanced formation of pores within the paper structure. As a result, the paper becomes more porous, which is reflected in the increased porosity values observed in Fig. 4 as the concentration of HCl solution rises.

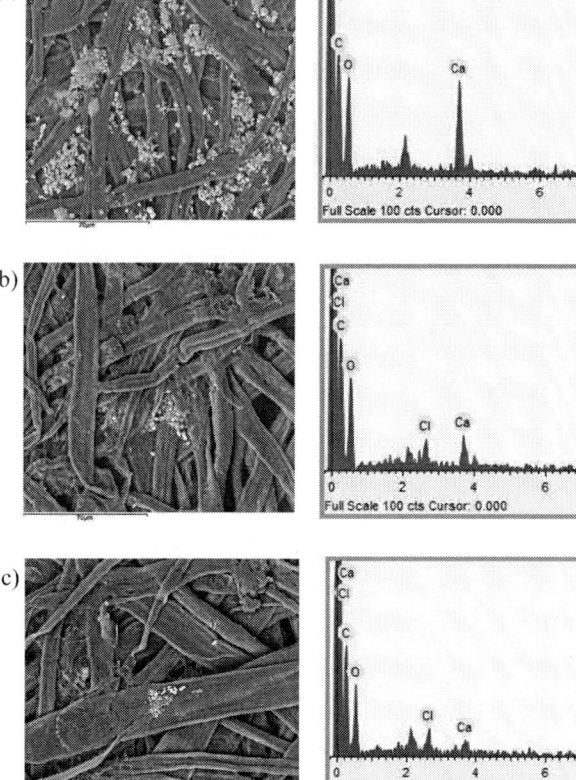

Fig. 3. SEM and EDX images of the sample (a) untreated paper (b) 0.1 mol HCl (c) 0.5 mol HCl

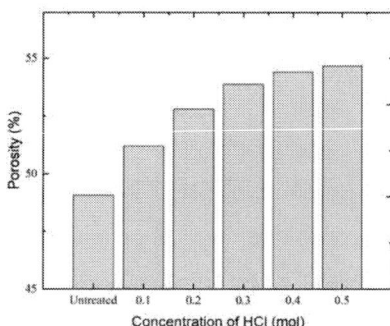

Fig. 4. Graph of porosity vs concentration of HCl solution

Based on Fig. 4, it can be seen that the porosity of 0.3 mol to 0.5 mol samples was almost similar. This could be due to a similar reaction ratio between the HCl solution and the surface of the paper, as represented in (2). Probably at 0.3 mol of HCl, the reaction between the acid and the paper fibers may reach a point where the available reactants become limited. Thus, no more changes happened with the increase in the concentration of HCl after that point. This can lead to a reduction in the rate of pore formation or a shift in the equilibrium of the reaction, resulting in diminished additional porosity gains.

IV. CONCLUSION

In conclusion, the immersion of printing paper samples in different concentrations of hydrochloric acid (HCl) solutions has significantly impacted the porosity. Several key findings have emerged through a combination of visual observation using SEM images and elemental analysis using EDX. Higher concentrations of HCl lead to more pronounced modifications in paper properties, including increased porosity. The porosity of the paper samples increases with higher concentrations of HCl. The paper immersed in 0.5 mol has the highest percentage of porosity at 56.5%. The results demonstrate the effectiveness of HCl treatment in modifying paper properties, with higher concentrations leading to greater enhancements in porosity and surface characteristics. These findings provide valuable insights into optimizing chemical treatment methods for tailoring paper properties to meet specific application requirements in humidity sensors.

ACKNOWLEDGMENT

This work was supported by the Universiti Teknologi Malaysia under UTM Fundamental Research Scheme (Q.J130000.3823.22H41). Authors also would like to express gratitude for the support from several laboratories in Faculty of Electrical Engineering, Universiti Teknologi Malaysia.

REFERENCES

[1] X. Zhang, D. He, Q. Yang, and M. Z. Atashbar, "Rapid, highly sensitive, and highly repeatable printed porous paper humidity sensor," Chemical Engineering Journal, vol. 433, p. 133751, Nov. 2021.

[2] X. Y. Guan et al., "Flexible humidity sensor based on modified cellulose paper," Sensors and Actuators B-chemical, vol. 339, p. 129879, Jul. 2021.

[3] C. Du, H. Li, B. Liu, J. Chen, H. Jian, and J. Zeng, "Effect of beating degree of fiber on the development of porosity in polyacrylonitrile-based activated carbon fiber paper," Diamond and Related Materials, vol. 128, p. 109228, Oct. 2022.

[4] F. P. Morais, A. M. Carta, M. E. Amaral, and J. M. R. Curto, "Micro/nano-fibrillated cellulose (MFC/NFC) fibers as an additive to maximize eucalyptus fibers on tissue paper production," Cellulose, vol. 28, no. 10, pp. 6587–6605, May 2021.

[5] A. M. Al-Qararah, A. Ekman, T. Hjelt, H. Kiiskinen, J. Timonen, and J. A. Ketoja, "Porous structure of fibre networks formed by a foaming process: a comparative study of different characterization techniques," Journal of Microscopy, vol. 264, no. 1, pp. 88–101, May 2016.

[6] A. Mautner and A. Bismarck, "Bacterial nanocellulose papers with high porosity for optimized permeance and rejection of nm-sized pollutants," Carbohydrate Polymers, vol. 251, p. 117130, Jan. 2021.

[7] C. Huang and N. L. Thomas, "Fabrication of porous fibers via electrospinning: strategies and applications," Polymer Reviews, vol. 60, no. 4, pp. 595–647, Nov. 2019.

[8] Y. Wang, "Cellulose fiber dissolution in sodium hydroxide solution at low temperature: dissolution kinetics and solubility improvement," 2008. [Online]. Available: http://smartech.gatech.edu/bitstream/1853/26632/1/wang_ying_200812_phd.pdf

[9] M. Y. Hashim, A. M. Amin, O. M. F. Marwah, M. H. Othman, M. R. M. Yunus, and N. C. Huat, "The effect of alkali treatment under various conditions on physical properties of kenaf fiber," Journal of Physics: Conference Series, vol. 914, p. 012030, Oct. 2017.

[10] "Precipitated Calcium Carbonate paper - Cales de Llierca," Cales De Llierca, Oct. 27, 2021. https://www.calesdellierca.com/applications/paper-industry-precipitated-calcium-carbonate/#:~:text=Precipitated%20Calcium%20Carbonate%20in%20paper,as%20Natural%20Carbonate%20(CC).

Effects of Growth Parameters on the Morphology of ReS$_2$ Nanoflakes Prepared by Chemical Vapor Deposition

M. F. M. Ruslan
Institute of Microengineering and Nanoelectronics (IMEN)
Universiti Kebangsaan Malaysia (UKM)
Bangi, Malaysia
m.farismusawwi@gmail.com

Muhammad Hilmi Johari
Institute of Microengineering and Nanoelectronics (IMEN)
Universiti Kebangsaan Malaysia (UKM)
Bangi, Malaysia
johari.hilmi@gmail.com

Syahira A. Hinayadullah
Institute of Microengineering and Nanoelectronics (IMEN)
Universiti Kebangsaan Malaysia (UKM)
Bangi, Malaysia
p106517@siswa.edu.ukm.my

P. Susthitha Menon
Institute of Microengineering and Nanoelectronics (IMEN)
Universiti Kebangsaan Malaysia (UKM)
Bangi, Malaysia
susi@ukm.edu.my

Abdul Rahman Mohmad
Institute of Microengineering and Nanoelectronics (IMEN)
Universiti Kebangsaan Malaysia (UKM)
Bangi, Malaysia
armohmad@ukm.edu.my

Abstract—The development of efficient ReS$_2$-based optoelectronic devices depends on the ability to synthesize high-quality nanoflakes with the desired morphology. In this work, we report the synthesis of ReS$_2$ nanoflakes with vertical and horizontal orientations via chemical vapor deposition (CVD) on SiO$_2$/Si substrates at atmospheric pressure. The ReS$_2$ samples were grown using rhenium trioxide (ReO$_3$) and sulfur precursors. It was found that growth temperature and ReO$_3$-substrate distance significantly influenced the nucleation density and morphology of the ReS$_2$ flakes, as observed by optical and field-emission scanning electron microscope images. High Re vapor environment favors the formation of vertical flakes. The growth then switches to a mixture of both vertical and horizontal flakes as the Re concentration decreases. Raman spectroscopy data show the characteristic E$_g$ and C$_p$ vibrational modes which belong to the ReS$_2$. The full-width-at-half-maximum of the 210.3 cm^{-1} peak is 8 cm^{-1} which is comparable to the high-quality ReS$_2$ flakes prepared by mechanical exfoliation. This work provides valuable insights on the effect of CVD parameters on the growth and morphology of ReS$_2$ nanoflakes.

Keywords—ReS$_2$, chemical vapor deposition, morphology

I. INTRODUCTION

Following the discovery of graphene, increasing interest has been shown in other types of layered materials such as transition metal dichalcogenides (TMDs) [1]. TMD is composed of transition metals (e.g. Mo, W) which are bonded to two chalcogen atoms (typically S, Se, or Te) [2]. Their unique crystal structure features weak van der Waals (vdW) forces between each atomic layer, enabling exfoliation to single-layer [3]. Unlike other 2D TMDs (e.g. MoS$_2$, WS$_2$), ReS$_2$ is attracting an increasing interest due to its direct bandgap, weak interlayer coupling (making its properties independent of thickness), and anisotropic behavior which is promising for optoelectronics devices [4]. When processed at nano-micro level, the structure exhibited improved optical properties, enhanced sensitivity and responsivity, high room temperature carrier mobility and large modulation band with

[5]. Due to the weak coupling between the layers, the bandgap of ReS$_2$ is relatively constant at ~1.55 eV regardless of the thickness [6]. It was also reported that the anisotropic behavior of ReS$_2$ originated from the zigzag Re-Re chains along the b-axis of the lattice which forms a stable distorted 1T phase [7].

ReS$_2$ nanoflakes have been prepared by several methods such as exfoliation techniques, the Bridgman method, and physical vapor deposition [8]. To date, chemical vapor deposition (CVD) is one of the promising synthesis techniques for producing large-area and high-quality ReS$_2$ [9]. Importantly, CVD is capable of producing 2D materials with low defect density, and large grain sizes, which are crucial for high-performance devices [10]. Furthermore, the compatibility of CVD with existing semiconductor fabrication processes may facilitate future integration of 2D materials into current device technologies [11]. Recently, the effect of CVD parameters (e.g. flow rate, growth temperature, and type of precursors) has been reported. In this work, we investigate the effect of Re vapor concentration by varying the separation distance between the rhenium precursor and the substrate. The evolution of the ReS$_2$ flakes morphology will be analyzed and discussed.

II. METHODOLOGY

A. Synthesis of ReS$_2$ nanoflakes

ReS$_2$ nanoflakes were prepared by a single heating zone CVD system, as demonstrated in Figure 1(a). The furnace was calibrated to obtain an accurate heating profile as shown in Figure 1(b). The calibration shows that the temperature across the furnace is not uniform, particularly close to the edges. Due to the lower melting temperature for sulfur (S) (115.21°C), it was placed near the edges of the furnace. In this experiment, 15 mg of ReO$_3$ powder and 150 mg of S powder were used as precursor materials while the substrate was SiO$_2$/Si. The position of the precursors and the substrate is also shown in Figure 1(a). Before growth, the furnace was purged for 20 minutes using argon (Ar) gas at a flow rate of 200 sccm to remove moisture and foreign molecules. The heating process

979-8-3503-7832-0/24 $31.00 © 2024 IEEE

was divided into two, (i) rapid temperature ramps between room temperature and 700°C in 30 minutes, and (ii) slow temperature ramps between 700 and 750 °C (growth temperature) in 15 minutes. Then, the temperature was maintained for 10 minutes. Finally, the CVD furnace was naturally cooled down to room temperature. In this study, the growth temperature of 650, 750 and 850 °C was used while the gas flow rate was fixed at 80 sccm. The distance between the ReO_3 and the substrate, d was varied between 8.4, 9.8 and 11.4 cm to obtain ReO_3 temperature of 655, 550 and 450 °C, respectively. The growth parameters are summarised in Table 1.

TABLE I.   VARIED PARAMETERS STUDIED IN THIS WORK

Sample	Distance between ReO_3 and substrate, d (cm)	Growth temperature (°C)	Gas flow rate (sccm)
A	8.4		
B	9.8	750	
C			80
D	11.4	850	
E		650	

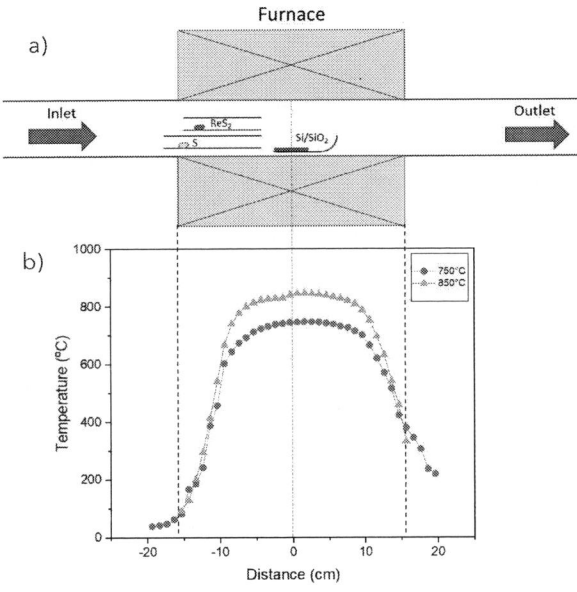

Fig. 1 (a) Schematic diagram of CVD set up to synthesize ReS_2 nanoflakes. (b) Temperature calibration of CVD furnace.

B. Material Characterizations

An optical microscope was used to observe ReS_2 deposition on SiO_2/Si substrate. Then, the morphology of the ReS_2 flakes was characterized using field-emission scanning electron microscopy (FESEM) at an accelerating voltage of 5.0 kV and emission current of 10 μA. To measure the vibrational properties of ReS_2, a confocal micro-Raman spectroscopy system was employed. This technique uses a laser beam with a wavelength of 532 nm and a spot size of around 1 μm. The laser power was kept low to avoid damaging the ReS_2 flakes. All data were calibrated to the Si reference peak at 520.7 cm^{-1}.

III. RESULTS & DISCUSSIONS

Figure 2, shows the optical images of sample A-E which were taken at the middle of the substrates. The optical images revealed that the density of ReS_2 deposition decreases when the ReO_3 precursor is placed further away from the SiO_2/Si substrate which was located at the center of the furnace. Based on the images, a relatively thick deposition was obtained for sample A. However, a mixture of greenish and vertical flakes was observed for sample B while sample C shows a high density of small greenish flakes. This indicates that the thickness of the deposited flakes reduces with the increase of d due to a lower concentration of Re vapor [12]. To reduce the

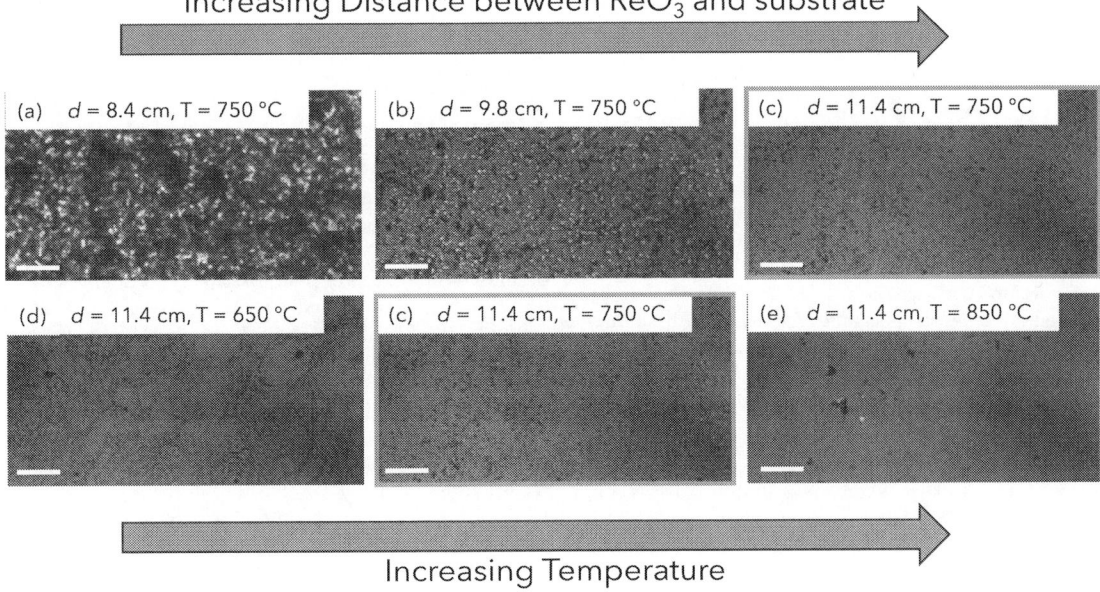

Fig. 2 (a) – (e) Optical images of sample A, B, C, E and D respectively. Scale bars: 10 μm.

979-8-3503-7832-0/24 $31.00 © 2024 IEEE   101

Increasing Distance between ReO₃ and substrate

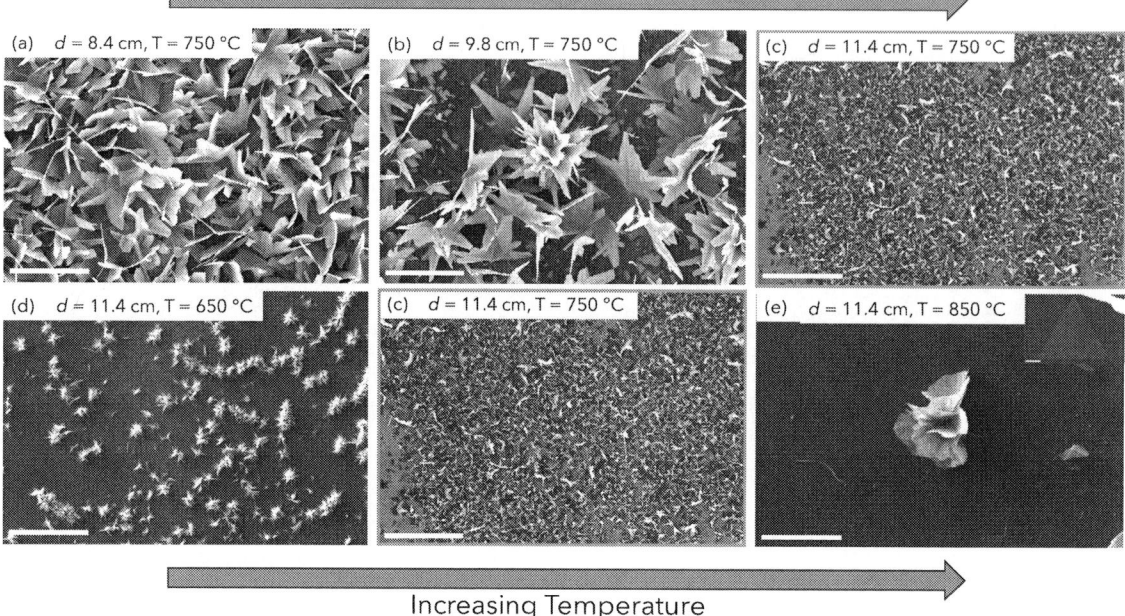

(a) *d* = 8.4 cm, T = 750 °C   (b) *d* = 9.8 cm, T = 750 °C   (c) *d* = 11.4 cm, T = 750 °C

(d) *d* = 11.4 cm, T = 650 °C   (c) *d* = 11.4 cm, T = 750 °C   (e) *d* = 11.4 cm, T = 850 °C

Increasing Temperature

Fig. 3 (a) – (e) FESEM images of sample A, B, C, E and D respectively. Scale bars: 5 μm. Inset shows a horizontal flake at different area on the same surface. Scale bar: 1 μm.

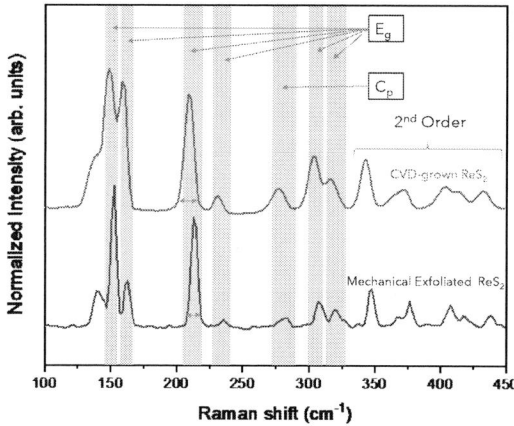

Fig. 4 Raman spectra of as-grown and mechanically exfoliated 2D ReS₂ flakes from 100 cm⁻¹ to 450 cm⁻¹

density of nucleation and increase the size of the flakes, the growth temperature was increased to 850°C. Fig 2(e) shows that a low density of horizontal and thin flakes with slightly larger dimensions was obtained for sample D, as expected. At lower growth temperature for sample E (650°C), darker shade of green is obtained which shows higher density of nanoflakes.

To further investigate the morphology of the deposited flakes, FESEM images were taken and analyzed, as shown in Figure 3(a)-(d). The FESEM image of sample A shows a high density of vertical ReS₂ flakes across the surface. For sample B, the density of vertical flakes decreased and horizontal flakes started to appear. Sample C also shows a mixture of both horizontal and vertical flakes but the size of the flakes reduced significantly. To promote more lateral growth, the

substrate temperature was increased by 100°C for sample D (Figure 3(d)). In addition to vertical flakes, horizontal flakes with triangular shapes were also obtained. When growth temperature is decreased, vertical flakes become dominant as shown in sample E.

To verify that the deposited flakes are indeed ReS₂, Raman spectra from the samples were measured, as shown in Figure 4. For comparison purposes, mechanically exfoliated ReS₂ flakes were also prepared and presented in the same plot. The vibrational modes from the CVD sample are consistent with the exfoliated ReS₂ flakes. For CVD-grown ReS₂, seven distinct Raman peaks were observed at 150.5, 160.1, 210.3, 231.5, 277.8, 304.8, and 316.3 cm⁻¹. It is important to note that, ReS₂ exhibits a significantly higher number of Raman peaks compared to other TMDs like MoS₂ and WS₂. This is due to the lower crystal symmetry of ReS₂, which allows for a greater number of distinct vibrational modes to be Raman-active. Raman modes at 150.4 and 160.1 cm⁻¹ correspond to the in-plane vibration (E_g) of Re atoms, while Raman modes at 210.3, 231.5, 304.8 and 316.3 cm⁻¹ correspond to the in-plane vibration (E_g) of S atoms. Lastly, Raman mode at 277.8 cm⁻¹ may be attributed to the in-plane and out-of-plane vibration of Re and S atoms (C_p). These characteristics peak assignments verify that deposited material is ReS₂ flakes [7]. Besides, the full-width-at-half-maximum (FWHM) of the 210.3 cm⁻¹ peak is 8 cm⁻¹ which is comparable to the high-quality ReS₂ flakes prepared by mechanical exfoliation (6 cm⁻¹). This shows that the CVD-grown ReS₂ flakes have a high crystal quality. The observed Raman peaks between 315-450 cm⁻¹ range can be attributed to the involvement of two phonons during light scattering, an indication of the presence of a second-order Raman process in ReS₂ [13].

979-8-3503-7832-0/24 $31.00 © 2024 IEEE

IV. CONCLUSION

This work investigates the influence of growth parameters on the morphology of ReS_2 flakes prepared by CVD. The ReS_2 nucleation density and layer thickness decrease with the increase of ReO_3-precursor substrate distance. Growth at high Re vapor concentration favors the formation of vertical flakes. However, the growth switches to a mixture of both vertical and horizontal flakes as the Re concentration decreases. Raman spectroscopy data confirms that the quality of the as-grown ReS_2 is on par with the mechanically exfoliated ReS_2 reference. These findings demonstrate the importance of understanding and optimizing the growth parameters to obtain high-quality ReS_2 flakes with the desired morphology.

V. ACKNOWLEDGEMENTS

This research was funded by Dana UKM (RR-2023-002) from Universiti Kebangsaan Malaysia.

VI. REFERENCES

[1] A. V. Kolobov and J. Tominaga, *Two-dimensional transition-metal dichalcogenides.* Springer, 2016.

[2] F. G. Aras *et al.*, "A review on recent advances of chemical vapor deposition technique for monolayer transition metal dichalcogenides (MX2: Mo, W; S, Se, Te)," vol. 148, p. 106829, 2022.

[3] R. Lv *et al.*, "Transition metal dichalcogenides and beyond: synthesis, properties, and applications of single-and few-layer nanosheets," vol. 48, no. 1, pp. 56-64, 2015.

[4] Y.-D. Cao, Y.-H. Sun, S.-F. Shi, and R.-M. J. R. M. Wang, "Anisotropy of two-dimensional ReS 2 and advances in its device application," vol. 40, pp. 3357-3374, 2021.

[5] G. Mohan Kumar *et al.*, "Effective modulation of optical and photoelectrical properties of SnS2 hexagonal nanoflakes via Zn incorporation," vol. 9, no. 7, p. 924, 2019.

[6] S. Tongay *et al.*, "Monolayer behaviour in bulk ReS2 due to electronic and vibrational decoupling," vol. 5, no. 1, pp. 1-6, 2014.

[7] M. Rahman, K. Davey, and S. Z. J. A. F. M. Qiao, "Advent of 2D rhenium disulfide (ReS2): fundamentals to applications," vol. 27, no. 10, p. 1606129, 2017.

[8] M. Hafeez, L. Gan, A. S. Bhatti, and T. J. M. C. F. Zhai, "Rhenium dichalcogenides (ReX 2, X= S or Se): an emerging class of TMDs family," vol. 1, no. 10, pp. 1917-1932, 2017.

[9] P. Hu *et al.*, "Lateral and Vertical Morphology Engineering of Low-Symmetry, Weakly-Coupled 2D ReS2," vol. 33, no. 13, p. 2210502, 2023.

[10] L. Tang, J. Tan, H. Nong, B. Liu, and H.-M. J. A. o. m. r. Cheng, "Chemical vapor deposition growth of two-dimensional compound materials: controllability, material quality, and growth mechanism," vol. 2, no. 1, pp. 36-47, 2020.

[11] Y. Chen, F. Fang, N. J. n. D. M. Zhang, and Applications, "Advance in additive manufacturing of 2D materials at the atomic and close-to-atomic scale," vol. 8, no. 1, p. 17, 2024.

[12] D. Bing, Y. Wang, J. Bai, R. Du, G. Wu, and L. J. O. C. Liu, "Optical contrast for identifying the thickness of two-dimensional materials," vol. 406, pp. 128-138, 2018.

[13] Z. Guo, A. Wei, Y. He, C. He, J. Liu, and Z. J. J. o. M. S. M. i. E. Liu, "Controllable growth of large-area monolayer ReS 2 flakes by chemical vapor deposition," vol. 30, pp. 15042-15053, 2019.

Impact of Varied Ag-GO Ratios on the Electrochemical Enhancement of Vitamin D Detection

Chan Kiki
Institute of Microengineering & Nanoelectronics (IMEN)
Universiti Kebangsaan Malaysia (UKM)
Selangor, Malaysia
P138066@siswa.ukm.edu.my

Nur Azura Mohd Said
Biotechnology & Nanotech Research Centre
Malaysian Agricultural Research & Development Institute (MARDI)
Selangor, Malaysia
nazurams@mardi.gov.my

Muhammad Aniq Shazni Mohammad Haniff
Institute of Microengineering & Nanoelectronics (IMEN)
Universiti Kebangsaan Malaysia (UKM)
Selangor, Malaysia
aniqshazni@ukm.edu.my

Mohd Hazani Mat Zaid
Duopharma Innovation Sdn Bhd
Selangor, Malaysia
Mohd-hazani@duopharmabiotech.com

Siti Nur Ashakirin Mohd Nashruddin
Institute of Informatics and Computing in Energy (IICE)
Universiti Tenaga Nasional (UNITEN)
Selangor, Malaysia
ashakirin@uniten.edu.my

Mohd Farhanulhakim Mohd Razip Wee
Institute of Microengineering & Nanoelectronics (IMEN)
Universiti Kebangsaan Malaysia (UKM)
Selangor, Malaysia
m.farhanulhakim@ukm.edu.my

P. Susthitha Menon
Institute of Microengineering & Nanoelectronics (IMEN)
Universiti Kebangsaan Malaysia (UKM)
Selangor, Malaysia
susi@ukm.edu.my

Abstract— Abstract— **Rapid detection of vitamin D levels in human serum presents significant benefits for diagnosing deficiency and managing treatments. This work reports the preparation and application of silver-graphene oxide (Ag-GO) composites for the development of vitamin D biosensors through chemical reduction method of silver nanoparticles (AgNPs) in aqueous suspension of graphene oxide (GO). Different concentrations of graphene oxide (GO) and silver (Ag) were tested to identify optimal electrochemical enhancements of screen-printed carbon electrodes (SPCE). By synthesizing Ag-GO composites and modifying SPCE, the electrochemical behaviour was investigated using differential pulse voltammetry (DPV). The Ag-GO composite prepared were characterised using Fourier-transform infrared spectroscopy (FTIR) and field-emission scanning electron microscopy (FESEM). AgGO synthesised from 2×10^{-3} mol dm^{-3} $AgNO_3$ and 1.0 g/L GO provided the most favourable conducting environment for electron transfer from the electrolyte to the electrode. The modified SPCE exhibited more than a threefold enhancement in peak current, reaching 37.98 μA, compared to an unmodified electrode.**

Keywords—graphene oxide-silver nanoparticles; screen-printed carbon electrodes; electrochemical sensor; biosensor; vitamin D detection

I. INTRODUCTION

Vitamin D deficiency is linked to numerous health issues, making its detection vital for early diagnosis and treatment [1]. Current methods, while effective, face limitations such as lack of sensitivity, high cost, and low accessibility [2]. The study aims to address these issues by developing an enhanced label-free opto-electrochemical sensor using silver nanoparticles-graphene oxide (Ag-GO)-modified screen-printed carbon electrodes (SPCE).

Prior research highlights the potential of graphene-based composites in biosensing due to their high surface area and excellent conductivity [3], [4]. However, the existing field of Ag-GO biosensors have yet to extend its application towards vitamin D detection. This significant gap presents a promising opportunity for further research and development in vitamin D detection.

This preliminary study serves as a vital foundation to the sensor development by optimizing the concentrations of GO and Ag. The modifications are expected to significantly elevate the sensor's performance by improving electron transfer and surface interaction dynamics.

II. MATERIALS & METHODS

A. Chemicals & Instruments

Graphene oxide (GO) was procured from Graphenea, Spain. Silver nitrate ($AgNO_3$), potassium ferrocyanide ($K_4[Fe(CN)_6]$) and potassium ferricyanide ($K_3[Fe(CN)_6]$) were bought from Sigma-Aldrich, Germany. Potassium chloride (KCl) and sodium borohydride ($NaBH_4$) were purchased from Merck, Germany. All solutions were prepared using Milli-Q grade water.

All electrochemical measurements were performed using 910 PSTAT mini potentiostat from Metrohm, Switzerland connected to a computer operated by PSTAT software. The SPCEs, DRP-110, were obtained from Metrohm DropSens, Switzerland and comprised of three electrodes. The working and auxiliary electrodes are made of carbon, while reference electrode is made up of silver. All electrochemical analysis was done using 5 mM $K_3[Fe(CN)_6]^{-3/-4}$ with 0.1 M KCl.

979-8-3503-7832-0/24 $31.00 © 2024 IEEE

B. Preparation and Modification of SPCE

In this preliminary experiment, Ag-GO was synthesized using a method adapted from a previous study [5], which involved a reduction process of silver nanoparticles in aqueous suspension of GO. Initially, solutions of 0.1%, 0.2%, 0.5%, and 1.0% (w/v) GO were prepared from a 2% w/v stock solution and ultrasonicated for 30 minutes. Following ultrasonication, 5 mL of 1 mM $AgNO_3$ was added to each GO solution and stirred for an additional 30 minutes. After this, 1 mL of 10 mM freshly prepared $NaBH_4$ was added dropwise with vigorous stirring. The reaction mixture was then stirred continuously for another 3 hours at room temperature to ensure complete reduction. This procedure was repeated with higher concentrations of $AgNO_3$, specifically 2 mM, 4 mM, and 8 mM, to evaluate the effects on the synthesis process and corresponding electrochemical performance. 10 μL of 0.1%, 0.2%, 0.5%, and 1.0% (w/v) GO and Ag-GO synthesised from 1 mM, 2 mM, 4 mM, and 8 mM $AgNO_3$ were electrodeposited onto the working electrode of the SPCEs using cyclic voltammetry (CV), conducted over a potential range of -1.5 to 1.0 V at a scan rate of 50 mV/s. The electrodes were washed with deionised water and let air dried.

III. RESULTS AND DISCUSSION

A. Electrochemical Study of GO and Ag-GO

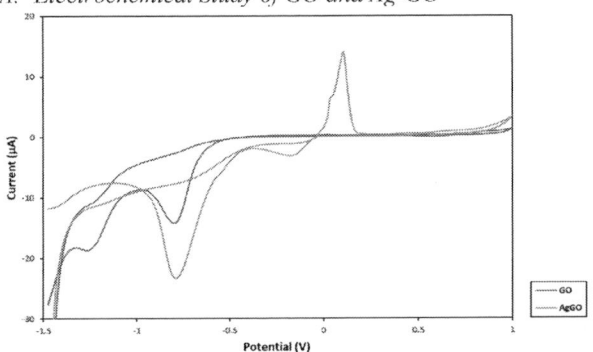

Fig. 1. Cyclic voltammograms of GO and Ag-GO electrodeposition.

Thin layer of GO and Ag-GO were deposited onto the SPCE through CV, as illustrated in Fig 1. During the electrodeposition process, the electroreduction peak attributed to the oxy-generated groups on GO surface was observed at -0.8 V [6]. This peak can also be seen for Ag-GO, indicating the retention of GO's functional groups. Additionally, Ag-GO also exhibited distinct anodic peak at 0.08 V due to the oxidation of AgNPs (Ag^0 to Ag^+) and a cathodic peak at -0.12 V. These additional peaks are believed to result from the oxidation-reduction reactions of the silver nanoparticles [7]. To further substantiate the successful deposition of GO and Ag-GO films on the SPCE, comparative voltametric analyses involving bare SPCE, GO-modified SPCE (GO/SPCE), and Ag-GO-modified SPCE (Ag-GO/SPCE) were conducted using both CV and differential pulse voltammetry (DPV). These comparative studies not only confirmed the deposition of thin films but also demonstrated the electrochemical changes attributable to the modified electrodes, supporting the effectiveness of the electrodeposition technique for surface functionalisation.

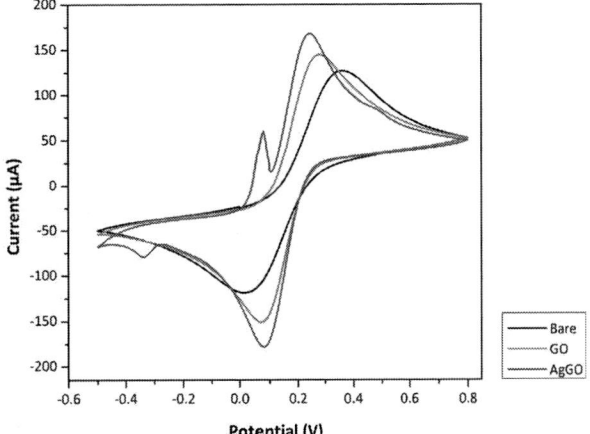

Fig. 2. Cyclic voltammograms of bare SPCE, GO/SPCE and Ag-GO/SPCE at scan rate of 100 mV/s.

The voltametric responses of bare SPCE, GO/SPCE and Ag-GO/SPCE were compared using CV in an electrolytic solution of 5 mM $K_3[Fe(CN)_6]^{-3/-4}$ with 0.1 M KCl. In Fig. 2, a marked increase in current was observed for the Ag-GO-modified electrodes compared to both the GO-modified and bare SPCEs. This enhancement in current response highlights the critical role of Ag-GO in amplifying the charge transfer rate across the electrode interface, thereby significantly improving the electronic conductivity of the modified electrode. This result demonstrated that Ag-GO plays an important role to enhance the charge transfer rate of the modified electrode which resulted in excellent electronic conductivity [8].

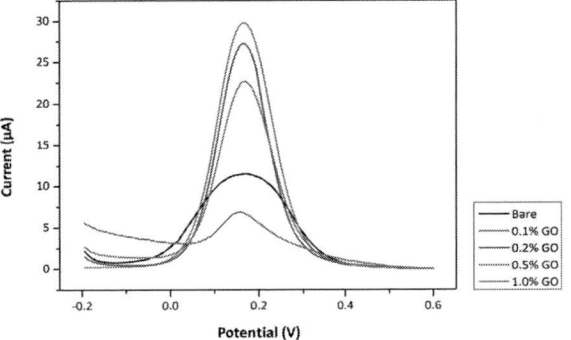

Fig. 3. Differential pulse voltammetry curves of bare SPCE and SPCE modified with 0.1%, 0.2%, 0.5%, and 1.0% GO at scan rate of 10 mV/s .

To determine the optimal concentration of GO for subsequent Ag-GO synthesis, electrodes modified with various GO concentrations (0.1%, 0.2%, 0.5%, 1.0% w/v) were evaluated using DPV. The DPV measurements were conducted across a potential range from -0.2 to 0.6 V, at a scan rate of 10 mV/s, with a pulse time of 25 ms, a pulse amplitude of 0.01 V, and a potential step of 0.005 V. As depicted in Fig 3, the electrode modified with 0.1% GO (1 g/L) exhibited the most pronounced current response, achieving a peak height of 29.44 μA. This represents a tripling in peak current compared to the bare unmodified electrode. At concentrations above 0.1%, GO tended to saturate on the electrode surface, forming a dense and insulating layer [9]. This layer impedes efficient electron transfer due to increased thickness and the presence of various oxygen-containing functional groups that is crucial for the immobilization of recognition molecules in the development of a vitamin D biosensor but adversely affect the electrode's conductive properties in higher concentration. Consequently, a 0.1% GO concentration was identified as optimal, striking a balance between maintaining functional group availability for molecular recognition and maximizing electrochemical performance. It was thus selected for further synthesis of Ag-GO.

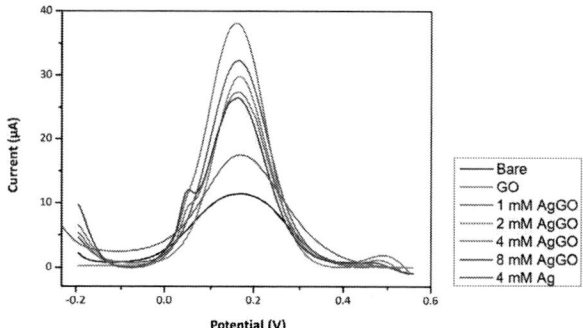

Fig. 4. Differential pulse voltammetry curves of bare SPCE, 0.1% GO-modified SPCE, and Ag-GO-modified SPCE with Ag of varying concentrations synthesised in 0.1% GO at scan rate of 10 mV/s.

Subsequently, the electrochemical performance of Ag-GO synthesised from 1 mM, 2 mM, 4 mM and 8 mM of AgNO₃ were studied using the same method (DPV) and parameters as GO. Fig 4 reveals that the reduction of 2 mM AgNO₃ in a 0.1% GO matrix resulted in the highest current response, exceeding that observed with 0.1% GO modification alone. This enhancement indicates a synergistic interaction between GO and AgNPs, which improves the electrochemical properties beyond those of the individual material. The 2 mM concentration of AgNO₃, which corresponds to approximately 1.67 mM of silver nanoparticles in the Ag-GO composite, was identified as the optimal concentration for achieving maximum electrochemical performance. At concentrations below 2 mM, the silver nanoparticles may not provide sufficient electroactive surface area for effective catalysis of redox reactions, thus diminishing the potential synergistic effects. This synergy is essential as it leverages both the conductive properties of graphene oxide and the catalytic capabilities of silver. Insufficient silver content may preclude the full realization of this combined effect, thereby limiting the enhancement of electrochemical reactions facilitated by GO. Conversely, at concentrations exceeding 2 mM, there is a tendency for silver nanoparticles to agglomerate [10]. Agglomerated nanoparticles might not expose as many active sites as well-dispersed nanoparticles, which decreases the overall electrochemical activity and thus lowers the current. With higher loadings of silver nanoparticles, the electrode surface can become overly congested, leading to diffusion limitations. Thus, maintaining the silver concentration at 2 mM optimizes the dispersion of nanoparticles and maximizes the electrochemical benefits of the Ag-GO composite.

B. FTIR, FESEM, and TEM

Fig. 5 shows the fourier transform infrared (FTIR) spectra of GO. The peak at 1634 cm⁻¹ is typically attributed to the C=C stretching vibrations in the aromatic rings, indicating the presence of unoxidised graphitic domains. This peak confirms the retention of graphitic structures within the GO. This is crucial as it suggests that the oxidation process used to create the GO has not disrupted the essential graphitic domains, which are vital for maintaining the conductive properties of the material. The peak at 3264 cm⁻¹ is assigned to the intermolecular bonded O-H stretching vibrations, representing the O-H stretching vibrations from hydroxyl groups within the graphene sheet.

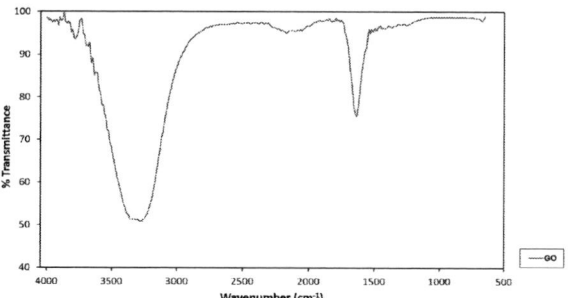

Fig. 5. Fourier transform infrared spectra of 0.1% GO.

The peak at 3778 cm⁻¹ represents the O-H stretching vibrations, typically indicating the presence of water molecules or surface hydroxyl groups associated with the GO. These peaks are indicative of the hydroxyl groups present on the GO surface. These functional groups are a direct result of the oxidation process and are critical for enhancing the solubility and processability of GO in various solvents, which can be particularly advantageous for composite materials and vitamin D detection [11].

Fig. 6. Field-emission scanning electron microscopy of the modified SPCE surface.

Field-emission scanning electron microscopy (FESEM) of the modified electrode surface (Fig. 6) revealed a roughened irregular texture compared to pristine unmodified electrode surface [12]. The silver nanoparticles scattered can be observed as bright specks dotting the graphene oxide landscape. The morphology of synthesized GO and Ag-GO was characterized using transmission electron microscopy (TEM). TEM images (Fig. 7) revealed that GO exhibited a smooth, flagstone-like morphology. In comparison, Ag-GO displayed a uniform dispersion of silver nanoparticles (AgNPs) across the graphene oxide sheets. The AgNPs, with diameters ranging from approximately 10 nm to 50 nm, were well-distributed, indicating a robust interaction between the AgNPs and the chemically active sites on the GO.

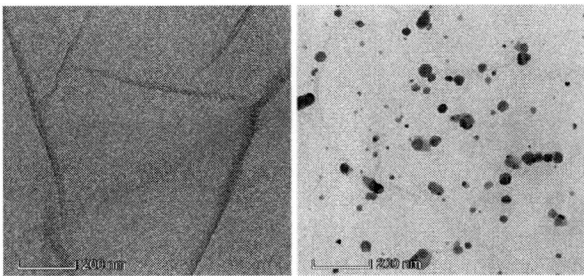

Fig. 7. Transmission electron microscopy image of GO (left) and Ag-GO (right).

This favourable interaction facilitates the deposition of AgNPs and contributes to their high dispersibility on the GO nanosheets [13]. The enhanced dispersibility of AgNPs on the GO nanosheets significantly increases the surface area, which is advantageous for improving the reactivity of the modified SPCE. This increase in surface area not only boosts the loading capacity for recognition molecules but also enhances the overall sensing performance, particularly in terms of sensitivity. Additionally, DPV data, as illustrated in Fig. 4, demonstrated a notable increase in the background current, which can be attributed to the increased surface area due to GO and AgNPs deposition. Moreover, it is worth noting that in optical applications, the preferred size of AgNPs typically ranges from 10 nm to 80 nm [14]. However, smaller nanoparticles are often more desirable in electrochemical applications, specifically those ranging from 10 nm to 50 nm [15]. These smaller nanoparticles offer a higher surface area to volume ratio, which enhances electrochemical activity by facilitating more effective electron transfer processes. Such characteristics are especially beneficial in sensor technologies, where increased surface area correlates with improved sensitivity and lower detection limits.

IV. CONCLUSION

This study successfully synthesised and determined the most effective concentrations of AgNPs and GO for electrochemical enhancement, highlighting the potential of Ag-GO composites for vitamin D detection. Specifically, the ratio of 2×10^{-3} mol dm^{-3} AgNO3 to 1.0 g/L GO was identified as optimal, significantly enhancing electron transfer capabilities and increasing the peak current more than threefold compared to unmodified electrodes in CV and DPV. Characterization via FTIR, FESEM, and TEM confirmed the successful synthesis of Ag-GO, maintaining essential graphitic structures and ensuring uniform nanoparticle dispersion within the particle size range of 10 nm to 50 nm, which enhanced electrochemical conductivity and potentially, optical activity. For future advancement, this preliminary work will be followed by biorecognition molecule implementation, and further development and optimization of an accessible and sensitive point-of-care vitamin D biosensor. Moreover, this study has the additional significance in its potential to pave the way for broader applications of Ag-GO composites in other biosensing fields, demonstrating versatility and promising routes for future interdisciplinary research.

ACKNOWLEDGMENT

The authors appreciate the technical support and facilities from the Institute of Microengineering & Nanoelectronics, Universiti Kebangsaan Malaysia. This research was funded by the Ministry of Higher Education via the Fundamental Research Grant Scheme (FRGS) with grant code of FRGS/1/2023/STG05/UKM/02/12.

REFERENCES

[1] N. Ali, 'Role of vitamin D in preventing of COVID-19 infection, progression and severity', J. Infect. Public Health, vol. 13, no. 10, pp. 1373–1380, Oct. 2020, doi: 10.1016/j.jiph.2020.06.021.

[2] B. Altieri et al., 'Vitamin D testing: advantages and limits of the current assays', Eur. J. Clin. Nutr., vol. 74, no. 2, Art. no. 2, Feb. 2020, doi: 10.1038/s41430-019-0553-3.

[3] Mohd Azerulazree Jamilan, Balqis Kamarudin, Dalila Rizuan, Mohd, and Fairulnizal Md Noh, 'Detection Of Vitamin D In Food Supplements Using a Simple Modification Of Electrochemically Reduced Graphene Oxide/Screen–Printed Carbon Electrodes', J. Smart Sens. Mater., vol. 1, no. 1, pp. 14–27, 2023.

[4] G. L. Radu, S. C. Liţescu, A. Enache, C. Albu, and S. A. V. Eremia, 'Synergism between Graphene and Molecularly Imprinted Polymers in Developing Electrochemical Sensors for Agri-Food and Environmental Analyses', Chemosensors, vol. 11, no. 7, p. 380, Jul. 2023, doi: 10.3390/chemosensors11070380.

[5] M. R. Das, R. K. Sarma, R. Saikia, V. S. Kale, M. V. Shelke, and P. Sengupta, 'Synthesis of silver nanoparticles in an aqueous suspension of graphene oxide sheets and its antimicrobial activity', Colloids Surf. B Biointerfaces, vol. 83, no. 1, pp. 16–22, Mar. 2011, doi: 10.1016/j.colsurfb.2010.10.033.

[6] M. H. Mat Zaid et al., 'PNA biosensor based on reduced graphene oxide/water soluble quantum dots for the detection of Mycobacterium tuberculosis', Sens. Actuators B Chem., vol. 241, pp. 1024–1034, Mar. 2017, doi: 10.1016/j.snb.2016.10.045.

[7] N. Elgrishi, K. J. Rountree, B. D. McCarthy, E. S. Rountree, T. T. Eisenhart, and J. L. Dempsey, 'A Practical Beginner's Guide to Cyclic Voltammetry', J. Chem. Educ., vol. 95, no. 2, pp. 197–206, Feb. 2018, doi: 10.1021/acs.jchemed.7b00361.

[8] S. Nur Ashakirin, M. Aniq Shazni M. Haniff, M. Hazani M. Zaid, M. Farhanulhakim M. Razipwee, and E. Mahmoudi, 'Urease immobilized electrodeposited silver reduce graphene oxide modified screen-printed carbon electrode for highly urea detection', Measurement, vol. 196, p. 111058, Jun. 2022, doi: 10.1016/j.measurement.2022.111058.

[9] W. Liu and G. Speranza, 'Tuning the Oxygen Content of Reduced Graphene Oxide and Effects on Its Properties', ACS Omega, vol. 6, no. 9, pp. 6195–6205, Mar. 2021, doi: 10.1021/acsomega.0c05578.

[10] N. Pauzi, S. C. Segaran, S. Mohamad, and S. S. Jamari, 'Silver nitrate concentration on silver nanoparticles formation attached on cellulose nanocrystal matrix', Mater. Today Proc., Apr. 2023, doi: 10.1016/j.matpr.2023.03.600.

[11] D. J. Joshi, J. R. Koduru, N. I. Malek, C. M. Hussain, and S. K. Kailasa, 'Surface modifications and analytical applications of graphene oxide: A review', TrAC Trends Anal. Chem., vol. 144, p. 116448, Nov. 2021, doi: 10.1016/j.trac.2021.116448.

[12] S. Chanarsa, J. Jakmunee, and K. Ounnunkad, 'A Bifunctional Nanosilver-Reduced Graphene Oxide Nanocomposite for Label-Free Electrochemical Immunosensing', Front. Chem., vol. 9, Apr. 2021, doi: 10.3389/fchem.2021.631571.

[13] A. Lange et al., 'Nanocomposites of Graphene Oxide—Silver Nanoparticles for Enhanced Antibacterial Activity: Mechanism of Action and Medical Textiles Coating', Materials, vol. 15, no. 9, p. 3122, Apr. 2022, doi: 10.3390/ma15093122.

[14] V. Sharma, D. Verma, and G. S. Okram, 'Influence of surfactant, particle size and dispersion medium on surface plasmon resonance of silver nanoparticles', J. Phys. Condens. Matter, vol. 32, no. 14, p. 145302, Apr. 2020, doi: 10.1088/1361-648X/ab601a.

[15] E. P. Medyantseva, D. V. Brusnitsyn, R. M. Varlamova, A. A. Maksimov, O. A. Konovalova, and H. C. Budnikov, 'Surface modification of electrodes by carbon nanotubes and gold and silver nanoparticles in monoaminoxidase biosensors for the determination of some antidepressants', J. Anal. Chem., vol. 72, no. 4, pp. 362–370, Apr. 2017, doi: 10.1134/S1061934817040086.

Piezoelectric Energy Harvesting from Thermal Vibrations Using Doped Graphene-MXene Heterostructure

Lijie Kou,
[1] Fuzhou Institute of Technology, Fuzhou 350506, People's Republic of China
[2] Institute of Microengineering and Nanoelectronics (IMEN), Universiti Kebangsaan Malaysia, 43600 Bangi, Selangor, Malaysia
P118051@siswa.ukm.edu.my

Poh Choon Ooi[*],
Institute of Microengineering and Nanoelectronics (IMEN), Universiti Kebangsaan Malaysia, 43600 Bangi, Selangor, Malaysia
pcooi@ukm.edu.my

Chang Fu Dee,
Institute of Microengineering and Nanoelectronics (IMEN), Universiti Kebangsaan Malaysia, 43600 Bangi, Selangor, Malaysia
cfdee@ukm.edu.my

Muhammad Aniq Shazni Mohammad Haniff,
Institute of Microengineering and Nanoelectronics (IMEN), Universiti Kebangsaan Malaysia, 43600 Bangi, Selangor, Malaysia
aniqshazni@ukm.edu.my

Abstract—Piezoelectric nanogenerators (PENG) can face challenges when integrated into high-temperature applications because of their temperature sensitivity. Heterostructures of specific 2D nanomaterials can potentially enhance PENG performance for practical applications at high temperatures. This study incorporated nitrogen-doped graphene (NGr) and $Ti_3C_2T_x$ MXene heterostructure nanofillers into the polyvinylidene difluoride (PVDF) matrix for energy harvesting in a high-temperature vibration environment. The solution processable nanogenerator device is achieved by optimizing the appropriate ratio of NGr to $Ti_3C_2T_x$. At room temperature, the nanogenerator showed an optimum output voltage of ~9.0 V and ~1.5 µA of current. Thereby, it raised to 24.0 V and 1.75 µA when the temperature increased to 90 ºC, obtaining a power density of 3.85 µW/cm². This outstanding performance is attributed to the designed NGr-$Ti_3C_2T_x$ quasi-3D heterostructure, where its rich interfacial features, excellent electrical conductivity, and localized elastic complexes synergistically promote the piezoelectric output of the energy harvester. Placing the device on the road could be used to collect the mechanical energy generated by the vibration of the car's movement and convert it into electrical energy, which opens up new development possibilities for addressing emerging energy issues.

Keywords—Nitrogen-doped graphene, $Ti_3C_2T_x$ MXene, Piezoelectric Nanogenerator, Quasi-3D Heterostructure, Thermal Property

I. INTRODUCTION

Integrating piezoelectric nanogenerators (PENGs) into portable electronic gadgets to facilitate miniaturization and ease of portability is an ongoing development. Generally, these portable electronics are driven by electrochemical batteries. Unfortunately, the battery life span is limited. As a result, additional costs of replacing them are incurred. Moreover, the replacement of old batteries causes excessive chemical waste. As such, efforts have been made to advance the implementation of energy harvesters as a self-power source for various portable electronic devices [1]. Therefore, integrating PENG into small electronic devices can be particularly useful in providing an uninterrupted power supply by harvesting abundant wasted energy without relying on traditional batteries while combating the battery waste issue and conserving the environment. PENG appeared to be an alternative sustainable green energy resource compared to environmentally dependent solar, tidal and wind turbine renewable energy. PENG can convert mechanical energy sources to electrical energy. Therefore, its ability to harvest mechanical energy from various mechanical movements, such as human motion, automotive engine vibration, and pressure, can be integrated with different types of renewable energy for multi-harvesting to ensure uninterrupted power.

Uninterrupted electrical energy supply relying on a green energy resource is a big challenge that requires an emerging solution. For example, solar panels can supply limited energy to the lighting of the highway network in a country ascribed to the sunlight radiation exposure hours and unpredictable weather conditions. The mentioned problem can be overcome using a multi-harvesting method, especially by combining the PENG device with the solar panel. In this instance, charges generated by PENG due to vehicle vibration can be stored in an electrical accumulator of solar panels. PENG generates higher charges when the heat deduced to the moving vehicle exceeds the room temperature (RT). Observing the frictional heat generated during driving and the actual road surface temperature makes it possible to predict the thermal state of the road. A study by Mustafa et al. showed that the maximum road surface temperature increases up to 80 ºC, which benefits the performance of the PENG device [2].

Polyvinylidene difluoride (PVDF) will be chosen for this study as it can withstand temperatures up to 200 ºC, enabling long-term stability under harsh temperature environments. While its copolymers may outperform in terms of long-term performance, thermal stability, and degradation [3], PVDF's unique properties make it a preferred choice. PVDF exhibits superior electroactivity and piezoelectric properties at the nanoscale compared to its copolymers. Additionally, PVDF can effectively accommodate the incorporation of various nanofillers, including heterostructures, without compromising its structural integrity or piezoelectric properties. Furthermore, PVDF can be easily processed into thin films, fibers, or other forms required for PENG fabrication, enabling the efficient integration of nanofillers and the development of high-performance PENG devices [4].

Introducing various nanofillers into the piezoelectric polymer host significantly improves the piezoelectric response of PVDF due to their physical and chemical properties. Generally, graphene, one of the first 2D nanomaterials, has been known widely and used as a filler to increase the content of β-phase in the polymer matrix and ultimately improve the electrical output of PENGs [5]. On the other hand, another reported 2D nanomaterial, transition metal carbides (MXene), has displayed high electrical conductivity, mechanical properties, and a large specific surface area. Titanium carbide ($Ti_3C_2T_x$) is a rapidly emerging MXene material discovered in 2011. It is primarily utilized as an energy storage material due to its unique layered structure and excellent chemical stability. The "T_x" in $Ti_3C_2T_x$ represents surface functional groups such as -F, -OH, and -O [6].

Therefore, in this study, the NGr-$Ti_3C_2T_x$ 2D/2D heterostructure nanofillers were dispersed in a PVDF polymer matrix to study the device's performance, as combining 2D materials to form heterostructures provides unprecedented opportunities for practical applications. The heterostructure formation can lead to synergistic effects, as each material may contribute distinct properties. Incorporating nanofillers into a PVDF matrix can significantly impact the material's electrical, mechanical, and thermal properties. Hence, thermal performance that mimics outdoor temperatures will also be carried out to investigate the behavior of the fabricated piezoelectric energy conversion devices based on mechanical vibration. This work will dissolve PVDF in a high-polar dimethylformamide solvent by introducing ionic salt to facilitate the dipole alignment. Then, the spin coating method will be used to fabricate the nanogenerator devices.

II. Experimental Details

A. Nanofillers and nanocomposite solutions preparation

This work purchased PVDF powder with an average $Mw\sim534\,000$, magnesium chloride ($MgCl_2$) powder, and N, N-dimethylformamide (DMF) solvent from Sigma Aldrich. All materials were used without further purification. $Ti_3C_2T_x$ MXene suspension was prepared by dispersing 10 mg of $Ti_3C_2T_x$ in 10 ml of DMF. Meanwhile, NGr suspension was prepared by adding 0.05 mg of NGr powder in 10 ml of DMF. Both suspensions were then ultrasonicated for 30 minutes at room temperature (RT). The polymer solution was prepared by dissolving 2 g of PVDF powder and 5 mg of $MgCl_2$ in 10 ml of DMF. The solution was magnetically stirred for 1.5 h at 80 °C.

Subsequently, NGr and $Ti_3C_2T_x$ nanofillers were mixed at different ratios before being incorporated with PVDF to obtain the optimal mixing ratio. The ratios investigated were 1:1, 1:2, 2:1, 1:3 and, 1:4 of NGr to $Ti_3C_2T_x$, and remarked as M1, M2, M3, M4, and M5, respectively. All the nanofiller mixtures were then ultrasonicated for 10 minutes at RT to achieve homogenous dispersion. After that, 1 ml of M1, M2, M3, M4, and M5 was then incorporated into a 4 ml PVDF polymer host, respectively to form the nanocomposite and labelled as PM1, PM2, PM3, PM4, and PM5. Similarly,

1 ml NGr and $Ti_3C_2T_x$ MXene suspensions were separately introduced into a 4 ml PVDF host for control samples purposes and indicated as PGr and PMXene, respectively. All the nanocomposites were also ultrasonicated for 15 minutes at RT.

B. Device fabrication

The 2.0 cm x 2.0 cm indium-tin-oxide (ITO) coated polyethylene terephthalate (PET) was used as the mechanically flexible substrates for fabricating the piezoelectric nanogenerator. Initially, the substrates were treated with oxygen plasma for 60 s to reduce the hydrophobicity. At first, the reference devices consisting of pure PVDF, PVDF incorporated with NGr, and $Ti_3C_2T_x$, were fabricated and labelled as RD1, RD2, and RD3, respectively. The main intention of the reference devices is to observe the improved performance of the heterostructure NGr-$Ti_3C_2T_x$. Afterwards, the prepared PM1, PM2, PM3, PM4, and PM5 were spin-coated on the ITO/PET substrates, respectively at 300 rpm for 5 s and followed by 1000 rpm for 60 s. The deposited pieces were carefully placed on an 80 °C hot plate for 2 minutes to form the piezoelectric layer (PEL). The same deposition procedure was followed to deposit PM1, PM2, PM3, PM4, and PM5 on another ITO/PET substrate to form the bistacked structure. After the heat treatment, the first and second PEL/ITO were attached carefully in the sequence of PM1-PM1, PM2-PM2, PM3-PM3, PM4-PM4, and PM5-PM5 to form the bistacked structure and named D1, D2, D3, D4, and D5 piezoelectric nanogenerators, respectively. Fig. 1 shows the depicted stacking structure diagram. The bistacked structure of D1, D2, D3, D4, and D5 was then further heat treated on the hot plate at 80 °C for 1.5 hours to dry the PEL effectively. TABLE 1 summarizes the prepared samples to ease the understanding.

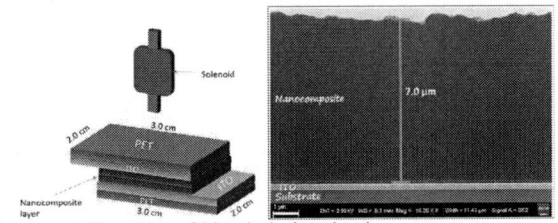

Fig. 1. (a) Illustration of bistacked piezoelectric nanogenerator structure. (b) Image of FIB-SEM cross-sectional structure.

TABLE 1. SUMMARY OF SAMPLE PREPARATIONS LABELLED AT DIFFERENT MIXING RATIOS OF NGR TO $Ti_3C_2T_x$.

Ratio of NGr: $Ti_3C_2T_x$	1:1	1:2	2:1	1:3	1:4	Reference Device		
Suspension	M1	M2	M3	M4	M5	-	NGr	$Ti_3C_2T_x$
Nanocomposite/ Polymer	PM1	PM2	PM3	PM4	PM5	PVDF	PNGr	PMXene
Device	D1	D2	D3	D4	D5	RD1	RD2	RD3

C. Structure and electrical characterization

The sample morphology characterizations were conducted using the field emission scanning electron microscope (FESEM) and the chemical structure characterization analyses included Fourier transform infrared spectroscopy (FTIR). The PENG device cross-section was characterized using Zeiss FIB-SEM. A 10 Hz and 5 N solenoid force impact was applied to identify the optimum performance sample. A 20 Hz vibrational tool was used to mimic vehicle moving vibration to harvest the vibrational mechanical energy at different temperatures. The device's open circuit voltage (V_{oc}) and short circuit current (I_{sc}) were recorded using a digital oscilloscope (Keysight DSO9404A) setup in differential mode. Meanwhile, I_{sc} was measured using a picoammeter (Keithley 6487) under the action of the external stimuli.

III. RESULTS AND DISCUSSION

Fig. 2 presents the cross-sectional FESEM image of NGr-$Ti_3C_2T_x$ MXene hybrid maintains the original layered configuration observed in the pure NGr film, with interspersed MXene layers.

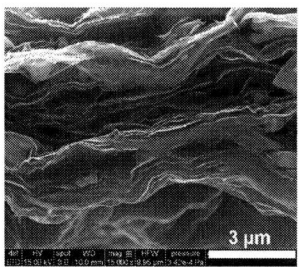

Fig. 2. Cross-sectional FESEM image of NGr- $Ti_3C_2T_x$ heterostructure in composition ratio of 70%/30%.

Fig. 3 shows that FTIR measurement was performed on the sample to assess crystallization behavior and compute phase content via molecular bond vibration. PVDF vibration peaks are caused by C-C bending, CH_2 swing, and CF_2 motion vibrations, which occur at 1071, 1176, and 1397 cm^{-1}, respectively [7]. The non-polar α-phase features distinctive peaks at 763, 795, and 976 cm^{-1}, whereas the all-trans polar β-phase has vibrational peaks at 840 and 1276 cm^{-1}. The γ-phase vibrations are minor and can be omitted in this study [8]. The peak at 878 cm^{-1} is the overlapping β and γ phases. The spectrum revealed that with the addition of a type of nanofiller, the organization of some non-polar molecules shifted to polar ordering, and the β-phase appeared. When NGr and $Ti_3C_2T_x$ MXene heterostructure nanofillers are added, the characteristic peak of the non-polar α-phase almost disappears, while the β-phase increases, indicating that NGr and $Ti_3C_2T_x$ MXene nanosheets effectively induce the PVDF molecular arrangement towards all-trans orientation. The β-phase content of samples PM1, PM2, PM3, PM4 and PM5 have been demonstrated electroactive content improvement of 87%, 90%, 84%, 79%, and 54%, respectively, according to the Equation in [9]. As aforementioned, it is expected that the PM2 sample attained the highest β fraction.

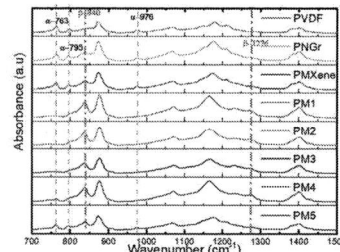

Fig. 3. FTIR of pure PVDF, PNGr, PMXene, PM1, PM2, PM3, PM4 and PM5, respectively.

Fig. 4 shows each sample's peak-to-peak (P-P) V_{oc} (in Fig. 4a) and I_{sc} (in Fig. 4b) achievement electrical characterization of the force impact induced by the solenoid during the press and release deformation of the PENG devices. The generated average P-P V_{oc} for RD1, RD2, RD3, D1, D2, D3, D4, and D5 is measured to be 4.5, 5.6, 5.4, 6.8, 9.1, 5.7, 6.7, and 4.9 V, respectively. Meanwhile, the same trend of average I_{sc} measurements was collected for RD1, RD2, RD3, D1, D2, D3, D4, and D5, which were 0.33, 0.70, 0.68, 1.22, 1.51, 1.01, 0.76, and 0.62 µA, respectively. Note that blending the NGr and $Ti_3C_2T_x$ at the appropriate ratio (1:2) further enhances the performance of the fabricated devices. The attained V_{oc} and I_{sc} of RD1 and RD2 are lower than D2, which indicates the synergistic effect of heterostructure nanofillers and can boost the device performance.

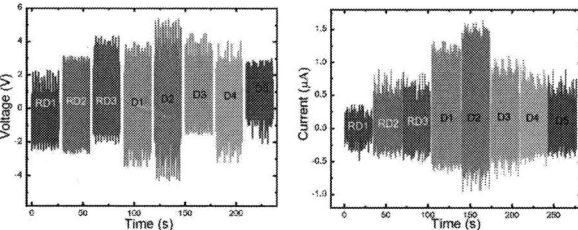

Fig. 4. (a) V_{oc} and (b) I_{sc} achievement of RD1, RD2, RD3, D1, D2, D3, D4, and D5 due to the solenoid force impact.

TABLE 2 compares the output performance of PVDF-based devices in this study to those reported in the literature, using other associated nanofillers at room temperature. Comparatively, the spin coating technique employed in this study offers cost-effectiveness and ease of operation. The PVDF/NGr/$Ti_3C_2T_x$ nanogenerator exhibited outperform achievement as recorded in Table 2, attaining relatively higher V_{oc} ~9.1 V and P_d ~3.1 mW/cm$_2$ when incorporating the different 2D nanomaterials in appropriate ratio to form quasi-3D heterostructure nanofillers.

TABLE 2. REPORTED OUTPUT PERFORMANCE COMPARISON OF THE PVDF-BASED PENGS WITH DIFFERENT NANOFILLERS AT RT.

Nanofillers in PVDF matrix	Fabrication process	V_{oc} (V)	I_{sc} (µA)	P_d (µW/cm^2)	Ref.
Graphene/BaTiO$_3$	Electrospinning	11.0	1.35	0.656	[10]
Ti$_3$C$_2$T$_x$	Electrospinning	3.15	0.134	0.021	[11]
rGO/MoS$_2$	Solution casting	2.4	0.68	0.81	[12]
NGr/Ti$_3$C$_2$T$_x$	Spin coating	9.1	1.51	3.1	This work

The thermal performance of D2 due to the vibrational tool was investigated to mimic the vibration caused by the moving vehicles at several temperatures. D2 was chosen because it demonstrated the outperforming device during the solenoid testing. The measurement was conducted by placing D2 on a heating film at different temperatures. Fig. 5

shows the harvested vibration V_{oc} thermal performance of D2 at RT, 70 °C, 80 °C, and 90 °C, respectively. The P-P V_{oc} measured were 9.1 V, 18 V, 21 V, and 24 V. The highest power density (P_d) of 3.85 μW/cm² at 3 MΩ is obtained at 90 °C.

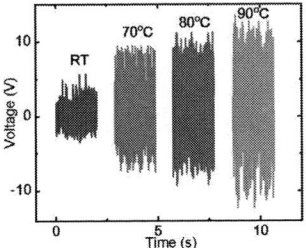

Fig. 5. Vibrational thermal performance of V_{oc} measurements for D2 device at RT, 70 °C, 80 °C, and 90 °C.

Fig. 6 depicts the schematic representation of the interactions. The construction of the quasi-3D interfacial interaction heterostructure designed in this work is derived from the excellent properties of the synthesized two different 2D nanomaterials with unique propertie. In detail, the strong electronegativity (δ⁻) of sp2 hybridized C atoms and pyridine N atoms in NGr, as well as the strong electronegative mutual repulsion of the $Ti_3C_2T_x$ surface functional groups (-F, -O, -OH), prevents MXene re-stacking and allows rich functional groups of Ti3C2Tx exhibit their high activities [13]. Consequently, quasi-3D interfacial interaction heterostructure ascribed to van der Waals force self-assembly formed. The strong interfacial contacts drive the all-trans arrangement of the PVDF molecular chains because the extremely electronegative surfaces of NGr and $Ti_3C_2T_x$ MXene function as nucleating agents, promoting β-phase formation. the reason for device D2 demonstrating an optimum piezoelectric output performance is that the dispersion concentration of MXene is lower than that of NGr, a higher proportion ratio is required to maintain the balance between these two nanofillers in the PVDF host.

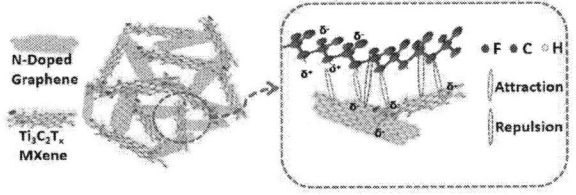

Fig. 6. Interfacial interaction mechanism at the room-temperature of NGr-$Ti_3C_2T_x$ MXene heterostructure.

IV. CONCLUSION

This study prepared 2D nanomaterials, NGr and $Ti_3C_2T_x$ MXene, using environmentally friendly electrochemical and chemical exfoliation method, respectively. Quasi-3D nano-heterogeneous configurations with promising electrical and mechanical properties were formed by interfacial coupling interactions between these two and used as nanofillers in the PVDF matrix. This led to a significant enhancement in the performance of the nanogenerator. The optimized nanofiller ratio at 1:2 for NGr and $Ti_3C_2T_x$ MXene synergistically increased the V_{oc} to 9 V at RT and 25 V at an elevated temperature of 90 °C, reaching an output power density of 3.85 μW/cm². The fabricated PENG device was perturbed by a vibrator simulating the mechanical vibration of a vehicle at high-temperature conditions. The device demonstrated the capability to scavenge the vibration into the electrical signals, suggesting the device's potential for practical applications. Upon heating, the enhanced interfacial coupling of the two nanomaterials constituent of the van de Waals interaction effectively promotes the formation of the electroactive β-phase in the PVDF molecular chain. It improves the performance of the PENG through synergistic electrical conductivity and localized mechanical elasticity, thus bringing about an enhanced piezoelectric energy output. This work provides a sustainable alternative for self-powered systems in areas such as electric energy harvest in vehicles and aerospace, paving the way for replacing conventional battery energy sources.

REFERENCES

[1] Han, Y.; Wu, F.; Du, X. Z.; Li, Z. H.; Chen, H. X.; Guo, D. X.; Wang, J. L.; Yu, H. Enhance vortices vibration with Y-type bluff body to decrease arousing wind speed and extend range for flag triboelectric energy harvester. *Nano Energy* 2024, *119*, 10.

[2] Mustafa, S.; Sekiya, H.; Hamajima, A.; Maeda, I.; Hirano, S. J. T. E. Effects of speeds and weights of travelling vehicles on the road surface temperature. 2021, *5*, 100077.

[3] Chai, M.; Tong, W.; Wang, Z.; Chen, Z.; An, Y.; Zhang, Y. J. J. o. H. M. Piezoelectric-Fenton degradation and mechanism study of Fe₂O₃/PVDF-HFP porous film drove by flowing water. *J. Hazard. Mater.* 2022, *430*, 128446.

[4] Wang, Y.; Zhu, L.; Du, C. J. M. Progress in piezoelectric nanogenerators based on PVDF composite films. *Micromachines* 2021, *12*, 1278.

[5] Wang, B.; Ruan, T. T.; Chen, Y.; Jin, F.; Peng, L.; Zhou, Y.; Wang, D. L.; Dou, S. X. Graphene-based composites for electrochemical energy storage. *Energy Storage Mater.* 2020, *24*, 22-51.

[6] Liu, Y.; Yu, J.; Guo, D.; Li, Z.; Su, Y. J. J. o. A.; Compounds Ti3C2Tx MXene/graphene nanocomposites: Synthesis and application in electrochemical energy storage. *J. Alloys Compd.* 2020, *815*, 152403.

[7] Mohamadi, S., Preparation and Characterization of PVDF/PMMA/Graphene Polymer Blend Nanocomposites by Using ATR-FTIR Technique. *Infrared spectroscopy-materials science* 2012, *1*, 213-232.

[8] Martins, P.; Lopes, A. C.; Lanceros-Mendez, S. Electroactive phases of poly[vinylidene fluoride]: Determination, processing and applications. *Prog. Polym. Sci.* 2014, *39*, 683-706.

[9] Gordon, H. R. Can the Lambert-Beer law be applied to the diffuse attenuation coefficient of ocean water? *Limnology Oceanography* 1989, *34*, 1389-1409.

[10] Shi, K.; Sun, B.; Huang, X.; Jiang, P. Synergistic effect of graphene nanosheet and BaTiO₃ nanoparticles on performance enhancement of electrospun PVDF nanofiber mat for flexible piezoelectric nanogenerators. *Nano Energy* 2018, *52*, 153-162.

[11] Zhang, J.; Yang, T.; Tian, G.; Lan, B.; Deng, W.; Tang, L.; Ao, Y.; Sun, Y.; Zeng, W.; Ren, X. Spatially confined MXene/PVDF nanofiber piezoelectric electronics. *Adv. Fiber Mater.* 2024, *6*, 133-144.

[12] Faraz, M.; Singh, H. H.; Khare, N. A progressive strategy for harvesting mechanical energy using flexible PVDF-rGO-MoS₂ nanocomposites film-based piezoelectric nanogenerator. *J. Alloys Compd.* 2022, *890*.

[13] Zhang, Y. M.; Li, Q. C.; Ye, X. W.; Wang, L.; He, Z. Y.; Zhang, T.; Wang, K. C.; Shi, F. Y.; Yang, J. Y.; Jiang, S. H.; Wang, X. R.; Chen, C. X. High-Performance Infrared Detectors Based on Black Phosphorus/Carbon Nanotube Heterojunctions. *Nanomaterials* 2023, *13*, 2700.

AI-based Image Processing Technique for Dielectrophoresis (DEP) in BioMEMs Applications

Clarence Augustine TH Tee
*College of Physics & Electrical Information Engineering,
Zhejiang Normal University
321004 Zhejiang
People's Republic of China*
catht@zjnu.edu.cn

En Hao Yu
*Photonics & Sensors Key Research Laboratory,
Zhejiang Normal University
321004 Zhejiang
People's Republic of China*
enhaotw@zjnu.edu.cn

W P Yeo
*Photonics & Sensors Key Research Laboratory,
Zhejiang Normal University
321004 Zhejiang
People's Republic of China*
svcat15@gmail.com

Burhanuddin Yeop Majlis
*Institute of Microengineering and Nanoelectronics (IMEN)
Universiti Kebangsaan Malaysia
43600 UKM Bangi
Selangor Malaysia*
burhan@ukm.edu.my

Muhamad Ramdzan Buyong
*Institute of Microengineering and Nanoelectronics (IMEN)
Universiti Kebangsaan Malaysia
43600 UKM Bangi
Selangor Malaysia*
muhdramdzan@ukm.edu.my

Ahmad Rifqi Md Zain
*Institute of Microengineering and Nanoelectronics (IMEN)
Universiti Kebangsaan Malaysia
43600 UKM Bangi
Selangor Malaysia*
rifqi@ukm.edu.my

Sheng Li
*Zhejiang Institute of Optoelectronics,
321004 Zhejiang
People's Republic of China*
shengli@zjnu.edu.cn

Le Song
*State Key Laboratory of Precision Measurement Technology and Instruments,
Tianjin University
300072 Tianjin
People's Republic of China*
songle@tju.edu.cn

Yelong Zheng
*State Key Laboratory of Precision Measurement Technology and Instruments,
Tianjin University
300072 Tianjin
People's Republic of China*
zhengyelongby@tju.edu.cn

Abstract—The integration of artificial intelligence (AI) methodology in the flexible electronics, BioMEMs and Dielectrophoresis (DEP) system provides an accurate, comprehensive, simple and cost-effective approach especially in the real time cell sorting, target identification, monitoring, biological entities separation, analysis and postulation. Here, a novel approach integrating AI algorithms with DEP technology for BioMEMs system has been demonstrated in real-time bacteria identification via detecting, identifying simultaneously, lock-in monitoring bacterial movement trajectories, analyzing, measuring and calculating bacteria's velocities of different frequencies. The AI-based bacterial recognition system creates real-time, fast, accurate, unlimited bacteria lock-in monitoring with high accuracy. Furthermore, the bacteria can be tracked, monitored, and analyzed using big data analysis and Deep Learning (DL) in real-time via a visual control interface system, as part of the Internet of Things (IoT) and intelligent information sharing between BioMEMS sensors and IoT networks.

Keywords— AI-based fractional-order differential algorithm, Dielectrophoresis (DEP), Image Processing, BioMEMs, Artificial Intelligence-Deep Learning (AI-DL)

I. INTRODUCTION

The existence of bacteria plays an important role in the ecological environment, but at the same time, it also poses a significant threat to human health. With the improvement of people's quality of life, health issues have become increasingly important. The emergence of antibiotics has solved this problem, but due to the widespread abuse of antibiotics, bacterial resistance to antibiotics has become a serious problem [1]. Therefore, solving the problem from the root has become one of the current solutions. Infection due to bacteria, could be coming from many sources. Of particularly common is from the existing food chain and

water supply. Water is the very existence of human-beings and in fact is one of the most important basic necessity ones must have in order to survive. Water is also indispensable for ones' daily chores and usage thus the ability to detect accurately, fast and cost effectively any harmful bacteria in drinking water is of paramount importance. One of the detection methodology is to identify and detect ESKAPE (an acronym comprising the scientific names of six highly virulent and antibiotic resistant bacterial pathogens including *Enterococcus faecium, Staphylococcus aureus, Klebsiella pneumoniae, Acinetobacter baumannii, Pseudomonas aeruginosa, and Enterobacter spp*) and enteric bacteria [1-3] in drinking water i.e., bacterial count in drinking water and whether meeting the water sanitary standard after going through a series of water treatments. Currently, the commercially available methods for detecting microorganisms are using fluorescence labeling detection and biochemical testing, after isolating live bacterial cells and culturing them. However, their detection cycle is long, with limitation of some compounds not able to be labeled with fluorescence. The testing equipment involved is huge and can only operate in high-end laboratory environment. All these shortcomings could be resolved with DEP approach [2-5]. DEP is an electrokinetic technique that has been proven to distinguish and isolate microorganisms based on their phenotype and physiological state as well as selectively manipulate cells organisms label-free. Dielectrophoresis' label-free technique does not affect living organisms. Due to different dielectric properties of each microorganism, when subjected to different electric fields, the polarities or charges on the microorganisms will exhibit different polarization phenomena [4]. For different dielectric properties of cells, they will be subjected to positive or negative electrophoretic forces. Sensors are manufactured using this principle can move microorganisms to different positions and locations

979-8-3503-7832-0/24 $31.00 © 2024 IEEE

under the influence of uneven electric fields [2-5]. Researchers [2-8] have incorporated DEP into biosensors and biological chips to perform multiple functions and improve bacterial detection significantly. Sensors using DEP technology have the advantages of high sensitivity, fast testing speed, and strong reliability [2, 4, 6-10]. Due to the large number, ever increasing mutation and minute size of microorganisms, it is very difficult to detect their presence with conventional methods, thus it is not uncommon that harmful microorganisms are being undetected in conventional methods [2-3, 7-9]. To solve this challenge, herewith, the proposed AI-based computer vision methodology using AI-DL algorithms for multi-points dynamic and real-time recognition and tracking of microorganisms. The AI tracking system could track and measure the movement as well as calculate and analyse the velocity of bacteria on the DEP with different frequencies condition. With these AI-driven functionalities and video analysis of bacteria movement, a more reliable, accurate and effective outcome for water quality safety measurement and assessment via the flexible BioMEMS sensors, are ensured.

II. METHODOLOGY & SYSTEM SET-UP

A. Bacterial detection and tracking methods

The proposed AI-DL algorithm is an open-source cross platform computer vision and machine learning software library which can perform fast, accurate and effective outcomes with the given parameters or requirements. The algorithms for mobile object detection are mainly based on four methods i.e., optical flow, deep learning, frame difference and background subtraction. Due to the large number and minute shape of bacteria, as well as the low clarity of the video outputs from conventional DEP imaging, background subtraction methodology has been chosen as the algorithm optimization for convenient, fast, and accurate outcomes. Background subtraction algorithm is the most suitable and intuitive approach for this case since by comparing the current frame with the background frame to achieve target detection, the background needs to be in a static state. Therefore, in this study, the background is a still image of the microscope field of view and the process requires collecting several frames of video images and average them to obtain a background model. The specific algorithm will read the current frame of the video and convert it into a grayscale image. The background subtraction algorithm is used to generate a foreground mask, and then a series of morphological operations are performed on the foreground mask to remove noise and fill in gaps. In order to prevent video jitter caused by hand shaking during output of DEP image, relevant filtering algorithms are being added.

After re-testing, it is necessary to detect and track the bacteria for their movement trajectory. The tracking of bacteria mainly involves two parts: bacterial detection and localization. In the DEP output video, bacteria are identified by the first few frames of the video, and the newly identified targets are located in subsequent video frames. The tracking algorithm mainly includes template based, feature based, density based, model based and deep learning. The main idea adopted in tracking algorithms is to initialize the tracker after identifying bacteria, add tracking algorithms to continuously update the position of bacterial changes in subsequent frames, and then optimize through templates and feature extraction to improve tracking accuracy.

B. Using Image Enhancement Algorithms to Improve Video Clarity

Due to the insufficient clarity of the video output from DEP, image enhancement is necessary for the video. Image enhancement is one of the important steps in image processing, which can mainly improve the visual quality of the image, making it more suitable for human observation or more conducive to subsequent image analysis and computer vision tasks. Image enhancement can be achieved through various methods, mainly by spatial domain methods and frequency domain methods. The commonly used methods for image enhancement include contrast adjustment, sharpening, de-noising, and edge enhancement. The most widely used technique in image enhancement is image sharpening, which enhances the edges and textures in achieving the best visual state. In order to better meet the requirements, traditional boundary scanning algorithms are being used for solving low image quality also. In this paper, due to the small size of the microorganisms, subtle differences are not easily detected. A novel AI-DL method with an improved algorithm based on fractional order differentiation theory has been applied to the DEP output videos. The optimization of this algorithm mainly uses ignored pixels in the new fractional order differentiation algorithm of the traditional Tiansi algorithm [11-14]. The advantage of this new algorithm is that it can avoid the pitfalls of traditional algorithm that only highlights image boundary information and ignores the processing process [12-14]. The AI-based fractional-order differential algorithm can perfectly adapt to and improve DEP output videos quality, sharpen microbial boundary information, improve its clarity and simplify subsequent operations. The specific operation and implementation process of the AI-based fractional-order differential algorithm are as follows:

i) Performing preliminary processing on the input image (blurred) to ensure that it is suitable for fractional order differentiation processing.

ii) Applying fractional order differential operators and use an improved 5×5 Tiansi template [12] to perform fractional order differential processing on the image. The formula for the calculation is:

$$I' = I + \lambda \cdot D^\alpha I$$

Where I is the original image, I' is the processed image, λ is the enhancement coefficient, and D^α is the fractional order differential operator.

iii) The subroutine for the operation can be represented as below:

```
def get_fractional_template(alpha):
    template = np.zeros((5, 5))
    for i in range(-2, 3):
        for j in range(-2, 3):
            template[i+2,j+2]=(-1)**(i+j)*
np.math.comb(alpha, abs(i+j))
    return template
```
Obtain Fractional Order Differential Template：
```
template = get_fractional_template(alpha)
```
iv) Performing subsequent processing on the processed image to reduce noise and enhance image details. The

weight of the template center pixel can be increased to suppress high-frequency noise.

v) Finally, adjust the fractional order and the weight of the template center pixel based on different scenes and images features. Setting the fractional order, enhancement coefficient, and sharpening coefficient for adjusting the sharpening of the images as below:

    alpha = 0.5
    lambda_ = 0.5
    beta = 1.5
    laplacian=cv2.Laplacian(enhanced_image,cv2.CV_64F)
    sharpened_image = enhanced_image - beta * laplacian

The application of AI-based approach here has significantly enhanced the video visual quality. Thus, enabling better accurate analysis of the video footage using DEP techniques. The flexibility of the approach can be further extended and working in parallel with various image enhancement technologies, optimization techniques and different video actual scenes conditions and needs with vast improvement.

C. Distance movement detection of bacterial and velocity

The methods for detecting the velocity/speed are mainly divided into real-time speed detection and average speed detection.

The real-time speed algorithm uses the ratio of the displacement difference between the previous and adjacent frames to the system frame change time to calculate the instantaneous speed of bacteria. Firstly, record the position of bacteria in the previous and current frames (x1, y1) and (x2, y2) Calculate the displacement difference between two frames as Eq. 1:

$$\Delta x = x_2 - x_1 \qquad (1)$$

Then use Eq. 2 to calculate the instantaneous velocity as below:

$$v = \frac{\sqrt{\Delta x^2 + \Delta y^2}}{\Delta t} \qquad (2)$$

Where Δt is the time interval between two adjacent frames.

The algorithm idea of average speed detection is calculated by recording the initial and final positions of target bacteria through pixel positions in the video, and using the Pythagorean equation, the movement distance of bacteria under the action of DEP force can be calculated. Record the initial and final positions of bacteria (x_i, y_i) and (x_f, y_f) then calculate the total displacement using Eq. 3 below:

$$d = \sqrt{\left(x_f - x_i\right)^2 + \left(y_f - y_i\right)^2} \qquad (3)$$

After that, calculate the average speed detection based on its corresponding time difference per Eq. 4:

$$v_{avg} = \frac{d}{T} \qquad (4)$$

Per DEP analytical rules, bacteria which escape above the 1st green line, l_1 (see Figs. 4 and 5) and below the 4th green line, l_4 (see Figs. 4 and 5) will not be accounted for the speed detection analysis. The actual gap between the two electrodes l_2 and l_3 is 20 µm. Likewise, the gap between the l_1 and l_2 lines is set at 20 µm, and the gap between the l_3 and l_4 lines is also set at 20 µm. Due to the use of pixel units in the program algorithm, conversion between video length and actual distance, or conversion between micrometers and pixels, is required. The retention of the action trajectory of bacteria tracking is recorded. For obtaining the x displacement, set the position of the 2nd line to l_2 and the position of the 3rd line to l_3, in the video per Eq. 5 below (which is also the distance between the electrodes):

$$\frac{l_3 - l_2}{20 \times 10^{-3}} = \frac{d}{x} \qquad (5)$$

where d is the bacterial movement distance obtained from the video using algorithms (with unit conversion of the unit, x in centimeters).

D. Data visualization analysis

Build a visual interface function that allows for direct video import and operation in multiple formats within the interface. Alternatively, a camera connected to a computer can be placed at the top of the microscope for real-time monitoring. By regularly refreshing data, the latest status of the data can be displayed in real time, allowing for a more intuitive observation of the real-time movement of bacteria. Through the output of DEP, various parameters can be analyzed and charts can be generated.

III. RESULTS AND DISCUSSION

The image enhancement has been performed on the original raw DEP video output (Fig. 1a) using the AI-based fractional-order differential algorithm (Fig.1b). The Fig.1b has improved tremendously with video image showing higher contrast ratio, clarity and less noise. It is clear that the novel approach has improved the image quality significantly and able to differentiate the microorganisms in the DEP and the background noise. The sharpened video would be used for herewith bacteria analysis work.

(a)   (b)

Fig.1. Comparison image before and after image enhancement

The simultaneous detection of unlimited bacteria targets with great accuracy has been demonstrated in a cost effective manner. Fig. 2. shows the real-time bacteria detection demonstration in progress. One of the bacteria was selected for demo displays with the algorithm operating at background (shown by a tracking-red-box). The dimension of the tracking-red-box is auto-adjusted in real-time with relation to the dimension of the tracked objects.

Fig.2. Multi bacterial detection tracking in progress

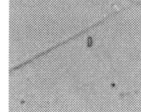

Fig.3. Zoom-in: Tracked bacterial (object)

Fig.4. shows the real-time detection of bacteria and its velocity respectively. Adding a velocity tracker (shown in Fig.5.) during the velocity analysis of a bacteria greatly improves its accuracy and prevents unrecognizable velocity error due to the bacteria spontaneous high speed movement in real-time.

Fig.4. Detection of bacterial velocity

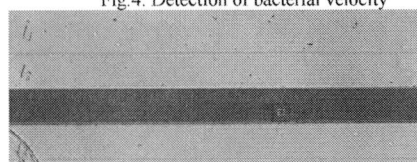

Fig.5.Velocity detection tracker for bacteria

It also minimizes tolerance differences error due to random and spontaneous bacteria movements across different frequencies conditions. The real-time velocity tracking of bacteria could be outputted to the AI-based image processing visualization interface as shown in Fig.6.

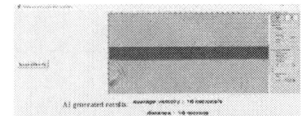

Fig.6. AI-based image processing visualization interface

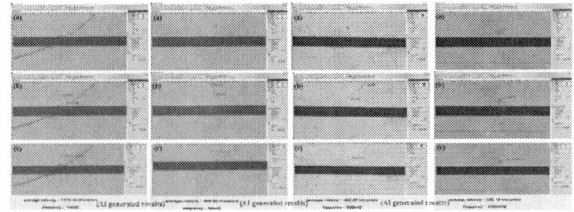

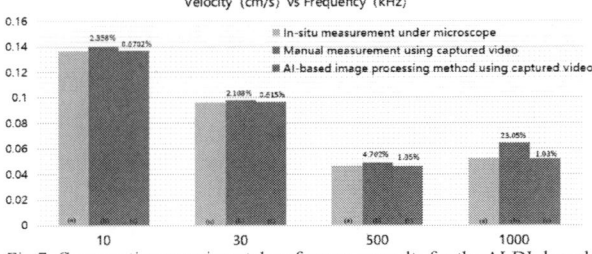

Fig.7. Comparative experimental performance results for the AI-DL based method

The accuracy performance results of the AI-based method, (c) have been compared experimentally with the actual in-situ measurement under a microscope, (a) and the manual calculation method based on captured videos, (b) (see Fig. 7). These 3 methods were compared experimentally in terms of error percentile deviated from the in-situ measurement under microscope for velocities versus frequencies of 10 KHz, 30 KHz, 500 KHz and 1MHz of *Staphylococcus aureus*, the ESKAPE bacteria [2-3]. The relative errors % benchmarked against the in-situ measurement under a microscope were calculated and demonstrated that the AI-based method had consistently performed much better than manual measurement calculation method with error % between 0.07% to less than 1%. Thus, the AI-based method can offer high accuracy comparable to in-situ measurement under a microscope while offering other superior features of measuring unlimited multiple bacteria simultaneously with ease and cost effective versus other methods.

IV. CONCLUSION

The feasibility of the novel approach integrating AI-DL algorithms with DEP in BioMEMs system has been demonstrated for the real-time simultaneous bacteria identification via detecting, identifying, lock-in monitoring bacterial movement trajectories, analyzing, measuring and calculating bacteria's velocities versus frequencies. The AI-based bacterial identification system creates a real-time, fast, accurate, unlimited simultaneous bacteria identifying mechanism with high accuracy yet cost effective.

ACKNOWLEDGMENT

The research work here has been supported by the research team (and international collaborators) within the Photonics & Sensors Key Research Laboratory as part of the cross-disciplined Research Project "Multi-Dimensional Nano-Microelectronics Functional Devices & System Engineering (MINDS)", an interdisciplinary and foresight collaborative key research initiative. The work is funded and supported by the China National High Talent Foreign Expert Grant under Grant G2021016010L, Zhejiang Normal University's Distinguished (Eminent) Professorship Grant and Zhejiang's "Pioneering and Leading Goose" R&D Programme Grant 2022C01030.

REFERENCES

[1] Christopher J L Murray, et. al, "Global burden of bacterial antimicrobial resistance in 2019: a systematic analysis", Lancet, Vol. 399, 629–55, February 12, 2022.

[2] Clarence Augustine TH Tee, et. al, "Electrode Isolation and Microfluidic Flow Effect Analysis on Dielectrophoretic Bioparticles Chaining for Flexible BioMEMS Application", IEEE Journal on Flexible Electronics, Volume: 2, Issue: 4, July 2023.

[3] W H M Wan Mohtar, et al., "Rapid detection of ESKAPE and enteric bacteria using tapered dielectrophoresis and their presence in urban water cycle", Process Safety and Environmental Protection 177, 427–435, 2023.

[4] R. Pethig, "Dielectrophoresis: An assessment of its potential to aid the research and practice of drug discovery and delivery", Adv. Drug Delivery Rev, 2013.

[5] J Cottet, et. al, "MyDEP: A New Computational Tool for Dielectric Modeling of Particles and Cells", Biophysical Journal 116, 1–7, January 8, 2019.

[6] M. Manvi, et. al, "Microelectronic materials, micro-fabrication processes, micromechanical structural configuration based stiffness evaluation in MEMS" Microelectron. Eng., vol. 263, Jul. 2022.

[7] Yunus FW, et. al, "Negative charge dielectrophoresis by using different radius of electrodes for biological particles." In 2017 IEEE Regional Symposium on Micro and Nanoelectronics, 84-87, 2023.

[8] Buyong MR, et. al, "Fabrication of thin layer membrane using CMOS process for very low pressure sensor applications", IEEE Int. Conf. Semiconductor Electronics, 363-369, 2008.

[9] R. Bashir, "BioMEMS: State-of-the-art in detection, opportunities and prospects," Adv. Drug Del. Rev, 6, 11, 1565–1586, 2004.

[10] PK Sekhar, et. al, "Review of sensor and actuator mechanisms for BioMEMS," Woodhead Publishing (Biomaterials), 46–77, 2012.

[11] Zhang S., et. al, "Fuzzy traffic video image enhancement based on fractional order differentiation", Optical Precision Engineering, 22 (3): 779-786, 2014.

[12] Yang Q, et. al, "Fractional calculus in image processing: a review", J. Fractional Calculus and Applied Analysis, 19(5):23-31, 2016.

[13] Chen X, et. al, "BioMEMS microfluidic chips for blood sample pretreatment", Sensing Technology, 19 (05B): 1991-1995, 2006.

[14] Zhang J, et. al, "Design of Real Time Video Surveillance System Based on WiMAX Technology", Television Tech, 32-3 91-92, 2008.

Enhancing Dielectrophoresis Analysis via Artificial Intelligence Integration

Muhamad Ramdzan Buyong
Institute of Microengineering and Nanoelectronics (IMEN)
Universiti Kebangsaan Malaysia
43600 UKM Bangi
Selangor Malaysia
muhdramdzan@ukm.edu.my

Muhammad Akmal Suhaimi
Institute of Microengineering and Nanoelectronics (IMEN)
Universiti Kebangsaan Malaysia
43600 UKM Bangi
Selangor Malaysia
p137669@siswa.ukm.edu.my

Arash Zulkaranain Ahmad Rozaini
Institute of Microengineering and Nanoelectronics (IMEN)
Universiti Kebangsaan Malaysia
43600 UKM Bangi
Selangor Malaysia
p124814@siswa.ukm.edu.my

Burhanuddin Yeop Majlis
Institute of Microengineering and Nanoelectronics (IMEN)
Universiti Kebangsaan Malaysia
43600 UKM Bangi
Selangor Malaysia
burhan@ukm.edu.my

Farahdiana binti Wan Yunus
Department of Physics
Xiamen University Malaysia 43900
Sepang Selangor Malaysia
farahdiana.wanyunus@xmu.edu.my

Noratiqah Yaakob
SilTerra Malaysia Sdn. Bhd.
Lot 8 Phase 2 Kulim Hi-Tech Park
09090 Kulim Kedah Malaysia
noratiqah_yaakob@silterra.com

Clarence Augustine TH Tee
College of Physics & Electrical Information Engineering
Zhejiang Normal University China
catht@zjnu.edu.cn

Céline Elie-Caille
Institut FEMTO-ST
UMR6174-CNRS
15B Avenue des Montboucons
25030 Besancon France
celine.elie@femto-st.fr

Abdullah Abdulhameed
Center Com. Systems & Sensing
King Fahd University
of Petroleum & Minerals,
Dhahran 31261
Saudi Arabia
dr.a.abdulhameed@gmail.com

Aminuddin Ahmad Kayani
Functional Materials &
Microsystems Research Group
School of Engineering
RMIT University
GPO Box 2476 Melbourne
Victoria 3001 Australia
amin.kayani@rmit.edu.a

Abstract—Dielectrophoresis (DEP) has emerged as a powerful technique for manipulating and analyzing particles based on their polarizability differences in electric fields. However, the analysis of DEP data often involves complex computational analysis methods and requires significant expertise. In this study, we propose a novel approach to enhance DEP analysis through the integration of artificial intelligence (AI) techniques. By leveraging AI algorithms such as image processing and velocity recognition, we aim to streamline the analysis process, improve accuracy, and enable real-time decision-making. This integration allows for automated classification of particles, identification of subtle patterns, and optimization of experimental parameters. the proposed method should revolutionize DEP analysis, paving the way for advancements in various fields such as biotechnology, nanotechnology, and medical diagnostics.

Keywords—Dielectrophoresis, artificial intelligence, Analysis, Integration

I. INTRODUCTION

Research into dielectrophoresis (DEP) at the Institute of Microengineering and Nanoelectronics (IMEN) began in 2010 and has continued since until now, yielding groundbreaking insights into its application across various fields, from environmental monitoring to medical applications [1-7]. This research follows a comprehensive methodology, starting with fundamental principles, numerical analysis, and simulation for modeling, and proceeds to experimental validation in various settings, including both laboratory and field tests. Biological specimens such as human blood cells, bacteria, and proteins are utilized as experimental samples. Currently, the technology readiness level (TRL) of DEP research at IMEN ranges from level 4, involving laboratory testing, to level 6, where field prototype systems are developed through collaborative multidisciplinary research endeavors. Leveraging insights from prior experimental experiences, we proposition a need to enhance DEP analysis through artificial intelligence integration for the resolution of interfacing and automatic control. For this drive, this manuscript outlines the artificial intelligence (AI) integration of subsequent actions and designs based on previous DEP research findings.

Convolutional Neural Network (CNN) is an AI technology used for image and video processing, recognition, segmentation and analysis. Since the framework of CNN used to create U-NET in 2015, the technology gives out a massive potential to be applied in biomedical industry. In 2022, a research from Thailand described the application of enhanced U-NET (ResU-NET and DenseU-NET) on the segmentation of kidney cell for potential cyst development to be highly successful at 87-95% accuracy [8]. They analysed 60 real patients' CT-scan images in implementing their technology. Rehman et al also published a study on segmentation of brain image using the U-NET framework in determining the presence of brain tumor. The showed an increase of detection performance at 1.6% compared to MRI technology [9]. There are also other studies involving U-

NET framework outside of biomedical application such as crack detection on road surface, robotics, and climate changes [10-12]. On the other hand, another AI and machine learning technology that could be considered would be the velocity detection method. Automated container terminal (ACT) uses artificial intelligence to analyze automated guided vehicle trajectory movement from port surveilance video. The publication was able to produce accurate vehicle kinematic data in real-time based on the trajectory and movement speed of these vehicles [13]. Furthermore, multiple object tracking technology (MOT), is currently being used in sport industry to track multiple players on the field. The technology could detect multiple players' position, movement and speed, and store them in the database. In automotive industry, the emergence of smart vehicles enquires the use of multiple cameras built together with smart computational systems. These systems use the basic of AI and machine learning to detect other vehicles' distance, speed and position in order to provide a safer driving environment. The use of AI had been implemented in biomedical purposes. Meanwhile, even tough there are examples mentioned above that listed the application of AI outside of biological world, the point to be highlighted is that the image processing technology and velocity detection technology could be tweaked and implemented together with DEP technology to further improve DEP efficiency.

Integration of DEP with AI has been an area of growing interest and ongoing research. General trends and potential developments that might have occurred since then advanced control systems. Researchers continue to explore AI-driven control systems for optimizing DEP-based processes. These systems leverage machine learning algorithms to adaptively adjust electric field parameters in real-time, leading to more efficient particle manipulation and enhanced performance of DEP devices. Predictive modeling, there is ongoing work in developing predictive models that combine computational simulations with AI techniques to accurately forecast the behaviour of particles in DEP systems. These models help in the design and optimization of DEP experiments, enabling researchers to achieve desired outcomes with greater precision. Automated experiment design, AI algorithms are being employed to automate the design of DEP experiments by efficiently exploring the parameter space and identifying optimal configurations. This automated approach accelerates the experimental process, allowing researchers to quickly iterate and refine DEP techniques for various applications.

The integration of DEP and AI holds great promise for advancing particle manipulation techniques and optimizing DEP-based applications across various fields. Continued research and development in this area are expected to lead to further innovations and practical implementations with significant societal and scientific impact.

II. DEP-AI INTEGRATION BENEFIT

DEP is a phenomenon in which a force is exerted on a dielectric non-conductive particle when it is subjected to a non-uniform electric field. This force arises due to the interaction between the induced dipole moment within the particle and the electric field gradient. DEP has various applications in fields such as biotechnology, microfluidics, and nanotechnology. Integration AI interfacing can be utilized for analyzing and enhancing DEP analysis processes via implementation of data analysis and pattern recognition.

AI can be employed to analyze the complex data generated from DEP experiments. By training machine learning algorithms on large datasets of DEP responses for different particles and field configurations, AI can recognize patterns and extract valuable insights. This analysis can help in understanding the behaviour of particles under varying electric field conditions, which is crucial for optimizing DEP-based applications. Process optimization of AI algorithms can be utilized to optimize DEP processes. By integrating AI models with DEP setups, researchers can develop closed-loop control systems that continuously monitor the DEP response and adjust experimental parameters in real-time to maximize particle manipulation efficiency. This can lead to improved sorting, trapping, and manipulation of particles, enhancing the overall performance of DEP based devices.

Predictive modeling, AI can aid in the development of predictive models for DEP. By combining computational simulations with machine learning techniques, researchers can create accurate models that predict the behaviour of particles in complex electric field environments. These predictive models can assist in the design and optimization of DEP devices, enabling researchers to achieve desired particle manipulation outcomes with higher precision. AI-driven predictive models are developed to simulate the behaviour of particles in DEP systems accurately. These models combine computational simulations with machine learning techniques to predict the trajectory, aggregation, and other properties of particles under varying electric field conditions.

Automated experiment design, AI interfacing can automate the design of DEP experiments. By leveraging AI-driven optimization algorithms, researchers can efficiently explore the vast parameter space of DEP configurations to identify optimal experimental conditions for specific applications. This automated approach accelerates the experimental process, enabling researchers to quickly iterate and refine DEP techniques for various practical tasks. The integration of AI interfacing with DEP offers promising opportunities for advancing particle manipulation techniques and optimizing DEP-based applications across diverse fields. By harnessing the capabilities of AI, researchers can unlock new insights, enhance performance, and accelerate innovation in DEP technology.

III. METHODOLOGY

A. DEP system

The DEP experiment was performed using a fluorescence microscope (Olympus BX50) equipped with a digital camera (Olympus DP22) and an adapter that enabled the image from the microscope to be viewed and focused on a computer system. The visual monitoring of the targeted particle is performed in bright field mode with the support of the Olympus cellSens imaging software. The input sinusoidal electrical signals from a function generator (Teledyne LeCroy-WaveStation, 2022; $\pm$ 20 VPP, 25 MHz) were directly connected to a prober with its source and ground terminals to supply a voltage of different frequencies to the microelectrode. A droplet of fluid suspension containing particles (5–10 μL) was pipetted onto the microelectrode and covered with micro-cover glass to ensure that the fluid was distributed evenly. The confinement of the sample droplet is an essential factor in making the fluid static for DEP manipulation while simultaneously controlling the particles within the small area. To control the environment, a small well was made with polyamide tape during each test on a microelectrode. After adding a droplet of particle, the fluid confinement tape well must cover the microelectrode region of interest (ROI). The cover glass slip is covered before the DEP experimental manipulation begins to make the fluid static and ensure a controlled environment. Microelectrode and particle identification were made by focusing on particles in ROI at 5×, 20×, and 50× magnifications until they were visible. The microelectrode length is initially calibrated in the software to the required gap. The manipulation of particles was done by applying an electric field to the DEP microelectrode. The particle's movement, trajectory and velocity were recorded in a time-lapse of 60 secs using the cellSens standard software.

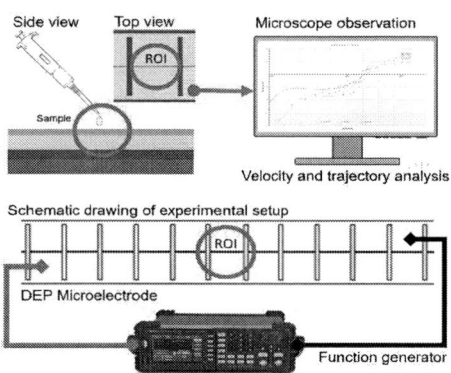

Fig. 1.   Schematic of experimental DEP system

B. Integration DEP and AI particle movement system

The AI system is used to extract particle movement details recorded by video. It is highly challenging if done manually, which can result in human errors. Therefore, an AI application capable of motion detection would greatly aid this research. The video recordings from the experimental day will be analyzed using AI image detectors that detect particle movement trajectory and measure the velocity of particle movement as the frequency increases from 100 KHz to 2 MHz. The analysis of particle movement, specifically regarding trajectory and velocity, is conducted for presentation in a polarization factor graph. This experimental polarization factor graph is also compared with numerical polarization factors graph. Furthermore, comparison of the velocity curve generated with the CMF curve demonstrates a minimal deviation, indicating the accuracy of the developed model after 3 to 5 iterations. The experimental results are validated through MyDEP numerical and COMSOL Multiphysics simulation, revealing a statistically minimal deviation between experimental and simulated data.

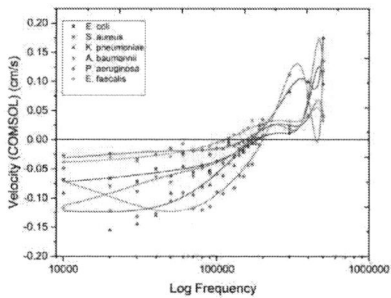

Fig. 2.   Simulation of DEP reaction for 6 types of bacteria

Figure above shows the COMSOL simulation of DEP reaction for 6 different type of bacteria. As shown in the figure, the velocity of all 6 bacteria at certain point in the frequncy range are easily distinguishable. By selecting the best frequency, a set of data can be fed to the AI system. The machine learning will eventually could determine what type of bacteria based on the velocity at the selected frequency based on multiple repetition data taken from laboratory experiment.

C. Integration DEP and AI image processing system

The integration of DEP and AI systems involves deploying AI interfacing to present dashboard results for the purposes of detection and quantification analysis. The input is based on polarization factors that resulted to trajectory positive DEP (PDEP), where is attracted to the highest field in region of interest (ROI), negative DEP (NDEP) where is repelled from the highest field to out from ROI, and crossover (fxo), where is no movement due to equivalent PDEP and NDEP forces. The polarization transition from PDEP to NDEP or vice versa detection methods, can be explored. Ultimately, the particle velocity magnitude such as PDEP, NDEP, can be effectively optimum characterized for detection purposes. A combination of these 7 parameters is ideally employed as the key criteria for achieving high-efficiency target detection. Furthermore, fit in all these parameters extracted from the experimental tests can verify numerical analysis and simulation modeling in closed-mode characterization and optimization. Then, quantification is carried out using the electrical capacitance method, which has been correlated and calibrated.

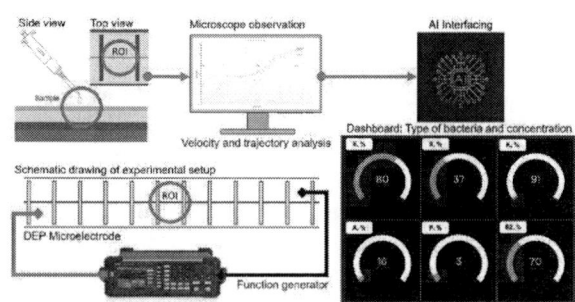

Fig. 3.   Integration DEP and AI system for DEP detection and quantification

979-8-3503-7832-0/24 $31.00 © 2024 IEEE   118

IV. RESULT AND DISCUSSION

DEP output refers to the results obtained from DEP experiments, which typically involve the manipulation, sorting, or trapping of particles using non-uniform electric fields. Analyzing DEP output involves understanding various parameters such as particle behaviour, movement, aggregation, or separation under different electric field conditions. An AI dashboard for analyzing DEP output would provide a user-friendly interface to visualize and interpret the experimental results.DEP dashboard function, data visualization the dashboard would display visual representations of DEP output data. This could include graphs, charts, or heatmaps illustrating parameters such as particle trajectories, velocity, concentration, or distribution. Users could customize the visualization based on their specific analysis needs, such as comparing different experimental conditions or tracking the behaviour of individual particles over time. Real time monitoring, the dashboard may offer real-time monitoring capabilities, allowing researchers to observe DEP experiments as they unfold. This could involve live streaming of microscopy images showing particle movement, along with updated data on relevant parameters. Real-time monitoring enables researchers to make immediate adjustments to experimental conditions and assess the effectiveness of particle manipulation strategies. Machine learning insights, AI algorithms integrated into the dashboard could provide insights derived from the DEP output data. For example, machine learning models could analyze patterns in particle behaviour to identify optimal electric field configurations for specific manipulation tasks. These insights could help researchers refine their experimental techniques and achieve better control over particle manipulation processes.

Statistical analysis, the dashboard could include tools for statistical analysis of DEP output data. Users could perform descriptive statistics, hypothesis testing, or regression analysis to quantify the effects of different experimental variables on particle behaviour. Statistical analysis helps researchers draw robust conclusions from their experiments and understand the underlying principles governing DEP phenomena. Customizable reporting, the dashboard may allow users to generate customizable reports summarizing key findings from DEP experiments. These reports could include visualizations, statistical summaries, and insights generated by AI algorithms. Customizable reporting facilitates collaboration, documentation, and communication of research results within interdisciplinary teams or with external stakeholders. Integration with experimental setup, in advanced setups, the dashboard could be integrated with the DEP experimental apparatus, enabling seamless control and synchronization of experimental parameters. Researchers could adjust electric field settings, particle types, or environmental conditions directly from the dashboard interface, streamlining the experimental workflow and ensuring consistency across experiments. An AI dashboard for analyzing DEP output provides a powerful tool for researchers to gain insights, optimize experimental protocols, and advance our understanding of particle manipulation using DEP techniques. By combining data visualization, real-time monitoring, machine learning insights, statistical analysis, and experimental control functionalities, such a dashboard empowers researchers to unlock the full potential of DEP technology for various applications.

ACKNOWLEDGMENT

The author acknowledges the Fundamental Research Grant Scheme (FRGS), grant number FRGS/1/2022/TK07/UKM/02/21 grant funded by Minister of Higher Education (MOHE). This work was performed in part at the Micro Nano Research Facility at RMIT University in the Victorian Node of the Australian National Fabrication Facility (ANFF). We acknowledge project funding from the Australian Research Council through IH210100040. The author would like to thank Professor Sharath Sriram, Professor Madhu Bhaskaran, and Professor Sumeet Walia, who is leading the Functional Materials and Microsystems Research Group at RMIT University for allocating resources and technical expertise in micro and nanofabrication activities for this project.

REFERENCES

[1] Buyong MR, Abd Aziz N, Majlis BY. "Fabrication of thin layer membrane using CMOS process for very low pressure sensor applications". In 2008 IEEE International Conference on Semiconductor Electronics 2008 Nov 25 (pp. 363-369). IEEE

[2] Rashid NF, Deivasigamani R, Wee MM, Hamzah AA, Buyong MR. "Integration of a dielectrophoretic tapered aluminum microelectrode array with a flow focusing technique". Sensors. 2021 Jul 21;21(15):4957

[3] Rahim MK, Buyong MR, Jamaludin NM, Hamzah AA, Siow KS, Majlis BY."Characterization of permittivity and conductivity for ESKAPE pathogens detection". In 2018 IEEE International Conference on Semiconductor Electronics (ICSE) 2018 Aug 15 (pp. 132-135). IEEE

[4] Wee MM, Buyong MR, Majlis BY. "Effect of microchannel geometry in fluid flow for PDMS based device". In RSM 2013 IEEE Regional Symposium on Micro and Nanoelectronics 2013 Sep 25 (pp. 391-393). IEEE

[5] Buyong MR, Ismail AG, Yunus FW, Abd Samad MI, Jamaludin NM, Rahim MK, Hamzah AA, Majlis BY, Abd Aziz N. "Tapered dielectrophoresis microelectrodes: Device, operation, and application." In 2019 IEEE Regional Symposium on Micro and Nanoelectronics (RSM) 2019 Aug 21 (pp. 172-175). IEEE

[6] Yunus FW, Hamzah AA, Norzin MS, Buyong MR, Yunas J, Majlis BY. "Dielectrophoresis: Iron dificient anemic red blood cells for artificial kidney purposes". In 2018 IEEE International Conference on Semiconductor Electronics (ICSE) 2018 Aug 15 (pp. 5-8). IEEE

[7] Yunus FW, Hamzah AA, Buyong MR, Yunas J, Majlis BY. "Negative charge dielectrophoresis by using different radius of electrodes for biological particles." In 2017 IEEE Regional Symposium on Micro and Nanoelectronics (RSM) 2017 Aug 23 (pp. 84-87). IEEE

[8] Sun, P., Mo, Z., Hu, F., Song, X., Mo, T., Yu, B., ... & Chen, Z. (2022). Segmentation of kidney mass using AgDenseU-Net 2.5 D model. Computers in Biology and Medicine, 150, 106223.

[9] Rahman, M. M., Sadique, M. S., Temtam, A. G., Farzana, W., Vidyaratne, L., & Iftekharuddin, K. M. (2021, September). Brain tumor segmentation using UNet-context encoding network. In International MICCAI Brainlesion Workshop (pp. 463-472). Cham: Springer International Publishing.

[10] Wang, L., & Ye, Y. (2020, August). Computer vision-based road crack detection using an improved I-UNet convolutional networks. In 2020 Chinese Control And Decision Conference (CCDC) (pp. 539-543). IEEE.

[11] Fu, L., & Li, S. (2023). A new semantic segmentation framework based on UNet. Sensors, 23(19), 8123.

[12] Singh, N. J., & Nongmeikapam, K. (2023). Semantic segmentation of satellite images using deep-unet. Arabian Journal for Science and Engineering, 48(2), 1193-1205.

[13] Zhang, J., Ioannou, P. A., & Chassiakos, A. (2006). Automated container transport system between inland port and terminals. ACM

ZnO and PZT Thin Film Piezoelectric MEMS Vibrational Energy Harvesters for Cardiac Pacemaker

Yap Jia Xui
Faculty of Engineering and Built Environment, Universiti Kebangsaan Malaysia (UKM),
Bangi, Malaysia.
jiaxui98@gmail.com

Avinash Kumaresan
Faculty of Engineering and Built Environment, Universiti Kebangsaan Malaysia (UKM),
Bangi, Malaysia.
P145160@siswa.ukm.edu.my

Jumril Yunas
Institute of Microengneering and Nanoelectronics (IMEN), Universiti Kebangsaan Malaysia (UKM),
Bangi, Malaysia.
jumrilyunas@ukm.edu.my

Seri Mastura Mustaza
Faculty of Engineering and Built Environment, Universiti Kebangsaan Malaysia (UKM),
Bangi, Malaysia.
seri.mastura@ukm.edu.my

Huda Abdullah
Faculty of Engineering and Built Environment, Universiti Kebangsaan Malaysia (UKM),
Bangi, Malaysia.
huda.abdullah@ukm.edu.my

Iskandar Yahya
Faculty of Engineering and Built Environment, Universiti Kebangsaan Malaysia (UKM),
Bangi, Malaysia.
iskandar.yahya@ukm.edu.my

Abstract— **MEMS piezoelectric energy harvesters (PEHs) have potential to convert mechanical energy of human heart vibrations into electrical energy to power the pacemaker. Challenges exist to match the resonant frequency of PEHs to heart wall low vibrational frequency and the pacemaker size constraint to fit the PEH. Moreover, the output performance of PEH with hybrid piezoelectric materials is not studied in literature. This work aims to design a MEMS PEH element with a resonant frequency <50 Hz, an output voltage of 2.5–2.8 V, a dimension of 40 mm by 6 mm, and optimise its output performance via a hybrid of ZnO and PZT. Correlation studies between the PEH design and output performance were performed to derive design strategies. The split-cantilever energy harvester (SCEH) structure was built using COMSOL Multiphysics. SCEH has resonant frequencies at 3.1 Hz and 4.9 Hz, with voltages of 36.21 V and 38 V respectively. The simulation results successfully meet the requirements for powering the pacemaker in terms of dimensions, resonant frequency, and generated voltage.**

Keywords— *MEMS, Piezoelectric Energy Harvester, Pacemaker, ZnO, PZT.*

I. INTRODUCTION

Battery powered pacemakers last for 5-8 years [1], which requires periodical replacement surgery that exposes the patients to a risk of serious complications. Piezoelectric energy harvesters (PEHs) are a viable solution as they can be used to convert mechanical energy from the heart into electrical energy to power the pacemaker. A standard cylindrical capsule leadless pacemaker has a dimension of 40 mm (l) by 6 mm (W), but different PEH designs proposed by other researchers are larger than this. In addition, the resonant frequencies of the proposed designs are not compatible with the human heart vibration signal, which is less than 50 Hz [2], [3], [4], [5]. Since the resonant frequency of PEH increases when it is scaled down, it poses a challenge to match the PEH resonant frequency to the human heart wall's low vibrational frequency. Besides, there are a variety of piezoelectric materials such as lead zirconate titanate (PZT) [6], aluminum nitride (AlN) [7], [8], and zinc oxide (ZnO) [9], [10], used in the PEH, but hybrid piezoelectric materials are yet to be ventured.

Thus, this work aims to design a ZnO-PZT hybrid MEMS PEH element in COMSOL Multiphysics software for pacemaker applications with a resonant frequency below 50 Hz, an output voltage of 2.5–2.8 V and dimensions below 40 mm in length and 6 mm in width. Both piezoelectric materials are chosen to be investigated in this work, as ZnO is easy to fabricate, abundant, and low cost while PZT exhibits a high piezoelectric coefficient. Hybrid configurations were considered as they have better output performance and higher energy conversion compared to PEHs with single type of piezoelectric material. The configuration of the 2-Dimensional (2D) PEH was fixed to a cantilever structure with tip mass attached at one end of the cantilever beam to reduce the resonant frequency of the PEH. Geometric parameters, types of substrates, and types piezoelectric materials were studied in the correlation study to find the optimal parameters for the PEH simulation. Once obtained, the parameters were applied to simulate a PEH that fulfils the requirement for the pacemaker application.

II. METHODOLOGY

The study commenced by conducting correlation studies to investigate the effect of geometric parameters based on the model shown in Fig. 1. The structure of the PEH model used for correlation studies in Fig. 1 is a bimorph cantilever beam with a tip mass. and a fixed mass. The substrate in the middle of the cantilever beam is sandwiched by two piezoelectric layers. Simulations were performed for 6 different parameters, which are beam length, beam thickness, piezoelectric layer thickness, ratio of piezoelectric layers to beam thickness, tip mass length and thickness.

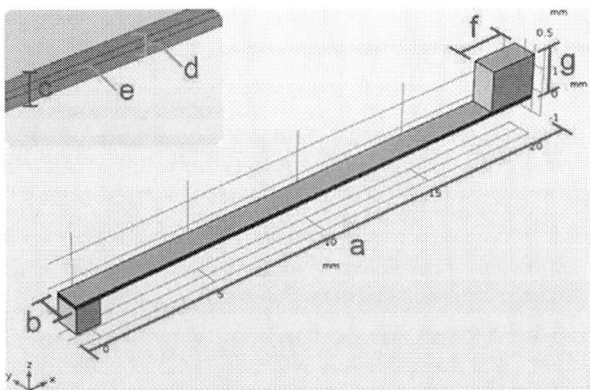

Fig. 1. PEH model used for correlation study: a) length b) width c) thickness d) piezoelectric top and bottom layers, e) substrate layer f) tip mass length g) tip mass thickness.

Their output resonance frequency, voltage and output powers were observed. The value that has the lowest resonance frequency and best voltage and power output were chosen as the optimum values. When one of the parameters is investigated, the values of the other parameters are fixed to ensure the accuracy of the simulation results obtained.

In the next correlation study, 4 configurations for PZT and ZnO (PZT-PZT, ZnO-ZnO, PZT-ZnO and ZnO-PZT), four substrates which are structural steel, aluminium, tungsten and silicon and their effect on resonance frequency, output voltage, power and efficiency were observed. Design strategies were finalized and implemented in modelling a split cantilever energy harvester (SCEH) PEH that fulfils the sizing and power requirements for pacemaker applications as a proof of concept. All simulations were carried out in COMSOL Multiphysics version 5.3, and it's used to prove the reliability and feasibility of a PEH for powering a pacemaker.

III. RESULTS

A. Geometric Parameter Correlation Study

TABLE 1 summarizes the optimum parameters for the PEH model shown in Figure 1, selected based on the lowest resonance frequency and highest output voltage and power from simulation results. Note that Fig. 2 focuses solely on the piezoelectric layer thickness parameter and the rest is not shown here.

TABLE I.   VALUES OF PARAMETERS FOR PEH MODEL IN FIG 1

Letter	Parameter	Values (mm)
a	Beam Length, l_b	76
b	Beam Width, w_b	14
c	Beam Thickness, t_b	0.16
d	Piezoelectric Layer Thickness, t_p	0.06
e	Substrate Layer Thickness, t_s	0.04
f	Proof Mass Length, l_m	4
g	Proof Mass Thickness, t_m	1.7

For beam length, f_n decreases with the increase of l_b. However, the low f_n of the model is limited to a range of

values, in which when l_b is further increased, f_n will remain in that range. The relationship between f_n and l_b can be explained by the Equation (1)[11]. Based on equation (1), f_n is inversely proportional to the square of l_b, and directly proportional to the square root of piezoelectric layer thickness, t_p and substrate layer thickness, t_s.

$$f_n \propto \frac{1}{l_b^2} \sqrt{t_p} \sqrt{t_s} \qquad (1)$$

The electrical output voltage and power decrease with increasing f_n and decreasing l_b. This situation can be explained by the fact that the tension and compression motion of the PEH decreases at higher f_n, as the cantilever beam has a shorter travel distance due to the time to recover before the next deformation cycle is constant, which in turn will reduce the output voltage and electrical power [12].

For beam width, f_n remains almost constant with increasing w_b. Based on Equation (1), w_b has no direct effect on f_n and the output voltage and electrical power are found to be proportional to w_b.

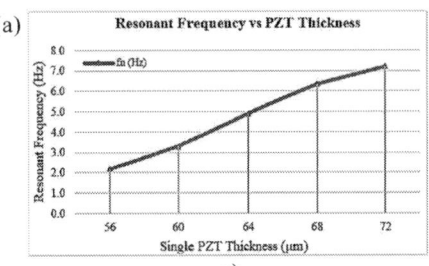

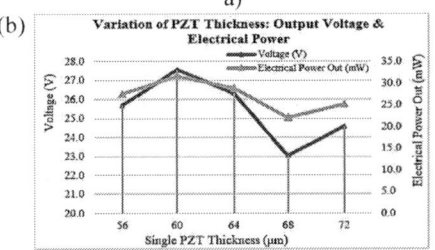

Fig. 2. Piezoelectric layer thickness a) Resonant frequency vs PZT film thickness b) Output voltage and power vs PZT film thickness.

For piezoelectric layer thickness and ratio of total piezoelectric layers to beam thickness (t), t_p and $\frac{t_p}{t}$ are proportional to f_n. Referring to Fig. 2 and TABLE 1, the chosen piezoelectric material thickness is 0.06 mm, as it is shown to have low resonant frequency and high voltage and power. The analysis of the obtained research results can be validate using Equation (1), as the bending stiffness of the piezoelectric layer decreases with the reduction of t_p. A low bending stiffness of the piezoelectric layer will generate a low f_n. The output voltage and electrical power are almost constant with the increase of t_p and $\frac{t_p}{t}$. It can be also noted that there is a max followed by a min feature in Fig 2b). Although the origin is unclear, it may be related to changes in output electrical polarization due to variations in thickness, thus affecting the bending pivot point. In conclusion, a thin piezoelectric layer is more suitable for the design of low f_n PEH while being able to generate a high output voltage and

979-8-3503-7832-0/24 $31.00 © 2024 IEEE

electrical power. However, due to the difficulty of fabricating very thin piezoelectric ceramics, the minimum thickness available for PZT piezoelectric ceramics is in the range of 10 μm to 100 μm [13].

For proof mass length, it is observed that increase of l_m results in a low f_n and this is due to the equivalent stiffness of the beam, k decreases with an increase in l_m. The corresponding relationship is explained by Equation (2)[11], where E is the stiffness of the material, I represent the moment of inertia of the beam cross-sectional area and ΣEI refers to the bending stiffness of the beam. As k decreases with increasing l_m, f_n also decreases with increasing l_m, based on Stoney's equation shown in Equation (3)[11], where m is the equivalent mass of the PEH [5], [11]. The output voltage and electrical power increases with increasing l_m but the increase in output power is much more significant compared to the increase in output voltage.

$$\kappa = \frac{12 \, \Sigma \, EI}{4 \, l_b^3 + 3 \, l_m l_b^2} \tag{2}$$

$$f_n = \frac{1}{2\pi} \sqrt{\frac{k}{m}} \tag{3}$$

For proof of mass thickness, as t_m increases, f_n decreases, the output voltage increases at an insignificant rate and the output power remains almost constant. The results of this simulation can be explained by Equation (2), where the stiffness of the beam, k does not change with the increase of t_m.

B. Substrate and Piezoelectric Layer Material Correlation Study

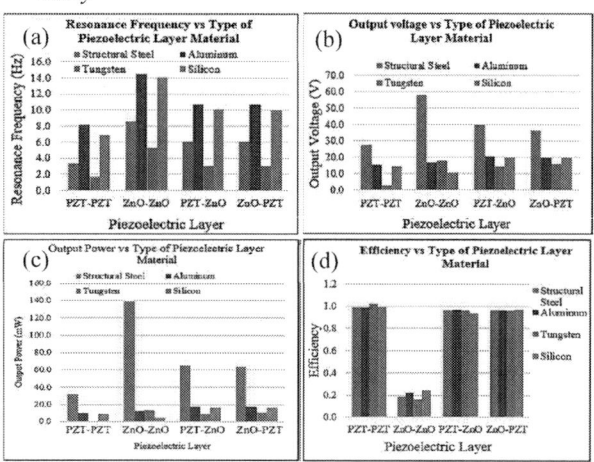

Fig. 3. (a) Resonant frequency (b) output voltage (c) output electrical power (d) efficiency of the PEH in response to the piezoelectric layer material and substrate.

Based on Fig. 3(a), ZT-PZT has the lowest f_n, but only generates low voltage and electrical power as compared to the other combinations. The efficiency of the PZT-PZT combination is the highest and is close to the value of 1.0 as shown in Fig. 3(d). As for ZnO-ZnO, its f_n is the highest compared to other combinations. The efficiency of the ZnO-ZnO is the lowest, approximately 0.3. For output voltage and power, ZnO-ZnO with structural steel and tungsten as

substrates produced the highest values, while for aluminium and silicon substrates, the output voltage and power produced was the second highest.

The hybrid combinations of PZT-ZnO and ZnO-PZT exhibit similar performance, with the second lowest natural frequency, f_n after PZT-PZT. Based on Fig 3(b) & (c), this hybrid combination achieves the second-highest output voltage and electrical power. However, due to material incompatibilities, a hybrid design for the bimorph cantilever is not feasible. As a proof of concept, a design with two beams was utilized and detailed in the following subsection.

In the substrate material correlation study, tungsten-based PEHs exhibit the lowest natural frequency f_n due to their high Young's Modulus and atomic mass. Conversely, aluminium, with the lowest atomic mass among substrates, has the highest f_n. Based on Equation (3), f_n is inversely proportional to the square root of the mass, m where an increase in m causes a decrease in f_n.

Even though the output voltage and power of PEH with structural steel substrates is the highest. However, the process to produce thin structural steel is difficult compared to other substrate materials because structural steel is a type of alloy. Thus, tungsten is a considered candidate because of its advantage to produce a low f_n PEH that matches the human heart wall's low vibration frequency while generating enough voltage to drive a pacemaker.

C. Split Cantilever Energy Harvester (SCEH) model simulation

Based on the data obtained above, a split cantilever energy harvester (SCEH) was simulated as a proof of concept. Each beam contains PZT and ZnO separately. The SCEH is shown in Fig. 4.

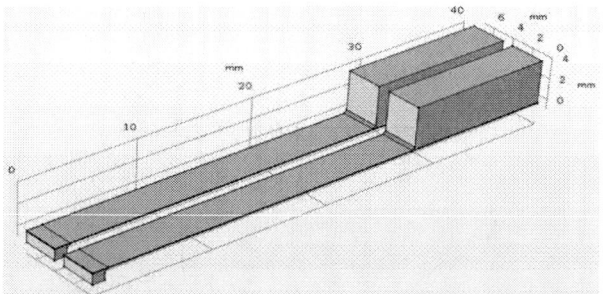

Fig. 4. 3D model of the Split Cantilever Energy Harvester.

The simulated SCEH model meets pacemaker requirements, with resonance frequencies at 3.1 Hz and 4.9 Hz—within the human heart wall vibrational frequency range (<50 Hz). The dual peaks in Fig 5a) and b) result from different materials used for each beam.

The output voltage generated by the SCEH is 36.21 V at the resonant frequency of 3.1 Hz and 38 V at the resonant frequency of 4.9 Hz. The efficiency curves depicted in Fig 5 (c) show a significant dip in efficiency at 4.9 Hz and this is due to destructive superposition interference during mechanical vibration of the cantilever beam. At resonance (4.9Hz), the cantilever's high vibration increases output voltage but also requires more mechanical power, resulting in lower efficiency. Beyond resonance, both output voltage and mechanical power drop, but efficiency increases due to the

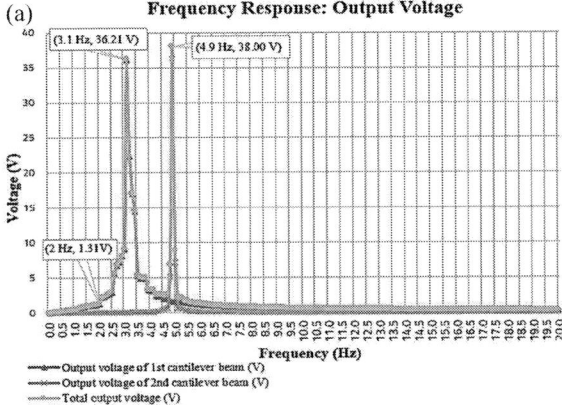

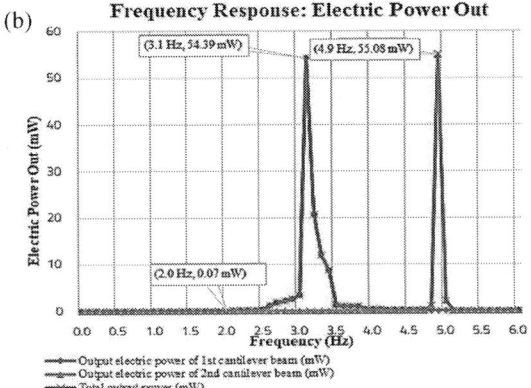

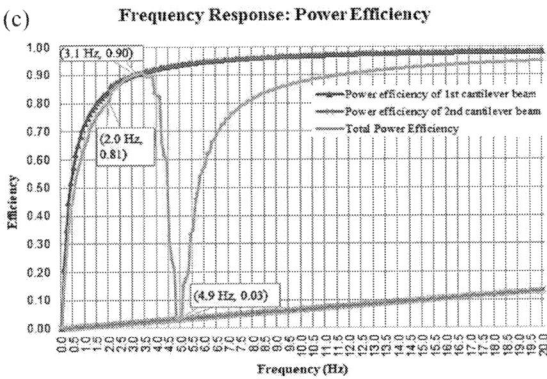

Fig. 5. a) resonant frequency b) output voltage c) electrical power of SCEH.

reduced input power needed. These simulations were conducted using an insertion body load with an acceleration of 1 g (9.8 m/s^2) on the entire PEH structure. A buck-boost converter can be used in the circuit to generate the voltage required by the pacemaker, either by producing an output voltage higher or lower than the output voltage of the PEH model. This output voltage is deemed to be sufficient to drive a standard pacemaker that normally operates in the 2.5 to 2.8 V voltage range.

SCEH reaches an efficiency of 90% at 3.1 Hz, indicating a high rate of stress-to-voltage conversion. However, this model does not work well at higher frequency components after 7 Hz as based on Fig. 5, the voltage and electric power generated

are nearly zero. Since SCEH is a combination of two single bimorph PEHs, if one of the PEH is damaged, it will affect the performance of the whole SCEH as these two PEHs are connected in series.

IV. CONCLUSION

In conclusion, a MEMS PEHs based on the SCEH structure was simulated as a proof of concept for pacemaker application with low resonant frequencies matching the human heart vibration frequency. Sufficient output voltage to power the pacemaker and small dimensions were also successfully developed in this work. SCEH has resonant frequencies at 3.1 Hz and 4.9 Hz, with output voltages of 36.21 V and 38 V. The requirements for powering the pacemaker have been achieved in terms of dimensions, resonant frequency, output voltage and power based on simulation analysis using COMSOL Multiphysics.

ACKNOWLEDGEMENT

This work was funded by Universiti Kebangsaan Malaysia (UKM) under the Geran Universiti Penyelidikan (GUP) scheme, grant code GUP-2022-006.

REFERENCES

[1] R. F. Padera and F. J. Schoen, *Cardiovascular Medical Devices*, Fourth Edi. Elsevier, 2020. doi: 10.1016/b978-0-12-816137-1.00067-2.

[2] S. Jay, M. Caballero, W. Quinn, J. Barrett, and M. Hill, "Characterization of piezoelectric device for implanted pacemaker energy harvesting," *J Phys Conf Ser*, vol. 757, p. 012038, Oct. 2016, doi: 10.1088/1742-6596/757/1/012038.

[3] A. Kumar, R. Kiran, S. Kumar, V. S. Chauhan, R. Kumar, and R. Vaish, "A Comparative Numerical Study on Piezoelectric Energy Harvester for Self-Powered Pacemaker Application," *Global Challenges*, vol. 2, no. 1, Jan. 2018, doi: 10.1002/gch2.201700084.

[4] A. Anand and S. Kundu, "Design of Mems Based Piezoelectric Energy Harvester for Pacemaker," in *2019 Devices for Integrated Circuit (DevIC)*, IEEE, Mar. 2019, pp. 465–469. doi: 10.1109/DEVIC.2019.8783311.

[5] A. Mostafa and L. Albasha, "Low Frequency Bimorph Cantilever Energy Harvester for Pacemaker Applications," in *2020 IEEE 5th Middle East and Africa Conference on Biomedical Engineering (MECBME)*, IEEE, Oct. 2020, pp. 1–4. doi: 10.1109/MECBME47393.2020.9265120.

[6] C. Dagdeviren *et al.*, "Conformal piezoelectric energy harvesting and storage from motions of the heart, lung, and diaphragm," *Proc Natl Acad Sci U S A*, vol. 111, no. 5, pp. 1927–1932, 2014, doi: 10.1073/pnas.1317233111.

[7] J. Nathan *et al.*, "Ultra-low Frequency PiezoMEMS Energy Harvester for a Leadless Pacemaker". IEEE, 2018. doi: 10.1007/978-3-319-67944-0_1.

[8] N. Jackson, O. Olszewski, C. O'Murchu, and A. Mathewson, "Powering a leadless pacemaker using a PiezoMEMS energy harvester," *Smart Sensors, Actuators, and MEMS VIII*, vol. 10246, p. 102460V, 2017, doi: 10.1117/12.2264437.

[9] A. Ashutosh and K. Sudip, Design of Mems Based Piezoelectric Energy Harvester for Pacemaker. 2019.

[10] S. Azimi *et al.*, "Self-powered cardiac pacemaker by piezoelectric polymer nanogenerator implant," *Nano Energy*, vol. 83, no. November 2020, p. 105781, 2021, doi: 10.1016/j.nanoen.2021.105781.

[11] Liu Chang, *Foundation of Mems* , 2nd ed. Prentice Hall, 2012.

[12] L. Dong *et al.*, "In vivo cardiac power generation enabled by an integrated helical piezoelectric pacemaker lead," *Nano Energy*, vol. 66, no. August, p. 104085, 2019, doi: 10.1016/j.nanoen.2019.104085.

[13] D. Wang *et al.*, "Fabrication and characterisation of substrate-free PZT thick films," *Ceram Int*, vol. 44, no. 12, pp. 14258–14263, 2018, doi: 10.1016/j.ceramint.2018.05.030.

Dielectrophoretic Rapid characterization of Antimicrobial resistance bacteria Escherichia coli

Arash Zulkaranain Ahmad Rozaini
Institute of Microengineering and Nanoelectronics (IMEN)
Universiti Kebangsaan Malaysia
43600 UKM Bangi
Selangor Malaysia
p145637@siswa.ukm.edu.my

Muhammad Akmal Suhaimi
Institute of Microengineering and Nanoelectronics (IMEN)
Universiti Kebangsaan Malaysia
43600 UKM Bangi
Selangor Malaysia
p137669@siswa.ukm.edu.my

Aminuddin Ahmad Kayani
Functional Materials & Microsystems Research Group
School of Engineering
RMIT University
GPO Box 2476 Melbourne
Victoria 3001 Australia
amin.kayani@rmit.edu.au

Abdullah Abdulhameed
Center Com. Systems & Sensing
King Fahd University
of Petroleum & Minerals,
Dhahran 31261
Saudi Arabia
dr.a.abdulhameed@gmail.com

Burhanuddin Yeop Majlis
Institute of Microengineering and Nanoelectronics (IMEN)
Universiti Kebangsaan Malaysia
43600 UKM Bangi
Selangor Malaysia
burhan@ukm.edu.my

Muhamad Ramdzan Buyong
Institute of Microengineering and Nanoelectronics (IMEN)
Universiti Kebangsaan Malaysia
43600 UKM Bangi
Selangor Malaysia
muhdramdzan@ukm.edu.my

Noraziah Mohamad Zin
Center for Diagnostic, Therapeutic and Investigative Studies Faculty of Health Sciences Universiti Kebangsaan Malaysia Jalan Raja Muda Abdul Aziz,
50300 Kuala Lumpur, Malaysia.
noraziah.zin@ukm.edu.my

Wan Hanna Melini Wan Mohtar
Faculty of Engineering and Built Environment, Universiti Kebangsaan Malaysia, 43600 UKM Bangi, Selangor Darul Ehsan, Malaysia
hanna@ukm.edu.my

Abstract— Rapid characterization antimicrobial resistance bacteria are crucial for critical solving of Antimicrobial resistance (AMR) increases around the world. Antibiotic resistance bacteria contribute to the numerous infections all around the world which lead to death over 4.95 million. Escherichia Coli is part of multidrug resistant bacteria and in this study, we present experimental of susceptible and resistant strain of Escherichia Coli on one single frequency point-based MyDEP simulation which is 0.47 MHz called crossover frequency to differentiate the dielectrophoresis (DEP) response between susceptible and resistant strain when exposed to it experimentally. This experimental finding show the reaction of susceptible and resistant type of E. coli using crossover frequency are different as both bacteria exhibit different polarization factor due to minimal different of physical properties. These support the rapid bacterial cell characterization to be an alternative aid in faster diagnosis of bacterial infections and benefit the clinical decision-making process for antibiotic treatment, addressing the critical issue of AMR.

Keywords— *Dielectrophoresis, Antimicrobial resistance, Escherichia coli, Characterization*

I. INTRODUCTION

Bacteria is one of major researched field since its discovered and the recent interest of bacteria and antimicrobial resistance (AMR) has shown alarming alert towards human's health. The major alarm of infectious disease caused by these AMR to human has been increasing due to the evolution of bacteria that rapidly changes through time especially on its resistance towards antibiotic. The growth of bacteria within short amount period of time changes with physical properties may causes various clinical manifestations with different antibiotic resistance. Infections are indeed challenging to be treated hence with the time-consuming, relatively expensive and increase of waste in terms of antibiotic misused on the gold standard technique which is microbial culture. This urges an innovative, rapid, accurate, and portable diagnostic tools characterize and manipulate multi-drug resistant strains based on its resistance by its physical properties. Applications of research on bacteria are essential for medicine, the pharmaceutical industry, biotechnology, bioremediation, and alternative energy production [1]. Developing these applications will utilize on controlling and treating the infectious disease caused by bacteria and will support on demand integration of many scientific disciplines.

II. DIELECTROPHORESIS (DEP) THEORY

A. DEP

Dielectrophoresis (DEP) is a kinetic technique that has been utilized to fulfil the requirement of real time detection and rapid characterization for biological particles especially bacteria. DEP is described as a motion of particles when they are exposed to non-uniform electric field based on three factors which are particles, medium and frequency input. As a label free technique, with non-additive chemical added for standard operating procedure, DEP had been explored and researched for its importance and advantages for real time detection and rapid characterization[1], [2]. The DEP response of a particle is determined by its intrinsic dielectric property which consist of conductivity and permittivity and physical properties of particle which are size, shape, thickness and arrangement that implies the characterization and detection of bacteria. DEP is the motion of particles in a non-uniform electric field that utilized for the particle's polarization[10,11]. DEP polarization is determined by the physical properties of the particles and the medium, such as permittivity and conductivity.

$$F_{DEP} = 2\pi r_{ext}^3 \varepsilon_0 \varepsilon_m \, Re[f_{CM}]\nabla E^2 \qquad (1)$$

Consequently, the connection of dynamic dielectric properties of particles, medium, and electric field are abbreviated using the Clausius Mossoti- Factor Re [FCM] in DEP force formulation as in Equation (1), where is the medium permittivity ε is the particle radius, and is the real part CMF of the particle, and defines the gradient of the external field magnitude square. is the Clausius-Mossotti factor of the particle in a medium and for a spherical particle. is the permittivity for vacuum, $8.854 \times 10-12$ F/m. CMF of particle are based of dynamic dielectric properties of the particle with the surrounding medium. CMF can have a value between 1 and -0.5 which consist of pDEP, nDEP and crossover frequency (fxo). CMF of positive values indicated for positive DEP (pDEP), and the particles attracted to higher electric field regions. The CMF of negative values showed negative DEP (nDEP), and the particle repulsed from higher electric field regions. At the value of 0, the pDEP and nDEP forces vanish and showed this static condition is known as the crossover frequency (fxo). The manipulation of these response can be applied for manipulation, separation and characterization of particles in medium especially biological particles.

B. DEP work

First stage of this study was carried out in prediction through simulation by the bacteria model parameter which is E coli. The simulation is achieved by MyDEP tool. MyDEP is a simulation software tool to plot the polarization factor of bacteria and medium against frequency using the dielectric particle response and medium input parameters to AC electric field by calculate the effective polarization and permittivity. Then, one single point is chosen which is crossover frequency , a static movement for particles when exposed to DEP on certain specific frequency for main experiment. Since every particle cover different physical properties especially bacteria, bacteria exhibit different crossover frequency that can be a baseline for detection on a single frequency point. Characterization of bacteria cells based on DEP response on microelectrode under a microscope were observed when exposed to DEP. Thus conclude the difference of DEP response for resistant and susceptible E coli on one single point crossover frequency. The physical properties of same species bacteria but different resistance can be proven and demonstrated using DEP experiment based on DEP response due to the minimal difference of physical properties on susceptible E coli and resistant E coli.

C. Bacteria model

The polarization frequency of AMR are based on the distinguishing features and differences in their properties especially between susceptible and resistant bacteria which includes thickness, size, cell wall, and charges, causing a

range of frequency responses. At low frequencies, the field is screened by a cell wall and membrane to cause DEP behaviour which cover outer region of bacteria. In contrast, at high frequencies, the electric field gradually leaks through the insulator shell, and cover inner region of the microbial cell's which is cytoplasmic to polarized. Negative gram bacteria due have low conductivity of cell wall due to their thinner cell wall (8-10 nm) compared to gram positive bacteria. The inclusion of the cell wall of AMR within the multiple shell model is essential for differential of bacteria based on its properties and explain the significant rise in conductivity of microbial suspensions within low-conductivity media.

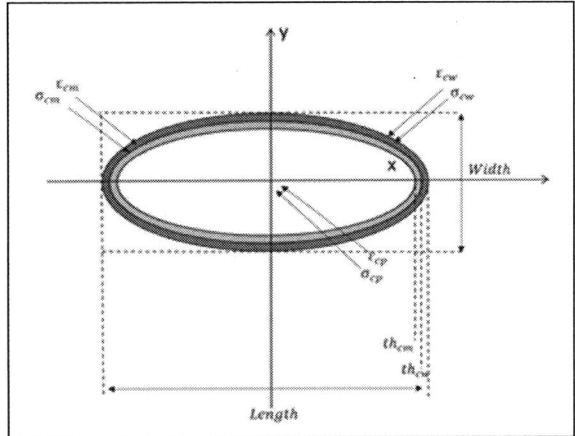

Fig. 1. Model simulation Gram Negative bacteria: Ellipsoid

The Re [f_{CM}] is the real part of CMF and can be formulated as in (1). For the double-shelled model such as bacteria, the first effective complex permittivity, ε*$_{e}ff_1$ and second effective complex permittivity ε*$_{e}ff_2$ needs to be calculated. The ε*$_{e}ff_2$ can be calculated by finding ε*$_{e}ff_1$, which depends on the permittivity of the bacteria core inside the cytoplasmic region and the permittivity of the cell membrane as described by Equation (2) and Equation (3) [3].

$$\varepsilon_{eff1}^* = \varepsilon_{cm}^* \frac{\left(\dfrac{r_{ext}}{r_{ext} - th_{cm}}\right)^3 + 2\dfrac{\varepsilon_{cp}^* - \varepsilon_{cm}^*}{\varepsilon_{cp}^* - 2\varepsilon_{cm}^*}}{\left(\dfrac{r_{ext}}{r_{ext} - th_{cm}}\right)^3 - \dfrac{\varepsilon_{cp}^* - \varepsilon_{cm}^*}{\varepsilon_{cp}^* + 2\varepsilon_{cm}^*}} \qquad (2)$$

$$\varepsilon_{eff2}^* = \varepsilon_{cw}^* \frac{\left(\dfrac{r_{ext}}{r_{ext} - th_{cw}}\right)^3 + 2\dfrac{\varepsilon_{eff1}^* - \varepsilon_{cw}^*}{\varepsilon_{eff1}^* - 2\varepsilon_{cw}^*}}{\left(\dfrac{r_{ext}}{r_{ext} - th_{cw}}\right)^3 - \dfrac{\varepsilon_{eff1}^* - \varepsilon_{cw}^*}{\varepsilon_{eff1}^* + 2\varepsilon_{cw}^*}} \qquad (3)$$

III. METHODOLOGY

A. E coli properties value

The bacteria's equivalent relative permittivity and conductivity and the frequency range were tabulated in table 1 and included in the MyDEP [3]–[5]. MyDEP was used to solve the FDEP exerted on the bacteria placed in the distilled water at different frequencies.

TABLE I. E COLI PROPERTIES

E coli properties	
Specifications	**Value**
Diameter of bacteria	3.35 μm
Width of bacteria	0.82 μm
Shell thickness Th_{cw}	15 nm
Cell wall conductivity σ_{cw}	0.5 S/m
Cell wall relative permittivity ε_{cw}	60
Cell membrane thickness Th_{cm}	8 nm
Cell membrane conductivity σ_{cm}	1.5E-6
Cell membrane relative permittivity ε_{cm}	10
Cytoplasm conductivity σ_{cp}	0.07
Cytoplasm relative permittivity ε_{cp}	70

B. Experimental setup

The electrodes are placed under a microscope OLYMPUS-BX53M as shown in Figure 2 and directly connected to an AC function generator WaveStation Function 2022 25 MHz /Arbitrary Waveform Generator through a prober. The function generator is set to supply an AC signal of 6 V peak to peak. The bacterial manipulation is monitored from the microscope by Olympus cell Sens standard software from top view towards the microelectrode [7]. The region of interest (ROI) of microelectrode covered with polyamide tape and placed in between the 20 μm gap and the bacteria sample will be dropped on polyamide tape and covered with a glass slide a static droplet (zero initial velocity) for movement of particles when applied with DEP.

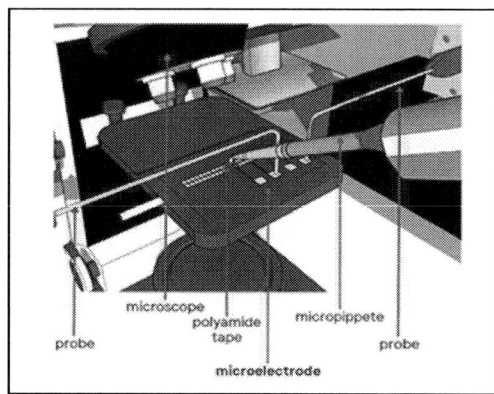

Fig. 2. Model DEP experimental set up

IV. RESULT AND DISCUSSION

A. DEP simulation

The simulation result of the CMF polarization is shown Figure 3 of E coli shows pDEP, nDEP and crossover frequency on distilled water (σ= 0.5 μS/m). The properties of E coli interpreted into electrical properties thus plot the CMF polarization. The polarization of cell wall E coli aureus begins with nDEP due to the low thickness and conductivity of cell wall. At frequencies beyond the nDEP level to gradually increase, become pDEP. The first fxo shows 0.47 MHz for E

coli when the blue line reach zero which static movement for bacteria. For second fxo, from pDEP, at higher frequencies, the polarization falls to fxo at 140 MHz and nDEP.

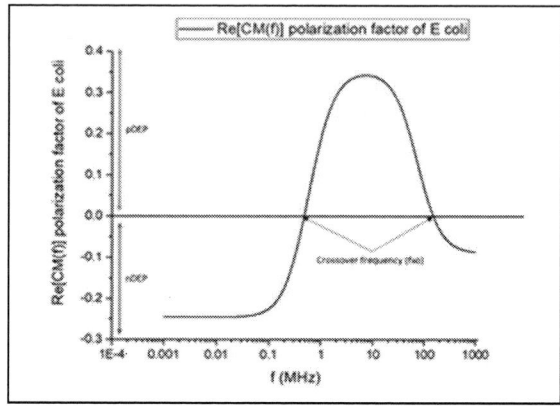

Fig. 3. MyDEP simulation: Model polarization factor efficiency of CMF E coli (blue line), crossover frequency at 0.47 MHz and 140 MHz for E coli

B. DEP experimental

The experimental DEP is tested to confirm DEP characterization and manipulation between two different strain of E coli which is susceptible and resistant strain. The crossover frequency in MyDEP which is 0.47 MHz and 140 MHz for isolation of E coli which could lead to deeper levels of analysis on specific factors. In this experimental, 0.47 MHz is used to differentiate the movement of both strain of bacteria. The result can be utilized as an indication for crossover frequency characterization and isolation for E coli thus proving differential physical properties based on different DEP response[7]. On this response, no bacteria movement were detected thus induced crossover frequency response. While detected bacteria movement either pDEP or nDEP show no crossover frequency.

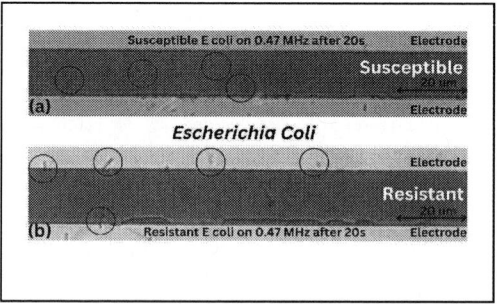

Fig. 4. (a) suscpetible E coli DEP response on 0.47 MHz : crossover frequency response, no bacteria movement after 20 seconfds exposed to DEP electroc field (b) resistant E coli on 0.47 MHz, pDEP response, bacteria attracted towards high electric field which is edge of electrode

The results of Figure 4 are demonstrated to be highly predictive for whole DEP characterization, the crossover frequency is potentially applicable to differentiate two different strain as susceptible strain show no movement which induced as crossover frequency [8]. Resistant strain of E coli show pDEP movement. The experimental result of microscopic image on Figure 4 shows constant movement of particle which indicate fxo on 0.47 MHz for susceptible E coli strain and resistant E coli strain show pDEP movement. These findings show by minimal different of physical properties and dielectric properties between similar species with different antibiotic profile could exhibit difference response on single crossover frequency point. These method which is rapid, label free and single platform no only enhance the quality of conventional method which require long processing time and expertise , but also giving the right detection and treatment for clinical and solution for AMR.

V. CONCLUSION

DEP technology is used to characterize biological particles in a medium based on their polarization factor. As the medium properties used are constant, the DEP response of the bacteria with a different antibiotic profile can be observed. DEP platform is rapid and selective for microbial characterization because of the different physical properties of bacteria in terms of shapes, sizes, wall thickness, and charges on the bacteria. The differences in the DEP responses of E coli susceptible and resistant are demonstrated through simulation, and experimentally. Using one single point of crossover frequency which is 0.47 MHz, two different response were observed for two different strain of E coli. In conclusion, the rapid characterization of bacteria using DEP differentiate bacteria strain due to different minimal physical properties Rapid assessment and characterization of bacteria are necessary for diagnosing and treating pathogenic bacteria diseases.

ACKNOWLEDGMENT

The author acknowledges the Fundamental Research Grant Scheme (FRGS), grant number FRGS/1/2022/TK07/UKM/02/21 grant funded by Minister of Higher Education (MOHE). This work was performed in part at the Micro Nano Research Facility at RMIT University in the Victorian Node of the Australian National Fabrication Facility (ANFF). We acknowledge project funding from the Australian Research Council through IH210100040. The author would like to thank Professor Sharath Sriram, Professor Madhu Bhaskaran, and Professor Sumeet Walia, who is leading the Functional Materials and Microsystems Research Group at RMIT University for allocating resources and technical expertise in micro and nanofabrication activities for this project.

REFERENCES

[1] M. K. Abdul Rahim, N. M. A. Jamaludin, J. Santhanam, A. A. Hamzah, and M. R. Buyong, "Rapid ESKAPE pathogens detection method using tapered dielectrophoresis electrodes via crossover frequency analysis," *Sains Malaysiana*, vol. 49, no. 12, pp. 2913–2925, 2020, doi: 10.17576/jsm-2020-4912-04.

[2] M. Khairulanwar, A. Rahim, N. Mas, A. Jamaludin, J. Santhanam, and A. Azlan, "Determination of Dielectrophoretic Unique Crossover Frequency by Velocity of Enterobacter Aerogenes Trajectory," vol. 13, pp. 33–44, 2020.

[3] J. Cottet, O. Fabregue, C. Berger, F. Buret, P. Renaud, and M. Frénéa-Robin, "MyDEP: A New Computational Tool for Dielectric Modeling of Particles and Cells," *Biophys. J.*, vol. 116, no. 1, pp. 12–18, Jan. 2019, doi: 10.1016/j.bpj.2018.11.021.

[4] R. Deivasigamani, N. N. M. Maidin, M. F. M. R. Wee, M. A. Mohamed, and M. R. Buyong, "Dielectrophoresis prototypic polystyrene particle synchronization toward alive keratinocyte cells for rapid chronic wound healing," *Sensors*, vol. 21, no. 9, 2021, doi: 10.3390/s21093007.

[5] R. Deivasigamani *et al.*, "A dielectrophoresis proof of concept of polystyrene particles and in-vitro human epidermal keratinocytes migration for wound rejuvenation," *J. Appl. Polym. Sci.*, vol. 139, no. 44, p. e53096, Nov. 2022, doi: 10.1002/APP.53096.

[6] A. Sanchis *et al.*, "Dielectric characterization of bacterial cells using dielectrophoresis," *Bioelectromagnetics*, vol. 28, no. 5, pp. 393–401, 2007, doi: 10.1002/bem.20317.

[7] W. H. M. Wan Mohtar *et al.*, "Rapid detection of ESKAPE and enteric bacteria using tapered dielectrophoresis and their presence in urban water cycle," *Process Saf. Environ. Prot.*, vol. 177, pp. 427–435, 2023, doi: 10.1016/j.psep.2023.06.088.

[8] A. Zulkarnain and A. Rozaini, "Dielectrophoresis microbial characterization and isolation of Staphylococcus aureus based on optimum crossover frequency," no. August 2022, pp. 1–14, 2023, doi: 10.1002/elps.202200276.

Electronic Properties of AB-Stacked Bilayer Graphene Nanoribbons with Zigzag and Armchair Orientations

Yuki Wong
Faculty of Electrical Engineering
Universiti Teknologi Malaysia
Skudai, Johor, Malaysia
yukiwong@graduate.utm.my

Nurul Ezaila Alias
Faculty of Electrical Engineering
Universiti Teknologi Malaysia
Skudai, Johor, Malaysia
ezaila@utm.my

Cheng Siong Lim
Faculty of Electrical Engineering
Universiti Teknologi Malaysia
Skudai, Johor, Malaysia
lcsiong@utm.my

Choon Min Cheong
School of Engineering
INTI International College Penang
11900 Penang, Malaysia
choonmin.cheong@newinti.edu.my

Michael Loong Peng Tan
Faculty of Electrical Engineering
Universiti Teknologi Malaysia
Skudai, Johor, Malaysia
michael@utm.my

Abstract—The electronic properties of AB-stacked bilayer graphene nanoribbons (AB-BGNRs) with zigzag and armchair orientations are investigated in this study using a nearest-neighbour tight-binding (NNTB) approach combined with the non-equilibrium Green's function (NEGF) method via MATLAB. To address the lack of clear and simple methodologies for solving energy-eigenvalue problems in AB-BGNRs, the final model is derived using justifiable simplifying assumptions, such as the application of basis functions, discretization of the Hamiltonian operator, and utilization of the plane wave approximation, all grounded in many-body theories and the modified Hartree simplification theory. The Hamiltonian matrices for both orientations are constructed with open boundary conditions to mimic real-world structures. Results show that zigzag-oriented AB-BGNRs exhibit metallic characteristics due to the absence of a bandgap, while armchair-oriented AB-BGNRs alternate between metallic and semiconducting properties. DOS analysis reveals an increase in degenerate low-energy states with increasing width for zigzag orientation and a lack of quantum states in the mid-band region for semiconducting armchair variants. This research contributes to our understanding of AB-BGNRs' electronic characteristics and lays a foundation for future advancements in graphene-based electronic devices.

Keywords—Bilayer Graphene, NNTB, NEGF, Dispersion Relation, DOS

I. INTRODUCTION

Over the last half-century, scientists have strived to unearth new materials capable of surpassing the constraints imposed by size limitations [1-3] to enable the efficient fabrication of semiconductor nanodevices [4, 5], satisfying the evolving needs of rapid development in the modern electronic industry. Recently, graphene has garnered significant attention as a promising candidate for nanoelectronics, stimulating extensive investigation owing to its distinctive physical characteristics [6-8] and diverse range of applications [9-11]. Graphene, derived from the exfoliation of layers from bulk multilayered graphite, can be regarded as an infinite two-dimensional lattice. This material demonstrates versatile fabrication capabilities, with bilayer graphene standing out as the most straightforward option among alternative methods for fabricating few-layer carbon systems [12]. This study

specifically delves into the unique characteristics of AB-BGNRs. Since edge shape and orientation dominantly affect electronic properties, this study investigates armchair and zigzag edges, which are 30° apart in orientation. Fig. 1 shows the crystal structure of AB-BGNRs.

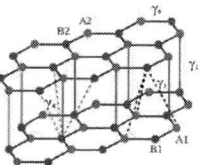

Fig. 1. Crystal lattice of AB-stacked bilayer graphene nanoribbon [13].

Since there is a scarcity of research that presents concise and easily understandable insights, this paper aims to share simple and replicable methods for solving energy-eigenvalue problems of AB-BGNRs by employing justifiable simplifying assumptions. The primary approach involves utilising the NNTB approximation, which is well-suited for modelling carbon allotropes such as AB-BGNRs to generate its resulting Hamiltonian operator, containing comprehensive information regarding the physical properties of the structure it describes. The electronic properties section defines and justifies the transport models used on the Hamiltonian operator to derive its energy spectrum, revealing electronic characteristics through dispersion relation and DOS visualisation. The modelling and simulation section implements these theories in MATLAB to analyse AB-BGNRs' electronic properties.

II. METHODOLOGY

A. Transport Models

Conventionally, semiclassical transport models relying on Boltzmann's Transport Equation (BTE) have been utilised to determine the electrical characteristics of devices, calculating an average value for the bulk system. However, quantum transport theories can provide a more accurate representation of the device, particularly at very small device lengths. Quantum transport theories, rooted in the Schrödinger equation [14] as shown in (1), are characterized by their ability to fully preserve the wave-like properties of electrons,

979-8-3503-7832-0/24 $31.00 © 2024 IEEE

attributed to the inherent wave function within the equation itself. This contrasts the semiclassical approach, where electrons are treated as classical particles subjected to classical Newtonian mechanics.

$$E\{\psi\} = [H]\{\psi\} \tag{1}$$

The solutions of the Schrödinger equation, obtained through either first-principles or semi-empirical methods, form the foundation for most quantum transport models, including the NEGF [15], which will be employed in this paper. The Hamiltonian operator, denoted as H in the Schrödinger equation, provides an accurate description of the physical structure of the device [16], indicating the positions of constituent particles and their interactions. Once defined, this operator, when combined with an appropriate wave function, ψ, enables the deduction of the available energy spectrums [17] for a given system, facilitating the determination of the electronic properties of the device.

B. Simplifying Assumptions for Device Modelling

Numerical analysis of Schrödinger's equation for all electrons within a device is laborious and feasible only for limited particles. Datta proposed the initial simplifying assumption for modelling devices, utilising only basis functions [18] representing electrons directly involved in transport. For instance, Fig. 2 shows a 1D system comprising three hydrogen atoms, where the hydrogen 1s electron orbital serves as the basis function.

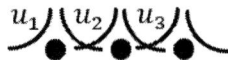

Fig. 2. 3 Hydrogen system, side-by-side with 1s wavefunction.

Extending this concept to more complex systems, such as a 2D graphene sheet, where only the π-electron of the 2pz orbital and its respective wave function are considered as the basis function. The overlap of the wavefunction of 1s orbitals solely with its directly adjacent counterpart, as depicted in Fig. 2, signifies the second significant simplifying assumption, which states that the wavefunctions are solely influenced by their nearest neighbour. Consequently, the NNTB model will be utilised throughout this paper.

Followed by the third simplifying assumption, the device Hamiltonian operator and its respective basis functions can be discretised and converted into matrix equations. This enables the solution for the device energy spectrum through computational methods utilising powerful matrix tools, justified by the discrete atomic structure of the lattice at the atomic level. With all these assumptions, the Hamiltonian matrix for the three hydrogen atom system in Fig. 2 can be derived as shown in (2), where E represents the energy eigenvalue of the system, ψ_1, ψ_2, and ψ_3 represent the wavefunctions corresponding to the 1s orbitals of the three hydrogen atoms, and u_1, u_2, and u_3 represent the 1s orbital potential function.

$$E \begin{Bmatrix} \psi_1 \\ \psi_2 \\ \psi_3 \end{Bmatrix} = \begin{bmatrix} u_{11} & u_{12} & 0 \\ u_{21} & u_{22} & u_{23} \\ 0 & u_{32} & u_{33} \end{bmatrix} \begin{Bmatrix} \psi_1 \\ \psi_2 \\ \psi_3 \end{Bmatrix} \tag{2}$$

In practical device modelling, terms on the diagonal, u_{11}, u_{22}, and u_{33}, denote self-interacting energies, effectively zero in this particular system as the 1s orbital electron lacks interaction with itself. However, subsequent sections of this paper will demonstrate how to adjust the self-interacting energy to extend this 1D model for higher-dimensional problems. Interacting energy functions between nearest neighbour wavefunctions, denoted as u_{12}, u_{21}, u_{23}, and u_{32}, are identical as they involve the same orbital electron of similar atoms. Conversely, u_{13} and u_{31} do not exist due to the second simplifying assumption, where the first and third hydrogen atoms do not interact with each other.

III. MODELLING AND SIMULATION

The most accurate solution to the Schrödinger equation is obtained through first principles but is only feasible numerically for a few particles [19]. Discretizing the equation into matrix form allows faster computational analysis, yielding energy eigenvalues. The basis function employed in graphene modelling consists of the wavefunction of the 2p$_z$ electron, which abstains from engaging in σ-bond formation between carbon atoms. Instead, it creates a freely moving delocalized π-orbital essential for charge transport and serves as the sole electron of interest for modelling purposes. In AB-BGNRs, each π-orbital interacts with three nearest-neighbour π-orbitals, which may vary in GNRs depending on unit cell definition. To ensure accuracy, the AB-BGNRs Hamiltonian will be formulated with open boundary conditions, while all dangling bonds at the edge carbon atoms are assumed to be passivated with hydrogen atoms [20], thereby not contributing to the electronic states near the Fermi level.

A. Generation of Alpha Matrix for Zigzag and Armchair AB-BGNRs

The smallest defined structure is referred to as a block, with a width of 1 unit cell, comprising 8 lattice points, including 4 from the top layer and an additional 4 from the bottom layer. Fig. 3(a) and (b) show the alpha unit cells with one block per unit cell width for zigzag and armchair AB-BGNRs, respectively.

(a)   (b)

Fig. 3. Alpha unit cell structures of (a) zigzag and (b) armchair AB-BGNRs with one block per unit cell width.

For the zigzag edge, the alpha matrix representing the unit cell in Fig. 3(a) is shown in (3) while the resulting alpha matrix for the armchair edge that represents the unit cell shown in Fig. 3(b) is depicted as (4). Parameter l_1, l_2, l_3, and l_4 denoting lower layer atoms, and u_1, u_2, u_3, and u_4 denoting upper layer atoms. Hopping integrals γ_0, γ_1, γ_3, and γ_4 describe their interactions. In MATLAB, the *diag* command is used to fill up both off-diagonals in order to construct this matrix.

$$\alpha = \begin{array}{c} \\ l_1 \\ l_2 \\ l_3 \\ l_4 \\ u_1 \\ u_2 \\ u_3 \\ u_4 \end{array} \begin{array}{c} \begin{array}{cccccccc} l_1 & l_2 & l_3 & l_4 & u_1 & u_2 & u_3 & u_4 \end{array} \\ \begin{bmatrix} 0 & \gamma_0 & 0 & 0 & \gamma_4 & \gamma_3 & 0 & 0 \\ \gamma_0 & 0 & \gamma_0 & 0 & \gamma_1 & \gamma_4 & 0 & 0 \\ 0 & \gamma_0 & 0 & \gamma_0 & \gamma_3 & \gamma_4 & 0 & 0 \\ 0 & 0 & \gamma_0 & 0 & 0 & \gamma_4 & \gamma_1 & \gamma_4 \\ \gamma_4 & \gamma_1 & \gamma_4 & 0 & 0 & \gamma_0 & 0 & 0 \\ \gamma_3 & \gamma_4 & \gamma_3 & \gamma_4 & \gamma_0 & 0 & \gamma_0 & 0 \\ 0 & 0 & \gamma_3 & \gamma_1 & 0 & \gamma_0 & 0 & \gamma_0 \\ 0 & 0 & \gamma_3 & \gamma_4 & 0 & 0 & \gamma_0 & 0 \end{bmatrix} \end{array} \tag{3}$$

979-8-3503-7832-0/24 $31.00 © 2024 IEEE   129

$$\alpha = \begin{array}{c} \\ l1 \\ l2 \\ l3 \\ l4 \\ u1 \\ u2 \\ u3 \\ u4 \end{array} \begin{bmatrix} \begin{array}{cccccccc} l1 & l2 & l3 & l4 & u1 & u2 & u3 & u4 \\ 0 & \gamma_0 & 0 & 0 & \gamma_4 & \gamma_3 & 0 & 0 \\ \gamma_0 & 0 & \gamma_0 & 0 & \gamma_1 & \gamma_4 & 0 & 0 \\ 0 & \gamma_0 & 0 & \gamma_0 & \gamma_0 & \gamma_3 & \gamma_4 & \gamma_3 \\ 0 & 0 & \gamma_0 & 0 & 0 & \gamma_4 & \gamma_1 & \gamma_4 \\ \gamma_4 & \gamma_1 & \gamma_4 & 0 & 0 & \gamma_0 & 0 & 0 \\ \gamma_3 & \gamma_4 & \gamma_3 & \gamma_4 & \gamma_0 & 0 & \gamma_0 & 0 \\ 0 & 0 & \gamma_4 & \gamma_1 & 0 & \gamma_0 & 0 & \gamma_0 \\ 0 & 0 & \gamma_3 & \gamma_4 & 0 & 0 & \gamma_0 & 0 \end{array} \end{bmatrix} \quad (4)$$

Based on (3) and (4), it is observed that the alpha matrix of the armchair edge is identical to that of the zigzag edge. This similarity arises from the consistent wavefunction couplings within the unit cells. The only distinction between the armchair and zigzag configurations with a width of one unit cell is their rotational orientation.

B. Generation of Beta Matrix for Zigzag and Armchair AB-BGNRs

The beta matrix describes the interactions between unit cells, identifying the interacting sites for π-orbitals between unit cells. Fig. 4(a) and (b) show the 1-block unit cells O, P, and Q lattice point π-orbital interactions for the zigzag and armchair edges of AB-BGNRs, respectively.

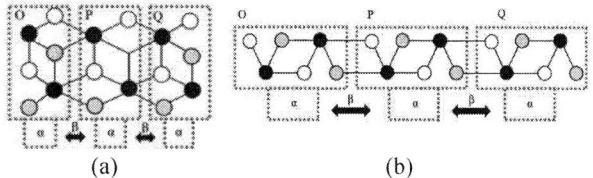

(a)   (b)

Fig. 4.   Interactions between lattice points O, P, and Q within a one-block unit cell for (a) zigzag and (b) armchair edges of AB-BGNRs.

Equations (5) and (6) illustrate the beta matrices for zigzag and armchair AB-BGNRs, respectively, showing the interaction of π-orbitals between unit cells P and O. The interactions of π-orbitals between unit cells P-Q and P-O are transposed of each other in the matrix representation.

$$\beta = \begin{array}{c} \\ pl1 \\ pl2 \\ pl3 \\ pl4 \\ pu1 \\ pu2 \\ pu3 \\ pu4 \end{array} \begin{bmatrix} \begin{array}{cccccccc} ol1 & ol2 & ol3 & ol4 & ou1 & ou2 & ou3 & ou4 \\ 0 & 0 & 0 & 0 & 0 & 0 & 0 & 0 \\ \gamma_0 & 0 & 0 & 0 & 0 & \gamma_4 & 0 & 0 \\ 0 & 0 & 0 & \gamma_0 & 0 & \gamma_3 & \gamma_4 & 0 \\ 0 & 0 & 0 & 0 & 0 & 0 & 0 & 0 \\ \gamma_4 & 0 & 0 & 0 & 0 & \gamma_0 & 0 & 0 \\ 0 & 0 & 0 & 0 & 0 & 0 & 0 & 0 \\ 0 & 0 & 0 & 0 & 0 & 0 & 0 & 0 \\ 0 & 0 & 0 & \gamma_4 & 0 & 0 & \gamma_0 & 0 \end{array} \end{bmatrix} e^{i\vec{k}(\overrightarrow{r_m}-\overrightarrow{r_n})} \quad (5)$$

$$\beta = \begin{array}{c} \\ pl1 \\ pl2 \\ pl3 \\ pl4 \\ pu1 \\ pu2 \\ pu3 \\ pu4 \end{array} \begin{bmatrix} \begin{array}{cccccccc} ol1 & ol2 & ol3 & ol4 & ou1 & ou2 & ou3 & ou4 \\ 0 & 0 & 0 & \gamma_0 & 0 & 0 & \gamma_4 & \gamma_3 \\ 0 & 0 & 0 & 0 & 0 & 0 & 0 & \gamma_4 \\ 0 & 0 & 0 & 0 & 0 & 0 & 0 & 0 \\ 0 & 0 & 0 & 0 & 0 & 0 & 0 & 0 \\ 0 & 0 & 0 & 0 & 0 & 0 & \gamma_0 & 0 \\ 0 & 0 & 0 & 0 & 0 & 0 & 0 & 0 \\ 0 & 0 & 0 & 0 & 0 & 0 & 0 & 0 \\ 0 & 0 & 0 & 0 & 0 & 0 & 0 & 0 \end{array} \end{bmatrix} e^{i\vec{k}(\overrightarrow{r_m}-\overrightarrow{r_n})} \quad (6)$$

C. Generation of Hamiltonian AB-BGNRs

Upon successful definition of both the alpha matrix and the beta matrix, the results can be merged to generate the final Hamiltonian matrix as shown in (7).

$$H = \begin{array}{c} \\ O \\ P \\ Q \end{array} \begin{bmatrix} \begin{array}{ccc} O & P & Q \\ \alpha & \beta & 0 \\ \beta' & \alpha & \beta \\ 0 & \beta' & \alpha \end{array} \end{bmatrix} \quad (7)$$

In MATLAB, the "kron" command is utilised to replace elements within a matrix with another matrix, simultaneously expanding the original size of the matrix.

IV. RESULTS AND DISCUSSION

A. Dispersion Relation

Fig. 5 shows the dispersion relations obtained for the zigzag orientation of AB-BGNRs.

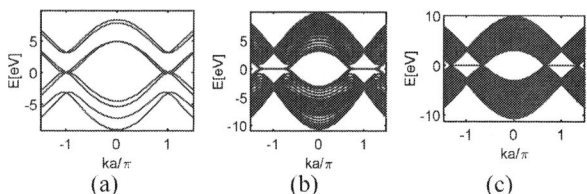

(a)   (b)   (c)

Fig. 5.   Zigzag edge dispersion relation for increasing widths: (a) 1, (b) 8, and (c) 20 blocks with a fixed length of 3 unit cells.

Based on Fig. 5, it is observed that the bandgap remained consistent at 0 eV across all widths of AB-BGNRs, indicating that all zigzag AB-BGNRs are metallic regardless of their width. Comparing the dispersion relations between this study and Ref. [21] as shown in Fig. 6, similarities in overall shapes and symmetry are noted, along with deviations in specific subbands.

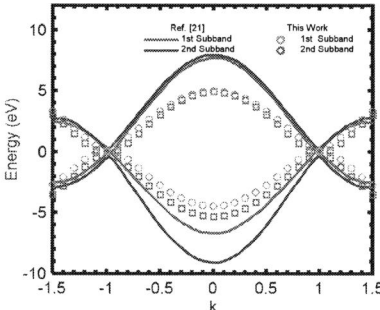

Fig. 6.   Comparison of zigzag edge dispersion relation between this study (dots) and previous work by Ref. [21] (solid lines).

These differences are attributed to varying tight-binding parameters and enhanced methodologies in the current study, which allow for more comprehensive parameter incorporation and precise energy level calculations. In particular, a detailed description of interactions within and between graphene layers is provided by this study, utilising four parameters compared to three in Ref. [21], thereby validating fundamental electronic characteristics and highlighting improvements over the previous work.

On the other hand, Fig. 7 shows the dispersion relations for the armchair orientation.

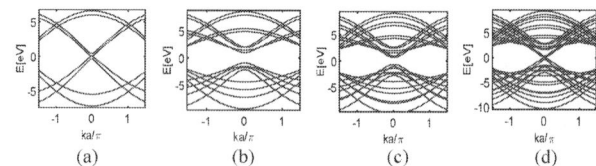

(a)   (b)   (c)   (d)

Fig. 7.   Armchair edge dispersion relation for increasing widths: (a) 1, (b) 2, (c) 3, and (d) 4 blocks with a fixed length of 3 unit cells.

Based on Fig. 7, the dispersion relation became denser and the bandgaps appeared closer when the width was increased. An oscillating bandgap was revealed, alternating between 0 eV (metallic) and non-zero eV (semiconducting) bandgaps depending on the width. As a function of the index n, this width exhibited a periodic relationship, as illustrated in (8).

$$f_{width}(n) = \begin{cases} semiconducting \ (3n-1), & \{n \in \mathbb{Z} | \ n \neq 0\} \\ semiconducting \ (3n), & \{n \in \mathbb{Z} | \ n \neq 0\} \\ metallic \ (3n+1), & \{n \in \mathbb{Z} \} \end{cases} \quad (8)$$

B. Density of States (DOS)

Fig. 8 and Fig. 9 show the DOS for zigzag and armchair orientations of AB-BGNRs, respectively.

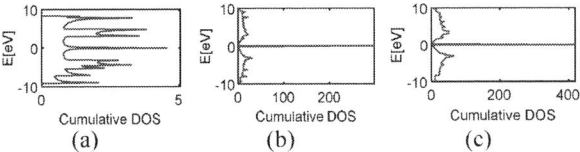

Fig. 8. Zigzag edge DOS for increasing widths: (a) 1, (b) 8, and (c) 20 blocks with a fixed length of 3 unit cells.

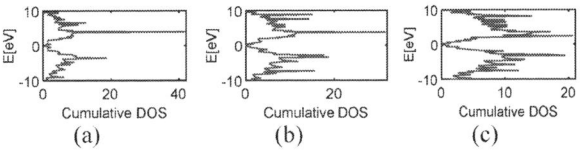

Fig. 9. Armchair edge DOS for widths of (a) 4 blocks (3n+1), (b) 5 blocks (3n-1), and (c) 6 blocks (3n) with a fixed length of 3 unit cells.

Based on Fig. 8, the results revealed numerous degenerate states in the lower energy bands of zigzag AB-BGNRs. The number of these states increased with width, sharply rising around the mid-band due to an abundance of similar energy states, showing the effect of widening. Based on Fig. 9, it is observed that the semiconducting variants lacked quantum states in the mid-band region, while the metallic variant displayed a low number of quantum states.

V. CONCLUSION

Quantum transport theories were simplified to study the electronic properties of AB-BGNRs using the NNTB model, validating modelling techniques. Zigzag orientation shows metallic properties with no band gap, while armchair orientation alternates between metallic and semiconducting states. DOS increases with width for both edge types. A detailed Hamiltonian structure was discussed. These findings lay a foundation for future advancements in graphene-based electronic devices, greatly benefiting real-time data acquisition and processing control boards in electronics. This progress paves the way for breakthrough research in interdisciplinary areas such as imaging applications, including optical coherence tomography [22, 23].

ACKNOWLEDGEMENT

This research was supported and funded by the Ministry of Higher Education under the Fundamental Research Grant Scheme, UTM, FRGS/1/2021/STG07/UTM/02/3 and UTM Fundamental Research (UTMFR), Q.J130000.3823.22H76. We express our sincere gratitude to Universiti Teknologi Malaysia (UTM) for their outstanding support and conducive research environment. We deeply appreciate the assistance from the Research Management Centre (RMC) and the Faculty of Electrical Engineering (FKE) at UTM. Yuki Wong acknowledges the financial support received as a Nexus Young Researcher of UTM.

REFERENCES

[1] E. Durgun, S. Tongay, and S. Ciraci, "Silicon and III-V compound nanotubes: Structural and electronic properties," *Physical Review B*, vol. 72, no. 7, p. 075420, 2005.

[2] S. Zhang, J. Zhou, Q. Wang, X. Chen, Y. Kawazoe, and P. Jena, "Penta-graphene: A new carbon allotrope," *Proceedings of the National Academy of Sciences*, vol. 112, no. 8, pp. 2372-2377, 2015.

[3] J. Zhao *et al.*, "Rise of silicene: A competitive 2D material," *Progress in Materials Science*, vol. 83, pp. 24-151, 2016.

[4] M. A. Kharadi, G. F. A. Malik, F. A. Khanday, K. A. Shah, S. Mittal, and B. K. Kaushik, "Silicene: From material to device applications," *ECS Journal of Solid State Science and Technology*, vol. 9, no. 11, p. 115031, 2020.

[5] K. Zhao, Y. Guo, and Q. Wang, "Contact properties of a vdW heterostructure composed of penta-graphene and penta-BN2 sheets," *Journal of Applied Physics*, vol. 124, no. 16, 2018.

[6] K. S. Novoselov, L. Colombo, P. Gellert, M. Schwab, and K. Kim, "A roadmap for graphene," *nature*, vol. 490, no. 7419, pp. 192-200, 2012.

[7] A. Scholz, T. Stauber, and J. Schliemann, "Dielectric function, screening, and plasmons of graphene in the presence of spin-orbit interactions," *Physical Review B*, vol. 86, no. 19, p. 195424, 2012.

[8] K. Wakabayashi, K.-i. Sasaki, T. Nakanishi, and T. Enoki, "Electronic states of graphene nanoribbons and analytical solutions," *Science and technology of advanced materials*, vol. 11, no. 5, p. 054504, 2010.

[9] A. K. Geim and K. S. Novoselov, "The rise of graphene," *Nature materials*, vol. 6, no. 3, pp. 183-191, 2007.

[10] P. Kumar, L. Panchakarla, and C. Rao, "Laser-induced unzipping of carbon nanotubes to yield graphene nanoribbons," *Nanoscale*, vol. 3, no. 5, pp. 2127-2129, 2011.

[11] Y. Zhang, Y.-W. Tan, H. L. Stormer, and P. Kim, "Experimental observation of the quantum Hall effect and Berry's phase in graphene," *nature*, vol. 438, no. 7065, pp. 201-204, 2005.

[12] A. V. Rozhkov, A. Sboychakov, A. Rakhmanov, and F. Nori, "Electronic properties of graphene-based bilayer systems," *Physics Reports*, vol. 648, pp. 1-104, 2016.

[13] G. Yang, L. Li, W. B. Lee, and M. C. Ng, "Structure of graphene and its disorders: a review," *Science and technology of advanced materials*, vol. 19, no. 1, pp. 613-648, 2018.

[14] E. Schrödinger, "An undulatory theory of the mechanics of atoms and molecules," *Physical review*, vol. 28, no. 6, p. 1049, 1926.

[15] S. Datta, "The non-equilibrium Green's function (NEGF) formalism: An elementary introduction," in *Digest. International Electron Devices Meeting*, 2002: IEEE, pp. 703-706.

[16] J. Baggott, *The meaning of quantum theory: a guide for students of chemistry and physics*. Oxford University Press, 1992.

[17] D. W. Oxtoby, H. P. Gillis, and L. J. Butler, *Principles of modern chemistry*. Cengage AU, 2016.

[18] S. Datta, *Quantum transport: atom to transistor*. Cambridge university press, 2005.

[19] M. P. Johansson, V. R. Kaila, and D. Sundholm, "Ab initio, density functional theory, and semi-empirical calculations," *Biomolecular Simulations: Methods and Protocols*, pp. 3-27, 2013.

[20] F.-P. Ouyang, L.-J. Chen, J. Xiao, and H. Zhang, "Electronic properties of bilayer zigzag graphene nanoribbons: First principles study," *Chinese Physics Letters*, vol. 28, no. 4, p. 047304, 2011.

[21] M. Junaid and G. Witjaksono, "Analysis of band gap in AA and Ab stacked bilayer graphene by Hamiltonian tight binding method," in *2019 IEEE International Conference on Sensors and Nanotechnology*, 2019: IEEE, pp. 1-4.

[22] R. Meleppat, M. Matham, and L. Seah, "An efficient phase analysis-based wavenumber linearization scheme for swept source optical coherence tomography systems," *Laser Physics Letters*, vol. 12, no. 5, p. 055601, 2015.

[23] K. Ratheesh, L. Seah, and V. Murukeshan, "Spectral phase-based automatic calibration scheme for swept source-based optical coherence tomography systems," *Physics in Medicine & Biology*, vol. 61, no. 21, p. 7652, 2016.

Reduction of Subthreshold Leakage Current in Resonant Gate Transistor

Mohamad Zain Azreen Ramli
PEC SE Asia Sdn Bhd
Penang, Malaysia
Zain.Azreen@peccorp.com

Arjuna Marzuki
School of Science and Technology
Wawasan Open University
Penang, Malaysia
arjunam@wou.edu.my

Masuri Othman
Institute of Microengineering and Nanoelectronics
Universiti Kebangsaan Malaysia
Bangi, Malaysia
masuri@ukm.edu.my

Rhonira Latif
Institute of Microengineering and Nanoelectronics
Universiti Kebangsaan Malaysia
Bangi, Malaysia
rhonira@ukm.edu.my

Abstract—The channel in a resonant gate transistor (RGT) transduces the resonator's mechanical movement into electrical signals. The device finds its application in acoustic sensing as a mechanoelectrical transducer. The channel region plays an eminent role to efficiently convert the resonator's vibration displacement into current flow between source and drain. In this paper, the subthreshold current for the channel is studied. The enhancement mode n-channel RGT and metal-oxide-semiconductor field-effect transistor (MOSFET) are fabricated. The channel width and length have been varied from 25 μm-540 μm and 10 μm-20 μm, respectively. The temperature to grow gate oxide and to anneal the ion-implanted source/drain have been altered from 950°C-1100°C. We have found that the subthreshold leakage current can be reduced by increasing the channel length, decreasing the channel width, decreasing the annealing and gate oxidation temperature and incorporating p-channel into the transistor device instead of n-channel. It is crucial for the subthreshold current to be small with no channel breakdown issue, so that the device can operate with high reliability and low power consumption.

Keywords—transistor, MEMS, gate oxide, channel region.

I. INTRODUCTION

The resonant gate transistor (RGT) was first introduced in 1967 by Nathanson et al. [1]. The structure of RGT is depicted in Fig. 1(a) which consists of a conductive resonator, positioned above a channel region of transistor with channel width W_c. The device's channel transduces the mechanical response of its resonating microstructure into electrical resonance. The electrostatically actuated device has gained significant research and commercial interest as filters, sensors and oscillators in many device applications. Coupled resonators have been used to create coupled bulk acoustic wave (BAW) microelectromechanical system (MEMS) resonators [2]. High quality factor of the micromachined device is on demand for smaller device structure, lower power consumption and better integration with microelectronics and circuits at die or package level.

The research was funded by Universiti Kebangsaan Malaysia, grant number GUP-2022-070

In [3], RGT has been used as resonant mass sensor for particle sensing application. Others have employed RGT as frequency-selective component in frequency and timing applications [4], or to study the dynamic pull-in effect in electrostatically actuated MEMS resonators for reliable and efficient operation conditions [5].

In our work, the resonant gate transistors have been fabricated into an array of RGTs to mimic the functionality of a cochlea. The mechanical performance of the resonators in RGT has been reported in detail elsewhere [6-9]. In Fig. 1(b), the cross-section view of RGT at the channel region shows the channel length L_c, source and drain regions, gate oxide and the spacing distance from resonator to gate oxide, d_i. Like the standard operation of a transistor, the gate-to-source voltage (V_{gs}) and drain-to source voltage (V_{ds}) are applied while the drain-to-source current (I_{ds}) is measured.

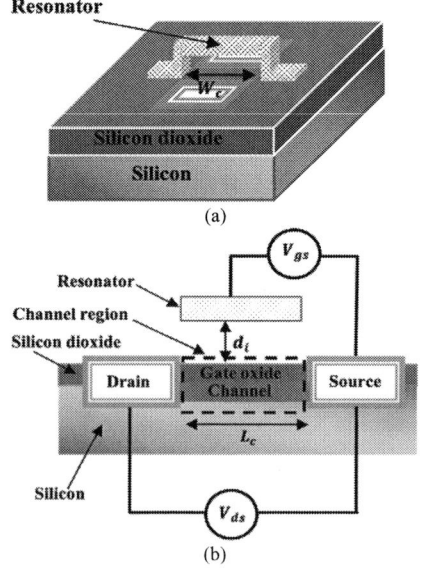

Fig. 1. (a) The schematic diagram of resonant gate transistor (RGT) with a micromachined resonator and W_c as the channel width. (b) The cross-sectional view at the channel region of length L_c with the application of gate voltage V_{gs} and drain voltage V_{ds}.

In this paper, we investigate the subthreshold leakage current behavior exhibited by the channel of RGT. The in-situ metal-oxide-semiconductor field-effect transistor (MOSFET) of identical channel's geometrical dimension with RGT is also fabricated as a test structure to investigate the channel region of RGT. The channel length, channel width and channel type have been varied. The influence of temperature during gate oxidation and source/drain annealing on device's subthreshold conduction is also examined. High subthreshold current reduces the strength in modulation of channel conductance with respect to the applied gate voltage, giving small transconductance value. Uncontrollably high subthreshold current can lead to channel breakdown and may cause high energy consumption, even when the device is in its off state. Hence, it is vital to minimize this leakage current in transistor devices.

II. FABRICATION

The fabrication process to create n-type channel for MOSFET and RGT involves the implantation of phosphorus ions onto boron-doped silicon substrate, forming source and drain regions of the device. Whilst in p-type channel MOSFET and RGT, boron ions are implanted onto phosphorus-doped silicon substrate to form source/drain. Annealing at 1100°C for 30 minutes in a furnace is performed to electrically activate the implanted ions. Aluminum 1% silicon thin film layer is employed as metal contacts for source/drain. The conductive gate/resonator is made of tantalum metal thin film layer. The channel region is insulated from gate by a thermally grown silicon dioxide layer called the gate oxide. Silicon dioxide of 70 nm thickness is attained by dry oxidation process at 1100°C for 30 minutes. Then, the source/drain regions and n-channel/p-channel are integrated with the resonator structure to form the n-channel/p-channel resonant gate transistor (n-RGT/p-RGT). Fig. 2 shows the fabricated RGT with the resonator length of 92 μm and an air gap distance from resonator to substrate of $d_i \sim 5$ μm.

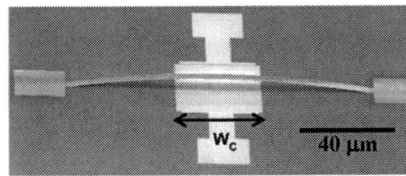

Fig. 2. The scanning electron micrograph (SEM) image of the fabricated resonant gate transistor with W_c as the channel width.

In the first iteration, n-channel devices have been fabricated with the size of $L_c = 10$ μm and $W_c = 25$ μm - 540 μm. Then, L_c has been increased to 20 μm and the temperature used for the annealing step of the implantation regions and for the gate oxidation process have been reduced to 950°C, in the second iteration. These two approaches have been taken to prevent channel breakdown within the channel of n-RGT and n-MOSFET. In the third iteration, the subthreshold leakage current (I_{th}) is minimized by changing the channel type from n-channel to p-channel. Compared to the fabrication process described previously for the n-channel devices, the implantations of source/drain regions for the p-channel devices have been performed after the growth of gate oxide layer due to the segregation of boron dopants. High furnace temperature during dry oxidation and annealing of the implantation regions depletes the surface concentration of boron as the boron dopants segregate from silicon substrate towards the grown silicon dioxide layer. Thus, the implantation of boron has to be performed after the gate oxidation process in order to minimize boron segregation effect at source/drain.

III. MEASUREMENT RESULTS

Direct current (DC) measurements for the fabricated devices are conducted using Hewlett Packard 4156B semiconductor parameter analyzer. The curves of drain current against drain voltage (i.e. I_{ds}-V_{ds} characteristics) for the four-terminal devices are obtained by varying the drain voltage and gate voltage. The source and substrate are grounded while the drain current is measured. The threshold voltage is the minimum gate voltage needed to be applied to the device to switch it on. I_{th} is the drain current I_{ds} that flows between source and drain when the device is in weak inversion region, by which V_{gs} is smaller than the threshold voltage. I_{th} can be found by switching off the gate terminal i.e. $V_{gs} = 0$ V. In our study, I_{th} is extracted from the transfer characteristics (i.e. I_{ds}-V_{gs} and $\sqrt{I_{ds}}$-V_{gs} plot) of the device operating in the saturation regime. At gate voltage V_{gs} much smaller than the threshold voltage, the average value of I_{th} is estimated.

In the first iteration, the n-MOSFET and n-RGT have been fabricated at 1100°C of gate oxidation temperature and source/drain annealing. Fig. 3 (a)(b) shows the measured I_{ds}-V_{ds} characteristics and the transfer characteristics of n-MOSFET with $L_c = 10$ μm and $W_c = 80$ μm. Clear distinction of the device's operation in linear and saturation regime is observed. The fabricated enhancement mode n-MOSFET is seen to be working in depletion mode instead of enhancement mode with the threshold voltage of ~ -1 V. The device is always on and requires $V_{gs} < -1$ V to be turned off. Other than the channel's depletion mode behavior, high subthreshold leakage current is measured from n-MOSFET. At $V_{gs} < -1$ V, an average of $I_{th} \sim 0.6$ mA is extracted from the transfer characteristics. There is a significant drain current flow even when the device is turned off.

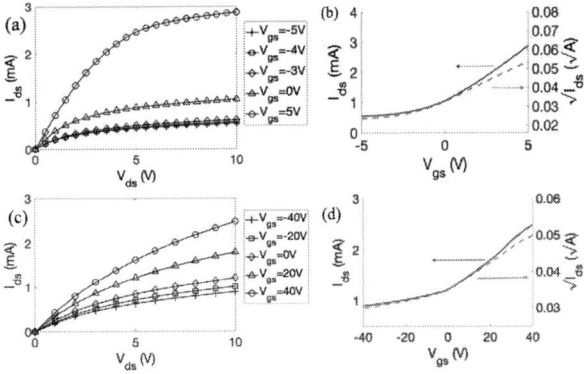

Fig. 3. DC measurement of the enhancement mode devices showing (a) I_{ds}-V_{ds} of n-MOSFET, (b) transfer characteristics of n-MOSFET, (c) I_{ds}-V_{ds} of n-RGT and (d) transfer characteristics of n-RGT with $L_c = 10$ μm and $W_c = 80$ μm.

Fig. 3 (c)(d) demonstrates DC measurement results for n-RGT that has equivalent channel size with n-MOSFET. Comparable range of drain current is measured for $V_{ds} = 0$ V-10 V. However, the operating resonator's voltage range (-40 V- 40 V) for n-RGT is one order higher than the gate

979-8-3503-7832-0/24 $31.00 © 2024 IEEE   133

voltage range used for n-MOSFET. Higher threshold voltage, ~ -20 V for n-RGT is due to the presence of ~ 5 μm air gap distance between the resonator and channel region. Thus, higher electric field induced by the resonator is required to turn the device on/off and modulate the channel conductance, compared to the situation of where the resonator is placed directly above the channel region.

The channel width W_c of n-MOSFET is increased from 25 μm to 540 μm for the first iteration in Fig. 4(a). The measured I_{th} increases and becomes relatively high, close to 1 mA. The device has been witnessed to be turned off with high subthreshold leakage current flowing between source and drain, leading to high energy consumption. Chanel breakdown occurred starting from W_c = 250 μm, whereby the gate loses control over channel and the drain current has been measured to be very dependent on V_{ds}.

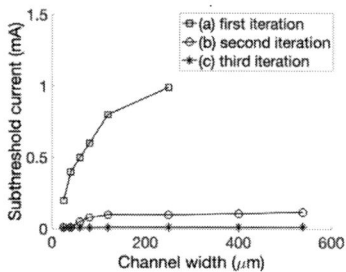

Fig. 4. Measured I_{th} with respect to W_c from n-MOSFETs fabricated at furnace temperature (a) 1100°C and L_c = 10 μm (first iteration) and (b) 950°C and L_c = 20 μm (second iteration). (c) Measured I_{th} from p-MOSFETs fabricated at furnace temperature 1100°C and L_c = 10 μm (third iteration) with respect to W_c.

Increase in L_c and decrease in temperature for gate oxidation and annealing process are presumed to reduce I_{th} and subsequently diminishes the possibility of devices experiencing channel breakdown. Therefore, temperature of 950°C and L_c = 20 μm are used in the second iteration. Fig. 4(b) demonstrates that the occurrence of channel breakdown has been eliminated. None of the fabricated n-MOSFETs have been observed to experience channel breakdown. At W_c = 80 μm, I_{th} has been reduced to 87% compared to the first iteration.

In the third iteration, p-MOSFETs have been fabricated with L_c = 10 μm and the temperature used to grow gate oxide and anneal the source/drain region is maintained at 1100°C. In Fig. 4(c), significant reduction of I_{th} to almost 98.6% compared to the first iteration is attained at W_c = 80 μm. Channel breakdown does not occur in the fabricated p-MOSFETs.

P-channel devices work in the opposite current and voltage polarities of the n-channel devices. DC measurement for p-channel devices in Fig. 5 shows the channel's enhancement mode characteristics. The device is normally off and requires gate/resonator voltage of higher than the threshold voltage to turn it on. Small average I_{th} of ~ -8.7 μA and I_{th} ~ -75 nA have been measured for p-MOSFET and p-RGT, respectively with the threshold voltage of -6.5 V and -20 V. There is an insignificant drain current flow when the device is turned off. In p-MOSFET, the drain current in the saturation regime increases from I_{ds} ~ 0 μA to -1.5 mA as the gate voltage increases from V_{gs} = -4 V to -15 V. Compared to n-MOSFETs, the measured drain current range from p-MOSFET is generally smaller, due to the smaller effective

carrier mobility for holes than electrons by a factor of 2 to 4 [10]. In p-RGT, the resonator operates at higher voltage range from -20 V to -24 V and higher threshold voltage compared to p-MOSFET of similar geometrical dimensions. Again, the measured discrepancies are related to the presence of air gap layer in RGT.

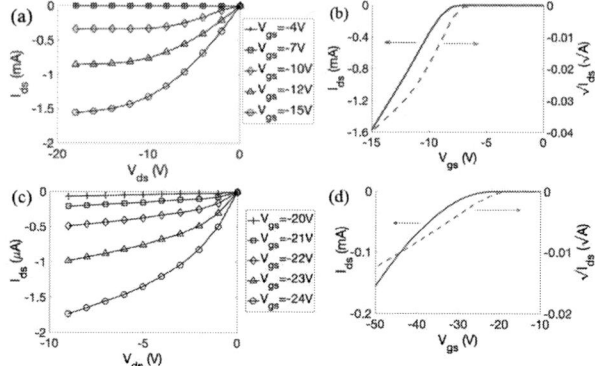

Fig. 5. DC measurement of the enhancement mode devices showing (a) I_{ds}-V_{ds} of p-MOSFET, (b) transfer characteristics of p-MOSFET, (c) I_{ds}-V_{ds} of p-RGT and (d) transfer characteristics of p-RGT with L_c = 10 μm and W_c = 80 μm.

IV. DISCUSSIONS

The observed depletion mode characteristics in enhancement mode n-channel devices indicates the presence of positive oxide charges (Q_{ss}) within the gate oxide layer. These trapped positive ion species within the gate oxide are located close to the gate oxide-silicon channel interface as shown in Fig. 6. Positive-charged Q_{ss} induces the formation of electron inversion layer in the channel. This means that the electron inversion layer can exist in the absence of any applied gate/resonator voltage, resulting in depletion mode behavior and negative threshold voltage in enhancement mode n-channel devices [10]. Boron segregation/light boron doping in silicon substrate/higher Q_{ss} instigates stronger electron inversion layer, leading to higher subthreshold leakage current.

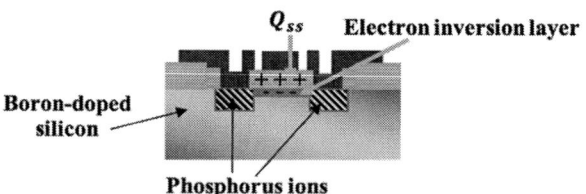

Fig. 6. The presence of positive oxide charges Q_{ss} at the gate oxide-silicon channel interface of the enhancement mode n-MOSFET. Q_{ss} induces the formation of electron inversion layer, causing the conduction of drain current at V_{gs} = 0 V [10].

The insertion of air gap spacing in RGT decreases the total capacitance between resonator and channel. The total capacitance is dominated by air capacitance and not the gate oxide capacitance anymore. Consequently, the strength of the perpendicular electric field from the resonator in RGT is smaller compared to the electric field exerted by gate in MOSFET, due to the presence of air gap layer. The depletion and enhancement of channel conductance by the resonator has become difficult whereby higher electric field is needed to turn the device on/off and modulate the channel

979-8-3503-7832-0/24 $31.00 © 2024 IEEE   134

conductance. This explains the higher operating voltage range and threshold voltage in RGT than in MOSFET.

The increase of subthreshold leakage current and the probability of channel breakdown with respect to the increase of channel width W_c have been speculated to be associated to the presence of high electric field within the channel, exerted by drain voltage. At larger W_c, the effect of electric field from drain voltage on the channel will be far greater and may tower over the perpendicular electric field from gate voltage, causing outcomes similar to short channel. Thus, the channel length can be designed to be adequately long to prevent short channel effects. However, with the application of V_{ds}, the depletion region around drain gets broaden. Thus, the distance between source and drain i.e. channel length decreases. The next approach will be to reduce the process temperature during device fabrication. Lower furnace temperature for the annealing of the implantation regions and gate oxidation process will assist to lessen the vertical and lateral diffusions of the phosphorus implanted source/drain regions i.e. making the effective channel length to be longer and the junction depths at source/drain to be shallower. The effect of electric field from drain voltage is predicted to be reduced as the drain becomes shallower and further away from source.

The influence of Q_{ss} that induces the formation of electron inversion layer within the silicon channel and contributes to high subthreshold current within the n-channel devices can be reversed by the use of p-type channel. The enhancement mode p-MOSFET and p-RGT require negative polarity of the gate/resonator voltage to form hole inversion layer and obtain drain current conduction. In enhancement mode p-channel devices, the electron inversion layer exists instantly in the channel due to the presence of Q_{ss}. Thus, larger negative gate/resonator voltage is needed to be applied to overcome this built-in electron inversion layer and create the hole inversion layer. Therefore, the existence of electron inversion layer in p-channel devices demonstrates the opposite effect by impeding drain current conduction from source to drain until hole inversion layer is formed.

V. CONCLUSION

Enhancement mode n-channel and p-channel MOSFETs and RGTs have been fabricated and the devices' subthreshold conduction have been studied. The occurrence of channel breakdown and high subthreshold leakage current are related to the high temperature processes involved during fabrication and high electric field effect on the channel. The device's electrical performance has been improved by increasing the channel length, decreasing the channel width and decreasing the annealing and gate oxidation temperature.

By employing p-channel instead of n-channel, the subthreshold leakage current has been drastically minimized and the occurrence of channel breakdown has been totally eliminated. The p-channel devices exhibit high modulation of channel conductance and can be turned off with considerably small subthreshold leakage current. This leads to low energy consumption and device heating due to the small off-state leakage current.

ACKNOWLEDGMENT

The authors would like to acknowledge various personnel involved from the Scottish Microelectronic Centre, The University of Edinburgh.

REFERENCES

[1] H. C. Nathanson, W. E. Newel, R. A. Wickstrom, and J. R. Davis, Jr, "The resonant gate transistor," IEEE Trans. Electron. Devices, vol. ED-14, no. 3, pp. 117-133, 1967.

[2] L. Wang, C. Wang, Y. Wang, A. Quan, M. Keshavarz, B. P. Madeira, H. Zhang, C. Wang, and M. Kraft, "A Review on coupled bulk acoustic wave MEMS resonators," Sensors, vol. 22, no. 10, 3857, 2022.

[3] C. -H. Weng, G. Pillai, and S. -S. Li, "A thin-film piezoelectric-on-silicon MEMS oscillator for mass sensing applications," IEEE Sensors Journal, vol. 20, no. 13, pp. 7001-7009, 2020.

[4] G. Wu, J. Xu, E. J. Ng, and W. Chen, "MEMS resonators for frequency reference and timing applications," Journal of Microelectromechanical Systems, vol. 29, no. 5, pp. 1137-1166, 2020.

[5] J. -H. He, D. Nurakhmetov, P. Skrzypacz, and D. Wei, "Dynamic pull-in for micro–electromechanical device with a current-carrying conductor," Journal of Low Frequency Noise, Vibration and Active Control, vol. 40, no. 2, pp. 1059-1066, 2021.

[6] T. B. Anak Ngelayang, B. Y. Majlis, M. A. Azam, F. Arith, R. Latif, "Platinum and Aluminium Microresonator Bridges for Artificial Basilar Membrane", Applied Mechanics and Materials, vol. 761, pp. 462-467, 2015.

[7] R. Latif, E. Mastropaolo, A. Bunting, R. Cheung, T. Koickal, A. Hamilton, M. Newton, and L. Smith, "Microelectromechanical systems for biomimetical applications", Journal of Vacuum Science and Technology B, vol. 28, no. 6, pp. C6N1-C6N6, 2010.

[8] R. Latif, E. Mastropaolo, A. Bunting, R. Cheung, T. Koickal, A. Hamilton, M. Newton, and L. Smith, "Low frequency tantalum electromechanical systems for biomimetical applications", Journal of Vacuum Science and Technology B, vol. 29, no. 6, pp. 06FE05/1-06FE05/6, 2011.

[9] E. Mastropaolo, R. Latif, T. Koickal, A. Hamilton, R. Cheung, M. Newton, and L. Smith, "Bimaterial electromechanical systems for a biomimetical acoustic sensor", Journal of Vacuum Science and Technology B, vol. 30, no. 6, pp. 06FD01/1-06FE01/7, 2012.

[10] R. Latif, "Microelectromechanical systems for biomimetical application", Ph.D thesis, The University of Edinburgh, 2012.

979-8-3503-7832-0/24 $31.00 © 2024 IEEE

Comparative Electrochemical Performance of Screen Printed Carbon, Gold and Graphene Electrodes

Mohammad Al Mamun
Nanotechnology and Catalysis Research Centre,
Universiti Malaya,
50603 Kuala Lumpur, Malaysia
Department of Chemistry,
Jagannath University,
Dhaka-1100, Bangladesh
zithrox@gmail.com

Yasmin Abdul Wahab
Nanotechnology and Catalysis Research Centre,
Universiti Malaya,
50603 Kuala Lumpur, Malaysia
yasminaw@um.edu.my

M. A. Motalib Hossain
Nanotechnology and Catalysis Research Centre,
Universiti Malaya,
50603 Kuala Lumpur, Malaysia
motalib123@yahoo.com

Hui Yin Nam
Nanotechnology and Catalysis Research Centre,
Universiti Malaya,
50603 Kuala Lumpur, Malaysia
huiyin@yahoo.com

Mohd Rafie Johan
Nanotechnology and Catalysis Research Centre,
Universiti Malaya,
50603 Kuala Lumpur, Malaysia
mrafiej@um.edu.my

Nurul Ezaila Alias
School of Electrical Engineering,
Universiti Teknologi Malaysia,
Johor Bahru, Malaysia
ezaila@fke.utm.my

Hanim Hussin
Faculty of Electrical Engineering,
Universiti Teknologi MARA,
Shah Alam, Malaysia
hanimh@uitm.edu.my

Maizan Muhamad
Faculty of Electrical Engineering,
Universiti Teknologi MARA,
Shah Alam, Malaysia
maizan@uitm.edu.my

Abstract—A comparative analysis of the electrochemical performance of screen printed carbon (SPCE), gold (SPGE), and graphene (SPGrE) electrodes is presented before integration into electrochemical biosensors. The electrodes were systematically examined based on key electrochemical parameters, including charge transfer kinetics, electrochemical reproducibility, and stability, considering $[Fe(CN)6]^{3-/4-}$ as a typical redox analyte using cyclic voltammetry (CV), differential pulse voltammetry (DPV), and electrochemical impedance spectroscopy (EIS). Results indicate that all the bare screen printed electrodes (SPEs) demonstrate significant irreproducibility (>10% of RSD) with poor stability of the electroactive surface. The graphene electrodes exhibit superior electrocatalytic properties with higher interfacial charge transfer rate constant (2.30×10^{-6} cms^{-1}) compared to the SPCE (1.40×10^{-6} cms^{-1}) and SPGE (1.72×10^{-6} cms^{-1}) surfaces. The findings provide valuable insights into the relative merits and drawbacks of SPEs, guiding the selection of suitable electrode materials for diverse biosensing applications.

Keywords—Electrochemical performance, screen printed electrode, cyclic voltammetry (CV), differential pulse voltammetry (DPV), Electrochemical Impedance Spectrosopy (EIS).

I. INTRODUCTION

Screen-printed electrodes (SPEs) stand out as one of the most extensively employed platforms for sensors and biosensors advancement due to their immense potential for miniaturization and commercialization [1, 2]. In the realm of biosensors development, the choice of electrode material plays a pivotal role in determining the device's sensitivity, stability, and overall performance. Among the myriad of materials available, screen printed carbon electrodes (SPCEs) have emerged as promising candidates due to their versatility, cost-effectiveness, and ease of fabrication [3, 4]. However, the quest for enhanced electrochemical performance has led

to considerable interest in exploring alternative materials that can surpass the capabilities of traditional carbon-based SPEs [5, 6].

Gold electrodes are favored in electrochemical sensing for their conductivity, stability, and biocompatibility. However, their high cost and limited surface area drive the search for better alternatives. Graphene, a two-dimensional carbon allotrope with exceptional properties, is gaining attention for biosensing platforms. Carbon-based SPEs (Screen Printed Electrodes) offer advantages over metal-based SPEs, such as cost-effectiveness and a wider potential window [7]. Yet, their surface inhomogeneity and irreproducibility pose challenges, hindering commercialization [5].

Before integrating biorecognition elements onto SPEs, it is essential to evaluate surface homogeneity and confirm the absence of interfering elements [5, 8]. Understanding SPE surface morphology and roughness is crucial, as they impact electrochemical performance. Different materials affect electrode conductivity, stability, and sensitivity. Ensuring surface suitability for functionalization is vital for specific analyte detection [9]. This characterization helps optimize sensor design for enhanced sensitivity, selectivity, and stability, establishing a baseline response in various conditions [6]. No previous investigations have compared SPEs performance before electrochemical biosensor construction.

This study aims to fill that gap by comparing the electrochemical performance of screen-printed carbon, gold, and graphene electrodes. Preferred methods like CV, EIS, and DPV will be used for screening. These methods are cost-effective and user-friendly. We will evaluate parameters such as electrochemical reversibility, peak separation potential, overpotential, charge transfer kinetics, reproducibility, and stability to determine the suitability of each electrode for biosensor integration, exploring their limitations.

II. MATERIALS AND METHODS

A. Chemicals and Reagentes

All reagents and solutions of analytical grade were made with deionized water. Potassium chloride (KCl), potassium ferrocyanide ($K_4[Fe(CN)_6]$), and potassium ferricyanide ($K_3[Fe(CN)_6]$) were obtained from Sigma-Aldrich (Germany) and used as received without further purification. An electrolytic medium of 0.1 M KCl-tris buffer (pH 7.2) was prepared using analytical-grade chemicals and used in the electrochemical cell. The commercial screen-printed electrodes (4 mm diameter) were supplied by Metrohm, DropSens, Switzerland.

B. Instrumentation and Electrochemical Measurements

Cyclic voltammetry (CV), differential pulse voltammetry (DPV), and electrochemical impedance spectroscopy (EIS) were conducted using a Metrohm AUTOLAB Potentiostat/Galvanostat controlled by NOVA 2.1.4 software. Screen-printed electrodes made of carbon, gold, and graphene served as the working electrode in a three-electrode configuration, with Ag/AgCl as the reference electrode and carbon as the counter electrode. The experiments were carried out at room temperature both in the presence and absence of $[Fe(CN)_6]^{3-/4-}$.

III RESULTS AND DISCUSSION

Electrochemical screening of Fresh SPEs

A low-cost, user-friendly method is essential to ensure SPE surface fitness before biosensor construction to avoid wasting expensive resources. Evaluating fresh SPCE, SPGE, and SPGrE surfaces using CV, DPV, and EIS (Fig. 1) after aqueous cleaning at ambient conditions can prevent biosensor rejection after identifying the unfit bare electrode surfaces.

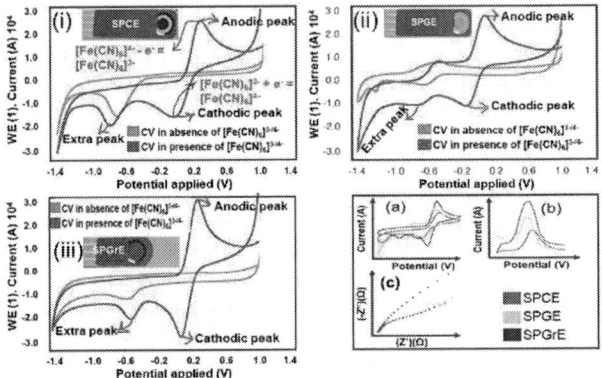

Fig. 1. Cyclic voltammetric analysis of (i) SPCE, (ii) SPGE, and (iii) SPGrE working electrode surfaces was conducted with and without redox species

(10 mM $[Fe(CN)_6]^{3-/4-}$) between 1.0 V and -1.5 V vs Ag/AgCl at a 100 mVs^{-1} scan rate. Comparative electrochemical behavior on SPCE (red), SPGE (yellow), and SPGrE (blue) surfaces was examined using (a) CV, (b) DPV, and (c) EIS in 10 mM $[Fe(CN)_6]^{3-/4-}$ prepared in KCl-tris buffer (pH 7.2). Note: Z' (real part) and Z'' (imaginary part) of impedance were measured at 10 mV amplitude and $10^5 – 10^{-1}$ Hz frequency.

From the cyclic voltammetry (CV) analysis for SPCE, SPGE, and SPGrE in Fig. 1(i), (ii), and (iiii) respectively, well-defined redox peaks were observed between -0.2 and 0.6 V with redox species (red curve) on all electrodes. A significant extra cathodic peak between -1.0 and -0.5 V was also noted on all surfaces, both with (red curve) and without (blue curve) redox species. This could be due to the supporting electrolyte or electroactive species on the SPE surfaces. Since an electrochemically inert electrolyte (KCl-Tris buffer) was used, it might be due to inherent electroactive functional groups on the SPE electrode surfaces [3]. Comparative CV (Fig. 1(a)), DPV (Fig. 1(b)), and EIS (Fig. 1(c)) analysis show the SPGrE electrode (blue curve) exhibited the highest electrocatalytic activity, with higher peak currents in CV and DPV, and consistent EIS results, compared to SPCE (red curve) and SPGE (yellow curve).

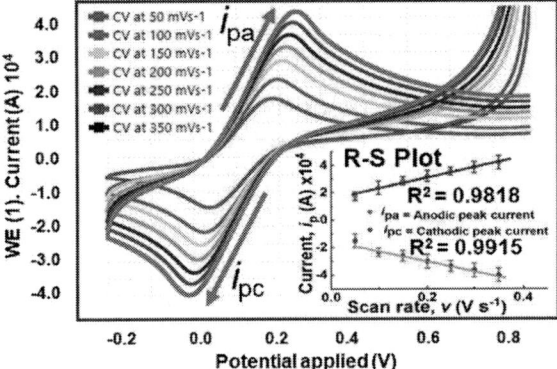

Fig. 2. Cyclic voltammograms (CVs) were recorded for a 10 mM solution of $[Fe(CN)_6]^{3-/4-}$ at various scan rates (50 mV/s, 100 mV/s, 150 mV/s, 200 mV/s, 250 mV/s, 300 mV/s, and 350 mV/s), spanning a potential range from -0.2 V to 0.8 V on a screen-printed graphene electrode (SPGrE) with Ag/AgCl reference. The corresponding Randles-Sevcik (R-S) plots, shown in the inset, illustrate both cathodic and anodic peak currents plotted against scan rate (V/s).

Effect of Scan Rate and Randless-Sevcik (R-S) Plot

The cyclic volatammograms of 10 mM $K_3[Fe(CN)_6]$ at various scan rates (mVs^{-1}) on different SPEs surface (as an example of Fig. 2 for SPGrE surface) between the specified potential window of 0.8 V and - 0.2 V vs Ag/AgCl were measured at ambient conditions. The Randless-Sevcik plots (peak current vs scan rate) (Fig. 2-inset) for both the anodic and cathodic peak current (i_p) were conducted following the Randles–Ševčík equation 1[10] :

$$i_p = 0.4463nFAC(nFvD/RT)^{1/2} \qquad (1)$$

In the given equation, i_p represents the maximum current measured in amperes (A), n stands for the quantity of electrons transferred, F denotes the Faraday constant (C mol^{-1}), A signifies the electrode's surface area in square centimeters (cm^2), D denotes the diffusion coefficient in cm^2

s^{-1}, C represents the concentration in mol L^{-1}, v indicates the scan rate in Vs^{-1}, R is the universal gas constant (J K^{-1} mol^{-1}), and T refers to the temperature in Kelvin (K).

The R-S plots of SPEs show the linear variation of peak current (R^2 = 0.9818 and 0.9915 for anodic and cathodic peak current, respectively) with the increase of scan rates (as illustrated in Fig. 2-inset), which satisfy the validity of Randless-Sevcik equation for the quasi-reversible charge transfer kinetics [3, 11]. We have also calculated the peak separation potential (ΔE) for all the SPEs surfaces (as tabulated in Table 1) using the equation 2.

$$\Delta E = (E_{pa}-E_{pc}) \qquad (2)$$

Where E_{pa} = anodic peak potential (V) , E_{pc} = cathodic peak potential (V). The lowest peak separation potential (0.160 V) was found on SPGE surface, which is still higher than the minimum peak separation potential 118 V [10]. This result indicate the quasi-reversible and heterogeneous electron transfer process through all the electrode-solution interfaces. To confirm this claim we have also calculated the peak current ratio (i_{pa}/i_{pc}) (Table 1) which was greater than the unity (for purely reversible charge transfer kinetics)[12]. Moreover, the rate of electron transfer for the anodic reaction (oxidation : $[Fe(CN)_6]^{4-}$ - e^- = $[Fe(CN)_6]^{3-}$) was significantly higher than the cathodic reaction (reduction: $[Fe(CN)_6]^{3-}$ + e^- = $[Fe(CN)_6]^{4-}$) through all the electrode-electrolyte interfaces [3]. However, this effect was more significant in SPGE surface ($i_{pa}>>i_{pc}$) compared to SPCE and SPGrE, as they are non-metallic carbon electrode surfaces. Because metal electrodes (such as SPGE) have a greater tendency for oxidation compared to carbon electrodes [11].

TABLE 1. CALCULATED PEAK SEPARATION POTENTIAL (ΔE), PEAK CURRENT RATIO (i_{pa}/i_{pc}), OVER POTENTIAL (η), CHARGE TRANSFER RATE CONSTANT (k), AND REPRODUCIBILITY (% RSD) FOR SPCE, SPGE AND SPGrE SURFACE AT THE SCAN RATE 100 mV s^{-1}.

Electrode ID	ΔE (V)	i_{pa}/i_{pc}	η (mV)	k (cm s^{-1}) $\times 10^6$	% RSD
SPCE	0.330	1.21	0.051	1.40	10.41
SPGE	0.160	1.62	0.080	1.72	9.77
SPGrE	0.223	1.15	0.062	2.30	12.67

On the other hand, since the peak potential separation is found to be higher than 118 mV (Table 1), the Tafel equation [13] can be expressed as follows:

$$i = i_o\, e\, (-\alpha\, n\, F \eta\, /\, R\, T) \qquad (3)$$

In the equation where the Tafel slope is defined as α n F/RT, α represents the transfer coefficient (ranging from 0 to 1), and η stands for overpotential (mV). Referring to Table 1 for the calculated overpotential described by equation (4), it is observed that the overpotential η exceeds 50 mV. This indicates that the redox process at the surface-electrolyte interface of the SPEs is not entirely reversible [13, 14].

$$\eta = (E_{pc} - E_o) \qquad (4)$$

where, $E_o = (E_{pc}+E_{pa})/2$ = Foraml potential (V)

To compare the charge transfer kinetics through the electrode solution interface we have determined the heterogeneous charge transfer rate constant (k) (Table 1) for all the screen printed electrode surfaces using the following equation 5 [10]:

$$i = nFAkC \qquad (5)$$

where n = 1, F = 96,584 C mol^{-1}, R = 8.314 J mol^{-1} K^{-1}, i denotes the exchange current (A), A represents the surface area of the SPE in cm^2, k is the electron transfer rate constant (cm s^{-1}) for the redox reaction $[Fe(CN)_6]^{3-}$ + e- = $[Fe(CN)_6]^{4-}$, and C stands for the solution concentration of $[K_4Fe(CN)_6]$ which is 10 mM. The computed values of k for the SPCE, SPGE and SPGrE were found as 1.4 x 10^{-6} cm s^{-1}, 1.7 x 10^{-6} cm s^{-1}, and 2.3 x 10^{-6} cm s^{-1} respectively. The significantly higher value of k for SPGrE surface indicates the electrocatalytic activity brought on by the conducting high surface area [13].

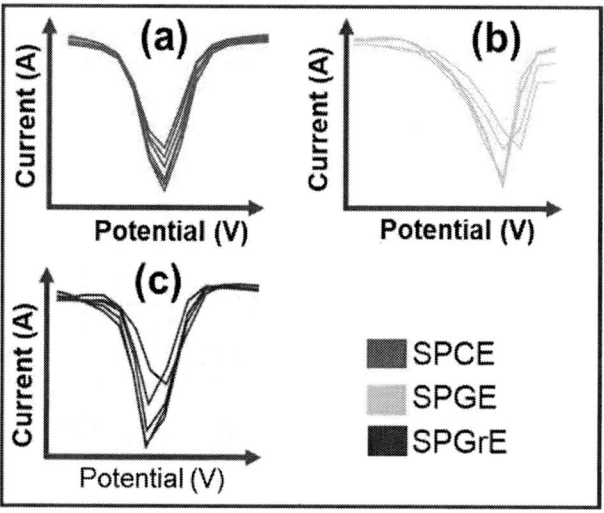

Fig. 3. (a) Reproducibility of the SPEs surface in DPV analysis using 10 mM $[Fe(CN)_6]^{3-/4-}$ in KCl-Tris buffer (pH 7.2) on SPCE (red),(b) SPGE (yellow), and (c) SPGrE (blue). The analysis was conducted within a potential window from +0.8 V to -0.2 V at a scan rate of 100 mV/s, modulation time of 0.05 s, and modulation amplitude of 0.025 V.

Reproducibility

We examined the reproducibility and stability of SPCE, SPGE, and SPGrE using DPV with 10 mM $[Fe(CN)_6]^{3-/4-}$ over five measurements in (Fig. 3). The % of RSD (relative standard deviation) is tabulated in Table 1. From the DPV analysis, we found that the peak potential on SPCE (Fig. 3a) remained constant, but the peak currents varied. On SPGE and SPGrE, both peak potential and current shifted. The relative standard deviations (RSD%) of peak currents for five consecutive measurements were 10.41% (SPCE), 9.77% (SPGE), and 12.67% (SPGrE). This indicates that the surfaces are irreproducible and unstable for redox species analysis, likely due to the present of electroactive species on the surface [3]. Because during the production and purification of carbon electrode materials, vigorous processing introduces inherent oxygenated functional groups into the carbon structure. In the case of SPGE, due to its limited potential window, the gold surface tends to oxidize at the higher potential range [11]. These oxygen functionalities are strongly bonded, making their removal challenging; the binding energy for C-OH is about 350 kJ/mol and for C=O is approximately 745 kJ/mol [15]. Traditional aqueous cleaning is insufficient to remove these groups from the surface of screen-printed electrodes (SPEs), affecting reproducibility and stability. Therefore, a cost-effective, simple, and sustainable method is essential for preparing reproducible and stable SPE surfaces for electrochemical biosensors.

III. CONCLUSION

In this comparative study, under identical experimental conditions, we aspire to provide valuable insights into the relative merits and weakness of screen printed carbon, gold, and graphene electrodes, thereby aiding researchers and developers in making informed decisions regarding the selection of electrode materials for diverse biosensing applications. The electrochemical behaviors of SPE surfaces exhibit quasireversible charge transfer kinetics with high peak separation potential and overpotential. Among them, the SPGrE electrode demonstrates the highest sensitivity with electrocatalytic activity in all CV, DPV, and EIS analysis. The electrochemical screening of the SPEs surfaces exhibit unexpected electrochemical peaks in the cathodic potential region which can influence the sensors electrochemical performance greatly due to the inhomogenety of the surface causes the irreproducibilty and unstability. By advancing and developing new methods for eliminating the interfering peaks from the negative potential window of the electrode surface, this study may endeavors to contribute towards the development of next-generation biosensing platforms that exhibit superior performance, reproducibility, stability and scalability in future.

ACKNOWLEDGMENT

The work is financially supported by the Ministry of Higher Education Malaysia (MOHE) via Fundamental Research Grant Scheme (FRGS/1/2022/TK09/UM/02/27). The authors also extend their appreciation to the Universiti Malaya (Grant No. ST055-2022 and ST095-2022) and Bangabandhu Science and Technology Fellowship, Bangladesh for the financial support during this research.

REFERENCES

[1] A. S. Calvo, C. Botas, D. Martin-Yerga, P. Alvarez, R. Menéndez, and A. Costa-Garcia, "Comparative study of screen- printed electrodes modified with graphene oxides reduced by a constant current," J. of The Electrochemical Society, vol. 162, pp. B282, August 2015.

[2] M. Al Mamun, Y. A. Wahab, M. M. Hossain, A. Hashem, and M. R. Johan, "DNA-Aptamer–Based Electrochemical Biosensors for the Detection of Thrombin: Fundamentals and Applications," in Functional Nanomaterials for Sensors: CRC Press, pp. 201-221, May 2023.

[3] M. Al Mamun, Y. A. Wahab, M. M. Hossain, A. Hashem, N. A. Hamizi, Z. Z. Chowdhury, S. F. W. Muhamad Hattad, I. A. Badruddin, S. Kamangar and M. R. Johan, "Differential Pulse Voltammetric Tuning of the Screen-Printed Carbon Electrode Surface to Enhance the Electrochemical Performance and Multiplex

[4] A. Hashem, A. R. Marlinda, M. M. Hossain, M. Al Mamun, M. Shalauddin, K. Simarani, and M. R. Johan., "A unique oligonucleotide probe hybrid on graphene decorated gold nanoparticles modified screen-printed carbon electrode for pork meat adulteration," Electrocatalysis, vol. 14, pp. 179-194, October 2023.

[5] M. Al Mamun, Y. A. Wahab, M. M. Hossain, A. Hashem, and M. R. Johan, "Electrochemical biosensors with aptamer recognition layer for the diagnosis of pathogenic bacteria: Barriers to commercialization and remediation," TrAC Trends in Analytical Chemistry, vol. 145, pp. 116458, October 2021.

[6] A. Hashem, M. M. Hossain, M. Al Mamun, K. Simarani, and M. R. Johan, "Rapid and sensitive detection of box turtles using an electrochemical DNA biosensor based on a gold/graphene nanocomposite," Beilstein J. of Nanotechnology, vol. 13, pp. 1458-1472, December 2022.

[7] A. J. S. Ahammad, T. Akter, A. A. Mamun, T. Islam, M. M. Hasan, M. Al Mamun, S. Faraezi, F. Z. Monira, and J. K. Saha, "Cost-effective electrochemical sensor based on carbon nanotube modified-pencil electrode for the simultaneous determination of hydroquinone and catechol," J. of The Electrochemical Society, vol. 165, pp. B390, June 2018.

[8] M. Al Mamun and A. J. S. Ahammad, "Characterization of carboxylated-SWCNT based potentiometric DNA sensors by electrochemical technique and comparison with potentiometric performance," J. of Biosensors & Bioelectronics, vol. 5, pp. 1, July 2014.

[9] A. J. Saleh Ahammad, M. A. Ullah, M. M. Hoque, M. Al. Mamun , M. K. Alam, A. N. Anju, M. N. Islam Mozumder, R. Karim, S. Sarker, and D. Min Kim, "Signal enhancement of hydroquinone and catechol on their simultaneous determination," International J. of Electrochemical Science, vol. 12, pp. 7570-7579, June 2017.

[10] A. J. Bard and L. R. Faulkner, "Student Solutions Manual to accompany Electrochemical Methods: Fundamentals and Applicaitons, 2e," 2002.

[11] M. Al Mamun, Y. A. Wahab, M. M. Hossain, A. Hashem, and M. R. Johan, "Scrap Gold Recovery: Recycling, Fabrication and Electrochemical Characterization of Low-Cost Gold Electrode," Malaysian Catalysis-An International Journal, vol. 2, pp. 1-20, October 2022.

[12] M Al Mamun, Y. A Wahab, M. M. Hossain, A. Hashem, M. R. Johan, N. E. Alias, H. Hussin, and "Determination of the Aptamer Probe Density by Double Layer and Redox Capacitance of CNT-Based Electrochemical DNA-Aptasensors," in 2023 IEEE Regional Symposium on Micro and Nanoelectronics (RSM), IEEE, pp. 74-77, November 2023.

[13] M. Al Mamun, Y .A. Wahab, M. M. Hossain, A. Hashem, K. A. Khan, M. R Johan, H. Hussin, and N. E. Alias "Electrochemistry of Green Ag Nanoparticles Modified Electrode Surface," in 2022 IEEE International Conference on Semiconductor Electronics (ICSE), IEEE, pp. 37-40, August 2022.

[14] S. Faraezi, M. S. Khan, F. Z. Monira, A. A. Mamun, T. Akter, M. Al Mamun, M. M. Rabbani, J. Uddin, and A. J. S. Ahammad "Sensitivity control of hydroquinone and catechol at poly (brilliant cresyl blue)-modified GCE by varying activation conditions of the GCE: an experimental and computational study," ChemEngineering, vol. 6, pp. 27, March 2022.

[15] S. Fletcher, "Screen-printed carbon electrodes," Electrochemistry of carbon electrodes, pp. 425-444, October 2015.

Evaluating the Impact of Upright and Inverted Pyramid Microstructures on the Optical Performance of Single Crystalline Silicon Solar Cells

Md. Yasir Arafat
Nanotechnology and Catalysis Research Centre,
Universiti Malaya,
Kuala Lumpur, Malaysia
yasir@um.edu.my

Yasmin Abdul Wahab
Nanotechnology and Catalysis Research Centre,
Universiti Malaya,
Kuala Lumpur, Malaysia
yasminaw@um.edu.my

Mohammad Aminul Islam
Faculty of Engineering,
Universiti Malaya,
Kuala Lumpur, Malaysia
aminul.islam@um.edu.my

Sharifah Fatmadiana Bt Wan Muhammad Hatta
Faculty of Engineering,
Universiti Malaya,
Kuala Lumpur, Malaysia
sh_fatmadiana@um.edu.my

Mohd Rafie Johan
Nanotechnology and Catalysis Research Centre,
Universiti Malaya,
Kuala Lumpur, Malaysia
mrafiej@um.edu.my

Nurul Ezaila Alias
School of Electrical Engineering,
Universiti Teknologi Malaysia
Johor Bahru, Malaysia
ezaila@fke.utm.my

Hanim Hussin
Faculty of Electrical Engineering,
Universiti Teknologi MARA
Shah Alam, Malaysia
hanimh@uitm.edu.my

Abstract— **This study examines the optical performance of single crystalline silicon solar cells with upright and inverted pyramid microstructures fabricated via Alkaline Chemical Etching and Metal Assisted Chemical Etching (MACE), respectively. Spectrophotometric and Finite Difference Time Domain (FDTD) analyses were used to evaluate light absorption and optical confinement. The weighted average reflectance results showed that inverted pyramids made with MACE had a significantly lower reflectance of 4.40% compared to 7.89% for upright pyramids, indicating superior light-trapping efficiency. This advantage is attributed to the favorable angular geometry and finer resolution of the MACE-fabricated inverted pyramids. These findings emphasize the importance of microstructural design and advanced fabrication techniques in enhancing the optical properties of photovoltaic materials, suggesting that tailored microfabrication strategies could significantly improve solar cell efficiency.**

Keywords— *Silicon Wafer Texturization, KOH Etching, MACE Process, Upright and Inverted pyramid microstructure, Surface Reflectance, Optical Performance, Single crystalline Silicon, FDTD, Solar Cell Optimization.*

I. INTRODUCTION

For many years, single crystalline silicon (sc-Si) and multi-crystalline silicon (mc-Si) solar cell, have dominated the worldwide solar energy industry [1]. Recently, the presence of sc-Si solar cells in the PV industry has been expanding quickly, attributed to their superior conversion efficiency and stability [2]. For researchers focused on solar cells, enhancing PV conversion efficiency through surface texturing of silicon wafers to improve light absorption is a primary goal. Texturing can lower the reflectivity of a planar Si wafer from over 45% to under 10%. [3,4]. Research into the texturing process facilitates the creation of a cost-effective and simplified fabrication approach, leading to enhanced efficiency in sc-Si solar cells.

KOH etching, an alkaline anisotropic chemical etching method, enhances photovoltaic efficiency by forming upright pyramid structures on silicon wafers that reduce reflectance and increase light absorption. The primary chemical reaction between hydroxide ions (OH⁻) and silicon (Si) can be represented as follows [4]:

$$Si + 2OH^- \rightarrow SiO_2 + 2H_2 + 2e^- \qquad (1)$$

The silicon initially reacts with hydroxide ions to form silicon dioxide (SiO_2) and hydrogen gas (H_2). The silicon dioxide is then dissolved by the KOH solution, which continues to expose fresh silicon for further reaction:

$$SiO_2 + 2H_2O + 2OH^- \rightarrow 2H_2SiO_3^{2-} \qquad (2)$$

The solubility of silicon dioxide in the KOH solution is critical, as it ensures the continuous removal of the reaction product, allowing the etching process to proceed efficiently.

MACE (Metal Assisted Chemical Etching) employs metal nanoparticles, including gold, silver, aluminum, copper, and nickel, as catalysts to create micro- or nanotextured surfaces on sc-Si. [5]. In this work, nickel (Ni) is used, and an optimized MACE technique is introduced that effectively reduced the surface reflectance. In the process of etching a Si substrate with a catalyst in an oxidizing solution, multiple reactions unfold, primarily at two distinct sites: the cathode and the anode. Specifically at the cathode, the sequence begins with the reduction of H_2O_2 at the metal surface, followed by the conversion of protons into hydrogen gas (H_2) [5]:

$$H_2O_2 + 2H^+ \rightarrow 2H2O + 2h^+ \qquad (3)$$

$$2H^+ + 2e^- \rightarrow H_2 \uparrow \qquad (4)$$

Additionally, at the anode, the etching reaction takes place where the Si substrate undergoes oxidation and dissolution:

$$Si + 4h^+ + 4HF \rightarrow SiF4 + 4H^+ \qquad (5)$$

$$SiF_4 + 2HF \rightarrow H_2SiF_6 \qquad (6)$$

Localized etching in the MACE process occurs through an electrochemical mechanism, where nanometer-sized metal particles serve as local cathodes [6]. A critical aspect of MACE involves the generation of holes (h+) from H_2O_2, facilitated by the reduction of oxidants like H_2O_2 or O_2, with the assistance of a metal catalyst. As a result, these holes are injected into the Si substrate at the metal interface. Metal nanoparticles, acting as catalysts, are surrounded by these holes due to electrostatic forces, leading to targeted etching of the Si directly beneath the catalyst [5].

II. EXPERIMENTAL PROCEDURE

In this section, two etching procedures are described: first, KOH anisotropic chemical etching, and then the MACE process. For both experiments, a P-type mono-crystalline Si (100) wafer with a thickness of 300±10 μm and a resistivity of 0.01-0.02 ohm-cm, measuring 2×2 cm², was used. Both experiments began with cleaning the Si (100) wafer using the RCA-1 and RCA-2 processes to effectively remove organic residues and metal ions from its surface. The subsequent steps are detailed in subsection A and B. The samples' morphological and optical properties were characterized using field emission scanning electron microscopy for surface morphology and compositional analysis, and a UV-vis-NIR spectrophotometer to measure surface reflectance within the 300-1200 nm wavelength range.

A. KOH Alkaline Chemical Etching

Typically, the KOH alkaline chemical etching process results in the formation of upright, randomly sized micro-pyramidal structures on the sc-Si surface layer. This section outlines the experimental procedure for the wet chemical anisotropic etching used in sc-Si fabrication. Different ratios of the KOH:IPA solution, as mentioned in Table I, have been tested to identify the optimum concentration of the etching solutions.

TABLE I.   SOLUTIONS BASED ON KOH: IPA MIXING RATIO

Component	Solution A	Solution B	Solution C	Solution D
KOH	1 g	3 g	3 g	5 g
IPA	5 ml	7 ml	5 ml	10 ml

After cleaning, the Si (100) wafer was treated to become hydrophilic using a 1:50 HF and deionized water mixture, followed by KOH alkaline chemical etching with IPA and KOH. Various quantities of KOH pellets (1g, 3g, 5g) and altered IPA volumes (5ml, 7ml, 10ml) were tested, maintaining constant deionized water at 125 ml, as detailed in Table I. The samples were etched for 10 to 30 minutes at a controlled temperature of 75 ± 5° C, and post-etching, immersed in a mixed solution of H_2O_2, HF, and deionized water at room temperature for 15 minutes.

B. Metal Assisted Chemical Etching (MACE)

Normally, the MACE method is divided into single-step and double-step processes [7]. In the two-step MACE used in this study, the 1st step includes metal deposition by one of three methods; for this work, deposition in an aqueous solution by calculating the molar ratio was chosen. The second step consists of etching in an HF: H_2O_2 solution.

TABLE II.   REAGENTS USED IN THE EXPERIMENT.

Action	Reagents Used
Si wafer cleaning	Hydrofluoric Acid HF, Hydrogen Peroxide H_2O_2, Hydrochloric Acid HCL, Ammonia $NH_3.H_2O$
Ni nanoparticle deposition	Ammonium Sulphate $(NH_4)_2SO_4$, Nickel(II) sulfate $NiSO_4(H_2O)_6$, Ammonium Floride NH_4F
Etching	Hydrofluoric Acid HF, Hydrogen Peroxide H_2O_2

After RCA-1 and RCA-2, followed by a hydrophilic treatment using a 50:1 mixture of HF acid and deionized water. The HF treatment was crucial for reducing surface hydrophobicity, preparing the substrate for deposition. Following this, the MACE process began with the deposition of Ni nanoparticles. A solution containing ammonium fluoride nickel (II) sulfate hexahydrate: $(NH_4)_3F$ $Ni(SO_4)_2(H_2O)_6$ was prepared as detailed in Table II. Subsequently, the Ni-coated Si wafers were etched in an HF: H_2O_2 solution mixed with deionized water at room temperature (25°C).

III. RESULTS AND DISCUSSION

In this section, the results of the samples are presented and analyzed. Fig. 1 (a), (b) displays the surface morphology of a sample that was etched using a KOH: IPA: H_2O mixture (1 g: 7 ml: 125 ml) for 20 minutes at a temperature of 75°C. Among the various parameters tested, this specific mixture achieved the most visually consistent textured surface. Consequently, the sample was chosen for characterization, and its results are discussed in this study. The UV-vis spectrometer reflectance results in Fig. 1 (c) show that wet chemical etching by KOH: IPA effectively reduces the reflectance, which is 23.34% less than the planar c-Si surface. The reflectance results prove that the experimental procedure successfully produced a low-reflective BSi surface layer.

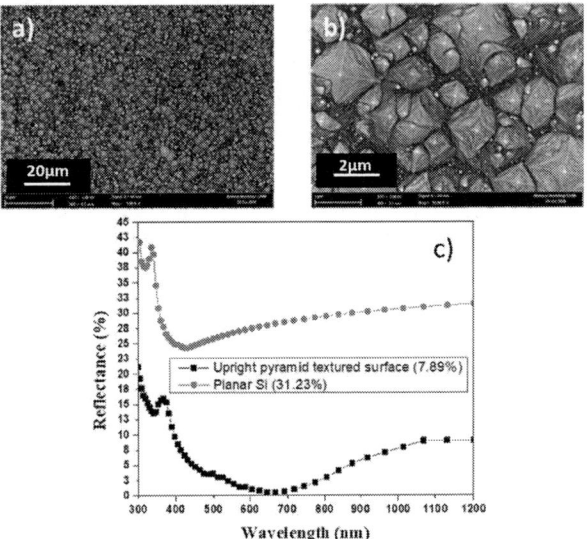

Fig. 1. (a) 20 μm and, (b) 2 μm zoom FESEM images show the effects of 20-minute anisotropic etching with KOH: IPA solution at 75°C, (c) Reflectance before and after etching.

979-8-3503-7832-0/24 $31.00 © 2024 IEEE

Each Si atom on the c-Si {100} plane has two dangling and two back bonds. During KOH etching, the alkaline etchant breaks these dangling bonds, initiating etching from the {100} plane toward the {111} plane [8]. The {111} plane, with one dangling bond and three back bonds, has higher activation energy and structural stability, making it resistant to etching. Consequently, etching predominantly occurs on the {100} plane, leading to the formation of upright micro-pyramidal structures as shown in Fig. 2 (a), and (b), with the cross-sectional FESEM image depicted in Fig. 2(c).

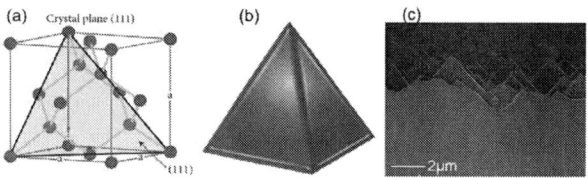

Fig. 2. (a) The {111} crystal plane, (b) a 3D schematic view of an upright pyramid, and (c) a cross-sectional FESEM image of the etched sample.

Fig. 3(a) and (b) display the surface and cross-sectional FESEM images of sc-Si wafers treated by Ni-MACE, revealing conspicuous variations in the size, depth, and morphology of the inverted pyramid structures. These structures form due to the corrosion from Ni nanoparticles. In the Ni-MACE process, Ni nuclei deposited at the Ni/Si interface cause the conduction and valence bands to bend downward, as the Fermi level of P-type silicon is lower than the work function of Ni ions. This results in electron transfer from Si to Ni^{2+}, promoting Ni nanoparticle growth and initiating silicon oxidation. The subsequent oxidation and HF dissolution precisely etch the silicon, creating inverted micro-pyramid structures on the wafer surface [6]. Figure 3(c) illustrates the substantial decrease in reflectance of the inverted pyramid structure when compared to the upright pyramid structure.

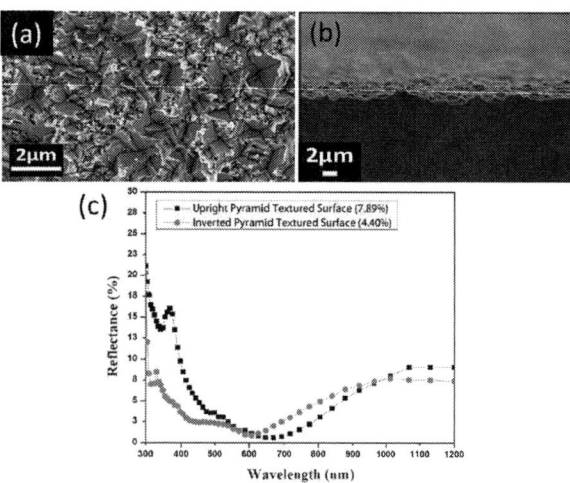

Fig. 3. (a) Upright view and, (b) cross-sectional view of the FESEM images of Ni-MACE etched sample. (c) Comparison of reflection results between the Ni-MACE sample and the KOH:IPA etched sample.

In the MACE process, metal nanoparticles catalyze the etching of Si {100} planes by focusing on dangling bonds,

leading to oxidation and selective etching. The {111} plane resists due to structural stability, with each electron supported by three back bonds and one dangling bond, resulting in higher activation energy and reduced susceptibility to etching. As Fig. 4(a) represents this preferential etching of the {100} planes produce inverted pyramid structures. Fig. 4(b) showcases these formations, while Fig. 4(c) presents a cross-sectional FESEM image of the inverted pyramids.

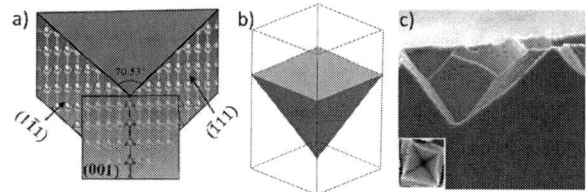

Fig. 4. (a) The {001} to {111} planes which forms the peak at a 70.53° angle. (b) 3d schematic of an inverted micro-pyramid, (c) Cross-sectional FESEM view of etched silicon, detailing the pyramid structures.

Based on the microstructural behaviour of the sample in terms of light incidence and reflectance, FDTD simulations have been conducted. The real surface microstructure dimensional data was used to construct the simulation model structures, as shown in Fig. 5(a) for the upright pyramid structure and Fig. 5(b) for the inverted pyramid structure. The XY views of these simulated models are also presented in Fig. 5(c) and Fig. 5(d), respectively.

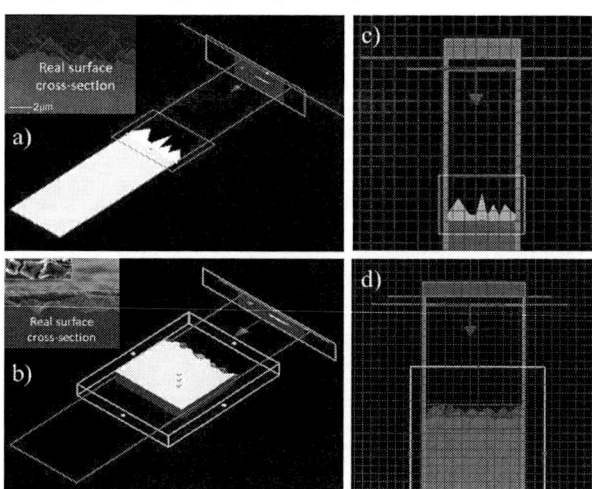

Fig. 5. FESEM cross-sectional and FDTD simulated perspective views of the (a) upright pyramid, (b) inverted pyramid microstructure model. The XY view of the FDTD simulated model structure of (c) upright pyramid, (d) inverted pyramid microstructure.

Advanced FDTD simulations were conducted to analyze the development of microstructure shapes, examining the pointing vector, electric field, and magnetic field distribution across different positions and as a function of frequency or wavelength. The field distributions for the upright pyramids in Fig.6 (a) and (b) represent the high-intensity interaction zones at the peaks, whether Fig. 6 (c) and (d) shows a concentrated field at the central base, suggesting a single, strong interaction zone of inverted pyramid microstructure. This configuration might be more effective for applications requiring focused field intensity, such as enhanced light trapping in photovoltaic cells. These visualizations provide insights into how different structural orientations—upright versus inverted—affect the behaviour of electric and magnetic fields at microscopic scales.

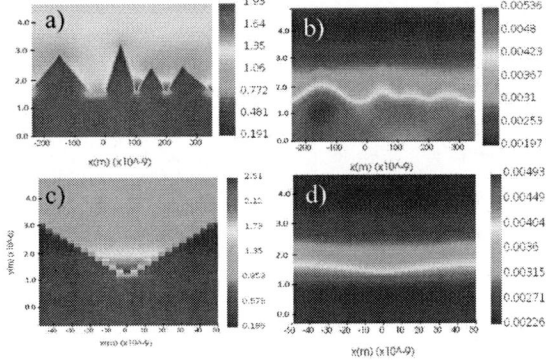

Fig. 6. (a) Electric field and, (b) Magnetic field data as a function of position and frequency/wavelength for upright pyramid structure, (c) Electric field and, (d) Magnetic field data as a function of position and frequency/wavelength for inverted pyramid structure.

The comparative analysis of reflectance results between real and simulated microstructural surface layers with upright and inverted pyramid structures reveals significant insights, as illustrated in Fig. 7. The inverted pyramid structure consistently shows superior performance in minimizing surface reflectance across both real and simulated models.

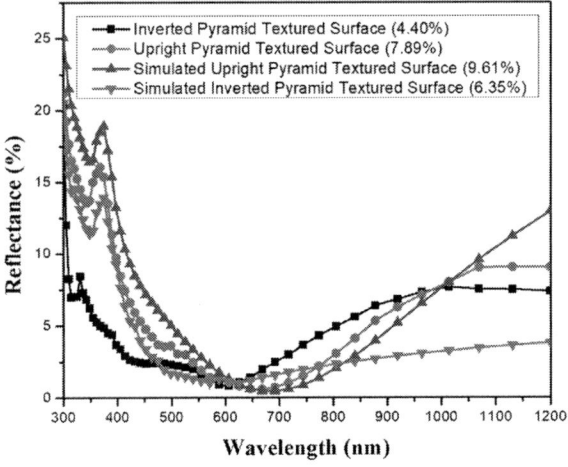

Fig. 7. Comparative reflectance results between real and simulated microstructural surface layers containing upright versus inverted structure surface layers.

In practical terms, the real inverted pyramid structure achieved a reflectance of 4.40%, significantly lower than the 7.89% observed in the real upright pyramid structure. This trend persisted in the simulated results, where the inverted pyramid structure exhibited a reflectance of 6.35%, compared to 9.61% for the upright pyramid structure. This data clearly supports the hypothesis that the inverted pyramid's design, by optimizing photon capture and minimizing reflective losses, enhances the optical properties of the surfaces.

IV. CONCLUSION

This study conducted a comparative analysis of upright and inverted pyramid microstructures on silicon wafer surfaces, textured via KOH and MACE etching methods. It confirmed that inverted pyramid structures significantly minimize surface reflectance, enhancing photon trapping and thereby improving photovoltaic efficiency. Real-world inverted pyramid structures achieved a reflectance as low as 4.40%, compared to 7.89% for upright pyramid structures. Similarly, simulated models reinforced these results, with inverted structures demonstrating lower reflectance values of 6.35% than their upright counterparts at 9.61%. These findings underscore the potential of advanced etching techniques in optimizing solar cell efficiency, providing a clear pathway for future enhancements in solar technology. This research highlights the importance of microstructural geometry in reducing reflective losses and increasing energy capture in photovoltaic applications.

ACKNOWLEDGMENT

The work is financially supported by the Ministry of Higher Education Malaysia (MOHE) via Fundamental Research Grant Scheme (FRGS/1/2022/TK09/UM/02/27). The authors also extend their appreciation to the Universiti Malaya (Grant No. ST055-2022).

REFERENCES

[1] P. Zhang et al., "An 18.9% efficient black silicon solar cell achieved through control of Pretreatment of Ag/cu mace," Journal of Materials Science: Materials in Electronics, vol. 30, no. 9, pp. 8667–8675, Mar. 2019.

[2] S. Sivaraj, R. Rathanasamy, G. V. Kaliyannan, H. Panchal, A. J. Alrubaie, M. M. Jaber, Z. Said, and S. Memon, "A comprehensive review on current performance, challenges and progress in thin-film solar cells," Energies, vol. 15, no. 22, p. 8688, 2022.

[3] X. Dai, R. Jia, G. Su, H. Sun, K. Tao, C. Zhang, P. Zhang, Z. Jin, and X. Liu, "The influence of surface structure on diffusion and passivation in multicrystalline silicon solar cells textured by metal assisted chemical etching (MACE) method," Solar Energy Materials and Solar Cells, vol. 186, pp. 42-49, 2018.

[4] D. Zhang, S. Jiang, K. Tao, R. Jia, H. Ge, X. Li, B. Wang, M. Li, Z. Ji, Z. Gao, and Z. Jin, "Fabrication of inverted pyramid structure for high-efficiency silicon solar cells using metal assisted chemical etching method with CuSO4 etchant," Solar Energy Materials and Solar Cells, vol. 230, p. 111200, 2021.

[5] H. Seidel, L. Csepregi, A. Heuberger, and H. Baumgärtel, "Anisotropic etching of crystalline silicon in alkaline solutions: I. Orientation dependence and behavior of passivation layers," Journal of the Electrochemical Society, vol. 137, no. 11, p. 3612, 1990.

[6] M. Y. Arafat, M. A. Islam, A. W. B. Mahmood, F. Abdullah, M. Nur-E-Alam, T. S. Kiong, and N. Amin, "Fabrication of black silicon via metal-assisted chemical etching—A review," Sustainability, vol. 13, no. 19, p. 10766, 2021.

[7] C. Bian, B. Zhang, Z. Zhang, H. Chen, D. Zhang, S. Wang, J. Ye, L. He, J. Jie, and X. Zhang, "Wafer-scale fabrication of silicon nanocones via controlling catalyst evolution in all-wet metal-assisted chemical etching," ACS Omega, vol. 7, no. 2, pp. 2234-2243, 2022.

[8] T. K. Adhila, H. Elangovan, K. Chattopadhyay, and H. C. Barshilia, "Kinked silicon nanowires prepared by two-step MACE process: synthesis strategies and luminescent properties," Materials Research Bulletin, vol. 140, p. 111308, 2021.

[9] M. A. Yasir, M. Aminul Islam, A. M. Wafi, F. Abdullah, T. K. Sieh, and N. Amin, "Study of black silicon wafer through wet chemical etching for parametric optimization in enhancing solar cell performance by pc1d numerical simulation," Crystals, vol. 11, no. 8, 2021.

A Flexible Framework Based on Finite-Element Method for Capacitance Extraction of 3-Dimensional Interconnects

Qiwen Zheng
School of Advanced Technology
Xi'an Jiaotong-Liverpool University
Suzhou, China
qiwen.zheng23@student.xjtlu.edu.cn

Ye Wu*
School of Advanced Technology
Xi'an Jiaotong-Liverpool University
Suzhou, China
ye.wu@xjtlu.edu.cn

Zichang Zhang
Electrical and Computer Engineering
Southern Illinois University Carbondale
Carbondale, USA
zichang@siu.edu

Abstract—**This paper proposed a capacitance extraction framework based on finite-element methods (FEM) to obtain the capacitance of three-dimensional (3-D) interconnect structures with multi-layer dielectrics. The proposed framework is demonstrated through its application to various 3-D interconnect configurations, revealing an error of no more than 8.03 % when compared to the commercial software Q3D, and 3.44 % compared to published papers. Notably, this framework exhibits enhanced flexibility in modeling and meshing, leading to a 1.5 % decrease in CPU time and a 40.7 % reduction in memory consumption compared to Q3D by optimizing the meshing strategy. Meanwhile, the proposed approach is based on the open-source software Gmsh and GetDP, ensuring high portability. Consequently, it can be adapted to various 3D interconnect structures, offering significant potential in designing very large-scale integrated circuits (VLSI) and enabling highly accurate parasitic extraction in post-layout simulations.**

Index Terms—**finite-element methods (FEM), Gmsh, GetDP, capacitance extraction, 3-D interconnects**

I. INTRODUCTION

With the scaling down of technology nodes, the reduction in the size of the device and circuit brings parasitic effects [1], which have negative effects on chip delay and power. Thus, precise parasitic extraction is crucial in the chip design process. 2.5-D methods offer rapid extraction speeds but an accuracy gap with 3-D field solvers [2]. 3-D field solvers are of high accuracy [3], the commonly used field solver for parasitic extraction include FEM [4], boundary element method (BEM) [5], floating random walk method (FRW) [6], etc. FEM has garnered significant interest due to its renowned accuracy and precision [4].

Traditional FEM is of high-accuracy but lack of speed, researchers made a lot of efforts to accelerate the FEM parasitic extraction, such as the matrix reduction [7], and the parallel technology [8]. However, FEM is complicated in extracting frequency-related parasitic parameters. In the past decade, people have proposed corresponding solutions: Ansys Q3D, the combination of FEM and BEM achieved high accuracy in low-frequency parasitic extraction [9]. In recent years, Stysch et al. focused on frequency-dependent parasitic extraction, for transient circuit parameters [10] and parasitic

impedance [11]. However, so far, commercial software such as Q3D and Raphael needs more flexibility in mesh generation and problem-solving.

In this paper, a flexible FEM capacitance extraction framework is proposed to obtain the capacitance of 3-D interconnects. Gmsh [12] is used to establish a 3-D model of interconnects structure; then GetDP [13] is used to extract capacitance from the 3-D structure. In addition, the obtained results are compared with commercial software results, and the proposed framework can achieve high accuracy.

II. RESEARCH METHOD

Capacitance extraction is obtain from the electrostatic potential, by solving Maxwell's equations. The potential distribution process is shown in Figure 1 (a). Using Gmsh to generate 3-D model and meshes, and GetDP to perform FEM on the nodes of meshes to obtain the potential. The capacitance between the master conductor and environment conductors will be calculated, as shown in Figure 1 (b). The potential of the master conductor is set to 1 V, and the potential of the environment conductors is set to 0 V.

It is worth mentioning that the setting of boundaries in FEM affects the accuracy of the results. The homogeneous Neumann boundaries and ground are set as shown in Figure 1 (b).

The boundary conditions of FEM in this paper are as follows, Dirichlet boundary conditions:

$$\begin{cases} \phi = 1, x \in \Gamma_m \\ \phi = 0, x \in \Gamma_{e,g} \\ \rho = 0, x \in \Gamma_c \end{cases} \tag{1}$$

Homogeneous Neumann boundary condition:

$$n \cdot \mathbf{D} = 0, x \in \Omega \tag{2}$$

The setting of the surface potential of the conductor and the ground potential are the Dirichlet boundary conditions in formula (1). The homogeneous Neumann boundary condition in formula (2) is also shown in Figure 2, which specifies the surface of the dielectric.

979-8-3503-7832-0/24 $31.00 © 2024 IEEE

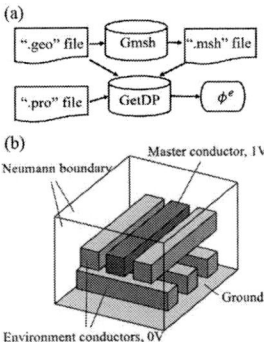

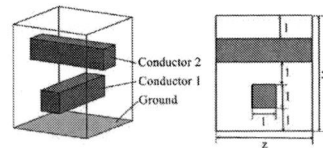

Fig. 1. (a) Flowchart of potential distribution process (including software used) (b) Schematic diagram of the interconnects model (3 × 3 crossover)

According to the Maxwell equation, the Poisson equation for scalar potential can be obtained

$$-\nabla \cdot (\varepsilon \nabla \phi) = \rho \tag{3}$$

At each element node, the basis function is

$$\phi^e = \sum_{j=1}^{K} \phi_j^e V_j^e \tag{4}$$

where K is the number of element nodes, V_j is the coefficient, the coefficient value can be obtained based on the nodes located on the boundary. Record the test function as φ, based on (3), obtain

$$\int_V \varepsilon \nabla \phi_j \cdot \nabla \varphi_j dV - \int_\Omega (\varepsilon \nabla \phi_j \cdot n) \cdot \varphi_j d\Omega = 0 \tag{5}$$

According to formulae (1) and (2), φ_j does not exist under the Dirichlet boundary condition, under the homogeneous Neumann boundary condition, $\varepsilon \nabla \phi_j \cdot n = 0$. Thus, the formula (5) is simplified as

$$\int_{V^e} \varepsilon^e \nabla \phi_j^e \cdot \nabla \varphi_j^e dV^e = 0 \tag{6}$$

GetDP will automatically use the values of boundary conditions to solve the basis function, thereby obtaining the potential ϕ on each element node. Substitute into $w = \frac{1}{2}\varepsilon(\nabla\phi)^2$ to obtain

$$W = \frac{1}{2} \sum_{i=1}^{K} \int_{V^e} \varepsilon^e (\nabla \phi^e)^2 dV^e \tag{7}$$

Finally, to calculate the capacitance of the master conductor.

$$C = \frac{2W}{\Phi^2} \tag{8}$$

where Φ is the potential difference between the master conductor and the environmental conductors.

III. NUMERICAL RESULTS

In this section, the errors of the proposed framework with published papers and commercial software for several different interconnects structures are presented, to validate the accuracy of the proposed framework.

Fig. 2. Schematic diagram of 1 × 1 crossover conductors (left) and its cross-sectional view (right). The units of the numbers marked in the figure are all micrometers.

A. 1 × 1 Crossover

As shown in Figure 2, the structure is composed of two $1 \times 1 \times z$ conductors perpendicular to each other. The dielectric constant is 1, and the bottom of the dielectric is the ground plane. The size of the dielectric is $z \times z \times 5$. The distance from conductor 1 to the ground, the distance between two conductors, and the distance from conductor 2 to the Neumann upper boundary are all 1. The length unit is micrometer.

Four cases with $z = 4, 5, 7$, and 10 are examined, and the results are compared with Raphael and Q3D. Table I presents the error between the proposed framework and Raphael, Q3D. The data indicates the proposed framework has a high degree of agreement with the Q3D and Raphael, with errors generally within 6.62 %. However, this structure is relatively simple and only contains one type of dielectric, more conductors with bends and multi-layer dielectric structure is conducted in the following experiment.

TABLE I
THE ERROR OF 1 × 1 CROSSOVER OF GETDP (COMPARE WITH Q3D AND RAPHAEL [5] RESPECTIVELY).

Conductor length z	C11		C22	
	GetDP/ Q3D	GetDP/ Raphael	GetDP/ Q3D	GetDP/ Raphael
4	-5.03%	-1.64%	-3.89%	-1.43%
5	-6.27%	-2.83%	-4.62%	-2.21%
7	-6.55%	-3.10%	-4.66%	-2.44%
10	-6.62%	-2.96%	-5.05%	-3.07%

B. 2 × 2 Crossover above Two Bends with Multi-layer Dielectrics

As shown in Figure 3 (a), the structure consists of 6 conductors, conductors 1 and 2 are bent, with a cross-sectional area of 1×1. The size of conductors 3, 4, and 5, 6 are both $1 \times 1 \times 13$. Their position is shown in the Figure 3 (b); the distance between the two conductors is 3. The length unit is millimeter. This structure consists of 7 layers of dielectrics, with thicknesses of 1, 1, 2, 1, 1, 1, and 1 from bottom to top. Each layer corresponds to a dielectric constant of 2, 3, 3, 4, 4, 5, and 5. The bottom of the entire structure is the ground plane, and the top is the Dirichlet boundary, with the potential set to 0.

The error comparison is presented in Table II. The maximum error between the proposed framework and Q3D is 8.03 %, while GetDP's results are in close agreement with DDM and SpiceLink, with a maximum error of only 3.44 %.

979-8-3503-7832-0/24 $31.00 © 2024 IEEE

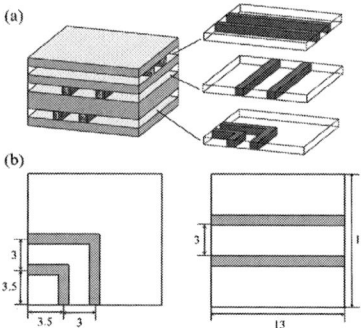

Fig. 3. (a) Schematic diagram of 2×2 crossover conductors above two bends with seven dielectric layers (b) Distribution map of the positions of two bend conductors and 2×2 crossover in the corresponding layers. The units of the numbers marked in the figure are all millimeters.

This demonstrates the proposed framework's effectiveness for interconnect structures with multi-layer dielectrics and bends.

TABLE II

THE ERROR OF 2×2 CROSSOVER ABOVE TWO BENDS WITH MULTI-LAYER DIELECTRICS OF GETDP (COMPARE THE RESULTS OBTAINED BY Q3D, DDM [14], AND SPICELINK [14] RESPECTIVELY).

Error	GetDP/Q3D	GetDP/DDM	GetDP/SpiceLink
C11	8.03%	0.74%	-0.90%
C22	1.27%	-2.48%	-2.48%
C33	1.34%	-3.44%	-1.50%
C44	1.69%	-3.03%	-1.69%
C55	-1.13%	-1.85%	-2.25%
C66	-0.86%	-1.69%	-2.09%

C. 3×3 Crossover with Multi-layer Dielectrics

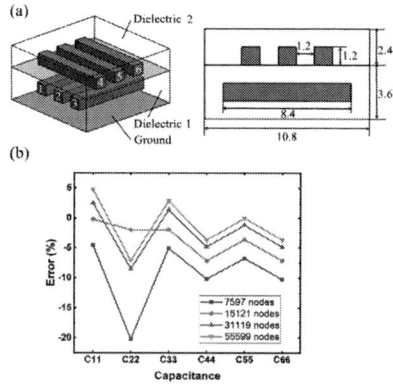

Fig. 4. (a) Schematic diagram of 3×3 crossover conductors with two dielectric layers (left) and its cross-sectional view (right). The units of the numbers marked in the figure are all micrometers. (b) The error of different mesh generation methods with the results obtain by Q3D in the test structure. The x-axis represents the six conductors of the structure.

The proposed framework has more flexibility in mesh generation. Reducing the number of meshes in areas with less potential changes will lead to decrease of CPU time and memory consumption while ensuring accuracy, relevant

experiments are conducted in this section. The impact of mesh generation strategy on accuracy, CPU time, and memory consumption is analyzed. As shown in Figure 4 (a), the test structure consists of two sets of perpendicular conductors, each containing three parallel conductors with a spacing of 1.2. The size of each conductor is $1.2 \times 1.2 \times 8.4$. The height of two dielectrics are 3.6 and 2.4, and the dielectric constant of 3.9 and 7.8, from bottom to top. The length unit is micrometer.

Gmsh is employed with four mesh generation strategies (with 7597, 15121, 31119, and 55599 nodes, respectively, mainly reduces the mesh in areas with less potential changes). Their errors compared to Q3D are displayed in Figure 4 (b). The errors for 15121, 31119, and 55599 nodes are acceptable, but overly dense meshes increase computational demands. The method with 15121 nodes is the optimal choice, reducing CPU time by 1.5 % and memory consumption by 40.7 % compared to Q3D. This demonstrates that a reasonable balance between mesh density and computational efficiency can be achieved without significantly compromising accuracy.

IV. DISCUSSION AND CONCLUSIONS

In this paper, a flexible framework based on FEM is proposed to obtain capacitance from 3-D interconnects. The results from the first two experiments demonstrate that the errors between the proposed framework and published papers are within 3.1 % and 3.44 %, respectively, while compared to the commercial software Q3D, the errors are within 6.62 % and 8.03 %, respectively. In the third experiment, a modified mesh generation strategy led to the error of -7.15 %, with a 1.5 % decrease in CPU time and a 40.7 % reduction in memory consumption compared to Q3D. Therefore, the proposed framework is of high-accuracy and flexibility, enabling the extraction of parasitic parameters for various interconnect structures.

As chip integration increases, the negative impact of parasitic capacitance on PPA indicators intensifies. The industry needs a parasitic extraction method of high accuracy. Therefore, a 3D field solver based on FEM is used to obtain high-accuracy parasitic parameters, helps to carry out more analysis of chip performance. The proposed framework can flexibly define the mesh, enhancing the generalization ability of the framework, which has great application prospects on chip-level parasitic extraction.

REFERENCES

[1] J. Chen, H. Peng, Z. Cheng, X. Liu, Q. Xin, Y. Kang, J. Wu, and X. Chu, "A Novel Power Loop Parasitic Extraction Approach for Paralleled Discrete SiC MOSFETs on Multilayer PCB," *IEEE Journal of Emerging and Selected Topics in Power Electronics*, vol. 9, no. 5, pp. 6370–6384, 2021.

[2] N. K. Karsilayan, J. Falbo, and D. Petranovic, "Efficient and Accurate RIE Modeling Methodology for BEOL 2.5D Parasitic Extraction," in *2014 IEEE 57th International Midwest Symposium on Circuits and Systems (MWSCAS)*, pp. 519–522, 2014.

[3] M. S. Abouelyazid, S. Hammouda, and Y. Ismail, "Accuracy-Based Hybrid Parasitic Capacitance Extraction Using Rule-Based, Neural-Networks, and Field-Solver Methods," *IEEE Transactions on Computer-Aided Design of Integrated Circuits and Systems*, vol. 41, no. 12, pp. 5681–5694, 2022.

979-8-3503-7832-0/24 $31.00 © 2024 IEEE

[4] S. Rajagopalan and S. Batterywala, "A 3-dimensional FEM Based Resistance Extraction," in *20th International Conference on VLSI Design held jointly with 6th International Conference on Embedded Systems (VLSID'07)*, pp. 565–570, 2007.

[5] W. Yu, Z. Wang, and J. Gu, "Fast Capacitance Extraction of Actual 3-D VLSI Interconnects Using Quasi-Multiple Medium Accelerated BEM," *IEEE Transactions on Microwave Theory and Techniques*, vol. 51, no. 1, pp. 109–119, 2003.

[6] M. Yang, W. Yu, M. Song, and N. Xu, "Volume Reduction and Fast Generation of the Precharacterization Data for Floating Random Walk-Based Capacitance Extraction," *IEEE Transactions on Computer-Aided Design of Integrated Circuits and Systems*, vol. 41, no. 5, pp. 1467–1480, 2022.

[7] A. Schade, F. Klotz, S. Jahn, and R. Weigel, "S-EHRFEM – Substrate Extraction using Highly Reduced FEM-meshes for Transient SPICE-simulation with Iterative Linear Solvers," in *2019 International Symposium on Electromagnetic Compatibility - EMC EUROPE*, pp. 812–819, 2019.

[8] G. Chen, H. Zhu, T. Cui, Z. Chen, X. Zeng, and W. Cai, "ParAFEMCap: A Parallel Adaptive Finite-Element Method for 3-D VLSI Interconnect Capacitance Extraction," *IEEE Transactions on Microwave Theory and Techniques*, vol. 60, no. 2, pp. 218–231, 2012.

[9] Ansys, "Ansys Q3D Extractor." (2023, Nov 28).

[10] J. Stysch, A. Klaedtke, and H. de Gersem, "Finite Element Extraction of Frequency-Dependent Parasitics," *IEEE Transactions on Magnetics*, vol. 58, no. 9, pp. 1–4, 2022.

[11] J. Stysch, A. Klaedtke, and H. De Gersem, "Broadband Finite-Element Impedance Computation for Parasitic Extraction," *Electrical Engineering*, vol. 104, no. 2, pp. 855–867, 2022.

[12] C. Geuzaine and J.-F. Remacle, "Gmsh: A 3-D Finite Element Mesh Generator with Built-in Pre- and Post-Processing Facilities," *International Journal for Numerical Methods in Engineering*, vol. 79, no. 11, pp. 1309–1331, 2009.

[13] P. Dular, C. Geuzaine, F. Henrotte, and W. Legros, "A General Environment for the Treatment of Discrete Problems and its Application to the Finite Element Method," *IEEE Transactions on Magnetics*, vol. 34, no. 5, pp. 3395–3398, 1998.

[14] Z. Zhu, H. Ji, and W. Hong, "An Efficient Algorithm for the Parameter Extraction of 3-D Interconnect Structures in the VLSI Circuits: Domain-Decomposition Method," *IEEE Transactions on Microwave Theory and Techniques*, vol. 45, no. 8, pp. 1179–1184, 1997.

Bandgap Modification of ZnO Nanorods for Enhanced Photocatalytic Application

Aini Ayunni Mohd Raub
Institute of Microengineering and Nanoelectronics
Universiti Kebangsaan Malaysia
Bangi, Malaysia
ainiayunni@ukm.edu.my

Siti Nur Ashakirin Mohd Nashruddin
Institute of Informatics and Computing in Energy (IICE), Department of Computing College of Computing & Informatics, University of Tenaga Nasional, Malaysia.
ashakirin@uniten.edu.my

Mohd Ambri Mohamed
Institute of Microengineering and Nanoelectronics
Universiti Kebangsaan Malaysia
Bangi, Malaysia
ambri@ukm.edu.my

Jumril Yunas
Institute of Microengineering and Nanoelectronics
Universiti Kebangsaan Malaysia
Bangi, Malaysia
jumrilyunas@ukm.edu.my

Abstract— In this paper, we report the bandgap modification of ZnO-based nanorods (ZnO NRs) through carbon-based nanocomposite (ZnO/rGO) and metal doping using Cu, Ni, Bi, Co, and Mg ions as the dopants. The hydrothermal method was used to grow the nanorod's structure by incorporating a metal doping solution during the growing process and rGO coating. The analysis of the synthesized materials shows that metal doping affects the bandgap of the ZnO NRs. The band gaps of pristine ZnO NRs, ZnO NRs/rGO, Cu, Ni, Bi, Co, and Mg-doped ZnO NRs are 3.25, 3.17, 3.07, 3.15, 3.11, 3.18, and 3.14 eV, respectively. Cu-doped ZnO NRs have the most significant reduction in bandgap energy compared to other metal-doped ZnO NRs. Meanwhile, the rGO coating causes bandgap reduction of ZnO NRs from 3.25 to 3.17 eV.

Keywords— *Cu doping, optical properties, reduced Graphene Oxide, and ZnO nanorods.*

I. INTRODUCTION

Photocatalysis using metal oxide semiconductor ZnO has garnered interest for its effectiveness in the photodegradation of various bio and non-bio-contaminants. Nonetheless, its utility is constrained because it can only absorb UV light (constituting 4% of the solar spectrum) and not visible light (comprising 43% of the solar spectrum). This limitation hampers the overall performance of the ZnO photocatalyst, exacerbated by a rapid recombination rate of photogenerated electron-hole pairs [1].

Metal doped-metal oxides constitute a significant class of materials that exhibit diverse and tailored properties, making them suitable for various applications. Metal ions into metal oxide lattices introduce new functionalities, impacting electronic and catalytic properties.

Doped metal oxides combine the intrinsic properties of metal oxides with the unique characteristics of metals/transition metals. Doping offers a versatile approach to tuning the electronic band structure and its catalytic activity of metal oxides [2]. The selection of transition metals for doping with elements such as Fe, Co, Ni, and Mn is commonly explored [3].

Alternatively, the heterostructure of ZnO and carbon materials can also enhance the electrical properties and catalytic performance of the nanocomposite of photocatalyst [4]. Reduced graphene oxide (rGO) gains significant attention owing to its similarities with pristine graphene in terms of its structural and functional properties. An effective synergy between ZnO and rGO was observed in waterborne Gram-negative and Gram-positive bacteria. This nanocomposite also proves to be a proficient photocatalyst for the degradation of Malachite Green dye [5].

Various synthesis methods incorporate transition metal dopants into metal oxides, including sol-gel, hydrothermal, and chemical vapour deposition [6, 7]. The synthesis method is crucial in controlling the morphology and distribution of dopants within the material.

Modified metal oxides have diverse applications, including catalysis, gas sensing, energy storage, and electronic devices [8, 9]. In conclusion, doped metal oxides represent a vibrant field of research with broad applications.

This study investigates the effects of doping using Cu, Ni, Bi, Co, and Mg as dopants and the integration of carbon on the bandgap of ZnO NRs. Cu-doped ZnO NRs have the most significant reduction in bandgap energy compared to other metal-doped ZnO NRs. Meanwhile, the rGO coating causes bandgap reduction of ZnO NRs from 3.25 to 3.17 eV.

II. EXPERIMENTAL METHOD

The remainder of this paper is arranged as follows. Section II discusses synthesising and characterising the metal-doped ZnO NRs and the ZnO NRs coated with rGO.

A. Synthesis

Hydrothermally, the pure and doped ZnO samples were grown on glass substrates. A magnetic stirrer was used to mix 20 mL of methanol with 0.01 M of ZnA, weighing 43.87 mg, at 60 °C for 30 minutes. Meanwhile, a 0.09 M NaOH solution weighing 72 mg was prepared by dissolving it in 20 mL of methanol, and the mixture was stirred for 30 minutes at room temperature. Subsequently, to produce stabilized ZnO NPs, the NaOH solution was gradually introduced into the ZnA solution while stirring with a magnetic stirrer at 60 °C for two hours. The resulting ZnO nanoparticle suspension solution was spin-coated onto clean glass surfaces, and then the substrates were annealed at 250 °C after coating.

An equimolar mixture of $Zn(NO_3)_2$ and HMTA in DI water was prepared and stirred for two hours to achieve a uniform solution. The substrates, covered with a ZnO nano seed layer, were inverted and placed in a beaker containing the precursor solution. Subsequently, the beaker was maintained at 95 °C for 4 hours. For Cu-doping, an equimolar solution of HMTA and zinc nitrate was prepared, and the $Zn(NO_3)_2$ solution was mixed with $Cu(NO_3)_3$ (4%Wt in the precursor solution). The substrates were cleansed with DI water and annealed at 250 °C for 1 hr. The same experiment was repeated using $Ni(NO_3)_2$, $Mg(NO_3)_2$, $Co(NO_3)_2$, and $Bi(NO_3)_3$.

For rGO coating, the ZnO NRs were treated with 50-watt oxygen plasma for 15 mins before GO solution (0.075 mg/mL) was spray-coated on top, and then the substrate was annealed for 1 hr. The details of the experiment can be referred to the previous report [10].

B. Characterization

Analysis techniques such as FESEM, UV-Vis, and XRD were employed. The crystal structure of the specimens underwent examination using a Bruker D8 Advance X-ray diffraction (XRD) instrument, utilizing CuKα radiation (λ = 1.54 Å). The morphological analysis was carried out using a Supra 55VP, Zeiss field emission scanning electron microscope (FESEM). UV-visible absorbance spectra were captured using a UV-visible spectrometer U39000-H, Hitachi.

III. RESULTS AND DISCUSSION

A. XRD analysis

All the XRD patterns of the samples show the wurtzite structure of ZnO (JCPDS No. 36-1451) demonstrated in Fig. 1. Due to the low concentration of doping ions and their increased likelihood of substituting zinc atoms, it is not possible to precisely determine the position of the metal. The intensity of the (100) peak is enhanced for Cu-ZnO and Bi-ZnO. The XRD examination reveals that the metals doped ZnO exhibits a hexagonal structure with high crystal quality.

Metals share a comparable atomic radius with Zn, facilitating their straightforward incorporation into the ZnO lattice. Additionally, cation doping ZnO NRs is a valuable tactic for increasing charge separation and decreasing the rate at which photoexcited electron-hole pairs recombine, which increases ZnO NRs' photoactivity.

Furthermore, cation doping of ZnO NRs is an effective strategy for improving charge separation and reducing the rate at which photoexcited electron-hole pairs recombine, which increases ZnO NRs' photoactivity.

For ZnO NRs/rGO, diffraction peaks for rGO can be observed at 24.5°.

B. Morphology analysis

From the morphology analysis of Fig. 2, introducing the metals increases the average diameter and decreases the height of ZnO NRs. This is due to the doping of metal ions affecting ZnO's crystalline nucleation and growth. In Fig. 2. (a) and (d), pristine ZnO has a height of 1139 nm and a diameter of 99.08 nm. In Fig. 2. (c) and (f), Cu-doped ZnO NRs has a height of 782.865 nm and a diameter of 99.31 nm.

Compared with the pristine ZnO NRs, the diameter increases at 23.58 %, and the height of the Cu-doped ZnO NRs decreases significantly, as much as 58.02 %. Metal atoms drastically affect crystalline growth and promote only

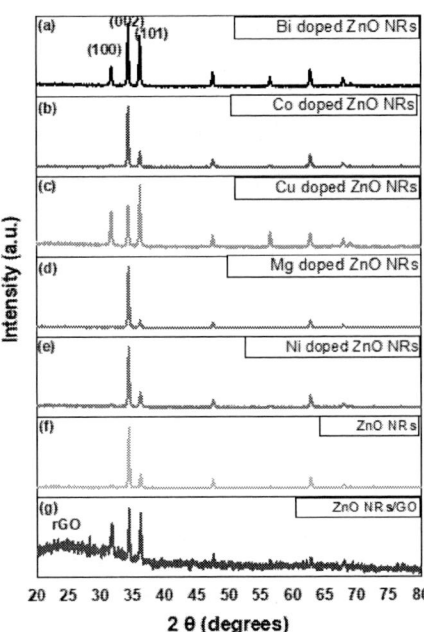

Fig. 1. XRD patterns of (a) Bi-doped, (b) Co-doped, (c) Cu doped, (d) Mg doped, (e) Ni-doped ZnO NRs, (f) ZnO NRs, and (g) ZnO NRs/rGO.

one-dimensional growth along the c-axis. However, due to the low doping percentage, the EDX can't detect the presence of metal elements.

In Fig 2. (b) and (e), the ZnO NRs/RGO wrinkled rGO layers cover the top of ZnO NRs. The rGO layers contribute to effective electron transfer from the ZnO CB to the rGO surface, enhancing photocatalytic activity.

C. UV-Visible spectroscopy

The pristine ZnO NRs show an absorption edge at a wavelength of 380 nm, which suggests a bandgap of around 3.25 eV. All metal-doped ZnO NRs demonstrate an improved sensitivity to visible light, as shown in Figure 3 (a). The absorption edge of all metal-doped ZnO NRs exhibits a redshift, indicating a decrease in the bandgap width following metal doping.

By projecting the straight line of the $(\alpha h\upsilon)^2$ against the photon energy $(h\upsilon)$ plot to the x-axis, the bandgap of the materials was determined using the Tauc Plot.

From the tauc plot, as shown in Fig. 3. (b), the band gaps of pristine ZnO NRs, ZnO NRs/rGO, Cu, Ni, Bi, Co, and Mg-doped ZnO NRs are 3.25, 3.17, 3.07, 3.15, 3.11, 3.18, and 3.14 eV, respectively as listed in Table 1. Cu-doped ZnO NRs have the most significant reduction in bandgap energy compared to other metal-doped ZnO NRs.

The introduction of metal ions resulted in a fundamental change in the bandgap of ZnO. Absorption measurements indicated a substantial decrease in the band gap of ZnO NRs after incorporating Cu ions. This phenomenon creates additional energy levels above the valence band (VB) and below the conduction band (CB).

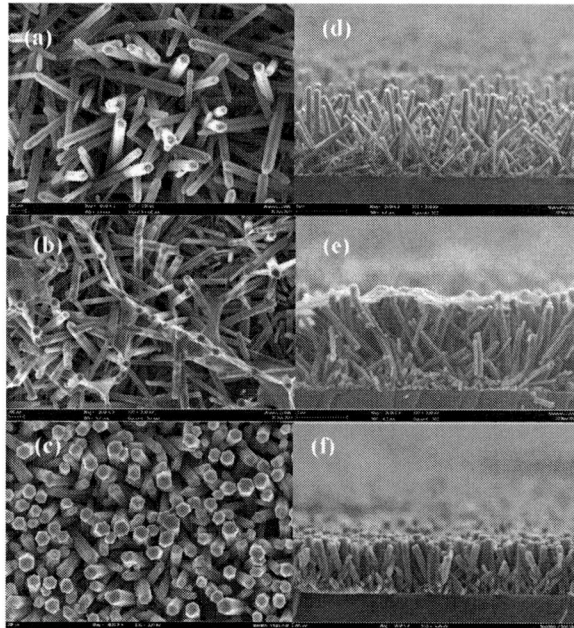

Fig. 2. FESEM images (a) morphology of pristine ZnO NRs, (b) ZnO NRs/rGO, (c) Cu-doped ZnO NRs, (d) cross-section of ZnO NRs, (e) ZnO NRs/rGO, and (f) Cu-doped ZnO NRs.

The doping of ZnO NRs with metal atoms is to enhance their ability to absorb visible light by decreasing the band gap. The presence of doped metal atoms modifies the arrangement of the Zn atom, adjusting the electronic properties of ZnO by increasing the highest valence band energy, decreasing the lowest conduction band energy, or introducing specific electronic energy levels within the range of energy levels between the valence and conduction bands [11], as depicted in Figure 4. Copper (Cu) acts as a p-type dopant by introducing a deep acceptor level below the conduction band [12]. Tauc's figure indicates that the bandgap energy is 3.07 eV.

The increased photocatalytic activity reported in the Cu-doped ZnO NRs can be attributed to the creation of more trapping sites, a greater surface area, and the effective generation and separation of electron-hole pairs [3].

Meanwhile, the increase in surface charge at the ZnO and rGO interface resulted in a shift of the optical band gap towards longer wavelengths. This phenomenon can account for a decreased ZnO NRs/rGO composite band gap. Consequently, the nanocomposite exhibited a reduced band gap of 3.17 eV, as demonstrated in Fig. 4. (b). The previous reports confirmed this finding [13, 14]. The Fermi level of rGO is -0.08 eV versus NHE. Therefore, the electron can be easily transferred from the ZnO CB to rGO [14].

Table 1. Calculated values of optical bandgaps for doped ZnO NRs.

Samples	Optical Bandgap (eV)
ZnO NRs	3.25
ZnO NRs/rGO	3.17
Cu doped ZnO NRs	3.07
Ni doped ZnO NRs	3.15
Bi doped ZnO NRs	3.11
Co doped ZnO NRs	3.18
Mg doped ZnO NRs	3.14

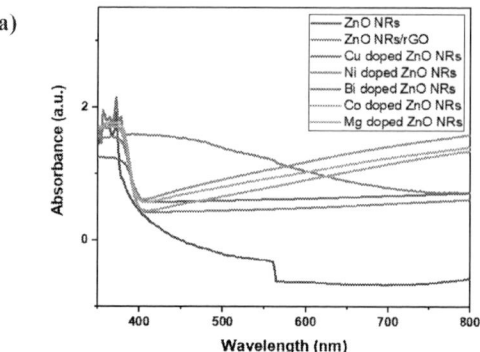

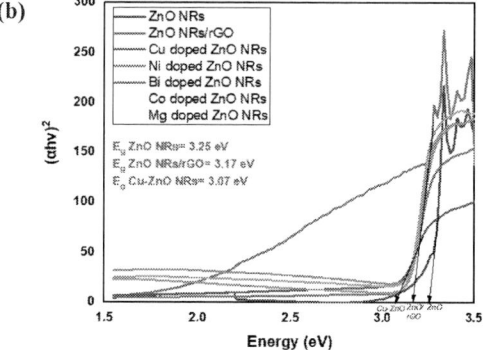

Fig. 3. (a) Absorbance spectra and (b) Tauc plot. and bandgap line analysis for Cu-doped and pure ZnO NRs.

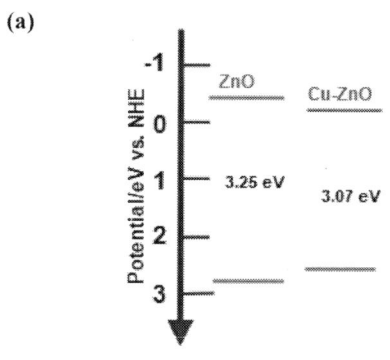

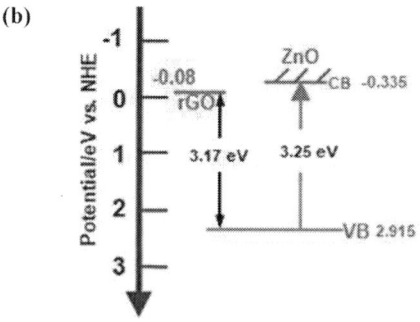

Fig. 4. Energy level diagram of (a) Cu-doped ZnO NRs, (b) ZnO NRs/RGO.

IV. CONCLUSION

The metals Cu, Ni, Fe, Mg, Co, and Bi were introduced as dopants to ZnO NRs to improve the photocatalytic activity of the metal oxide via a hydrothermal process. Compared to the pristine ZnO NRs, doped ZnO NRs and carbon-based ZnO composite show enhanced visible light absorption and energy bandgap reduction. The decreased energy bandgap and enhanced photon absorption in the visible spectrum resulting from the interaction between ZnO NRs and rGO could be attributed to the effective transfer of charges. The metal ions doping and rGO layer, which act as electron traps, lower the recombination rate of photogenerated carriers, thus enhancing the photocatalytic activity of the synthesized materials. In future work, photocatalytic activity for the photodegradation of the dye of each synthesized material will be investigated to prove the enhanced performance.

ACKNOWLEDGEMENT

This research was supported by Universiti Kebangsaan Malaysia under Project DIP-2023-003, Nanofabrication of Vertical Aligned ZnO/Graphene Nanorods using Lithography Nanopatterning Process, and by the Ministry of Higher Education Malaysia under Project Grant FRGS/1/2023/TK09/UKM/02/4, Antibacterial Photocatalytic Mechanism on Doped ZnO/Graphene Oxide Nanorods as Reusable Antibacterial Agent. The authors would also like to thank all UKM staff and technicians from the Institute of Microengineering and Nanoelectronics who contributed to this study.

REFERENCES

[1] A. Serrà et al., "Hybrid Ni@ZnO@ZnS-Microalgae for Circular Economy: A Smart Route to the Efficient Integration of Solar Photocatalytic Water Decontamination and Bioethanol Production," *Advanced Science*, vol. 7, no. 3, p. 1902447, 2020.

[2] P. Karuppasamy, N. Ramzan Nilofar Nisha, A. Pugazhendhi, S. Kandasamy, and S. Pitchaimuthu, "An investigation of transition metal doped TiO2 photocatalysts for the enhanced photocatalytic decoloration of methylene blue dye under visible light irradiation," *Journal of Environmental Chemical Engineering*, vol. 9, no. 4, p. 105254, 2021.

[3] K. Qi et al., "Transition metal doped ZnO nanoparticles with enhanced photocatalytic and antibacterial performances: Experimental and DFT studies," *Ceramics International*, vol. 46, no. 2, pp. 1494-1502, 2020.

[4] N. A. F. Al-Rawashdeh, O. Allabadi, and M. T. Aljarrah, "Photocatalytic Activity of Graphene Oxide/Zinc Oxide Nanocomposites with Embedded Metal Nanoparticles for the Degradation of Organic Dyes," *ACS Omega*, vol. 5, no. 43, pp. 28046-28055, 2020.

[5] P. Rajapaksha et al., "Broad spectrum antibacterial zinc oxide-reduced graphene oxide nanocomposite for water depollution," *Materials Today Chemistry*, vol. 27, p. 101242, 2023.

[6] J. Y. Jaenudin Ridwan, Akrajas Ali Umar, Aini Ayunni Mohd Raub, Azrul Azlan Hamzah, Jamal Kazmi, Asep Nandiyanto, Roer Eka Pawinto, and Ida Hamidah, "Vertically Aligned Cu-Doped ZnO Nanorods for Photocatalytic Activity Enhancement," *International Journal of Electrochemical Science*, 2022.

[7] Z. Han et al., "Transition metal elements-doped SnO2 for ultrasensitive and rapid ppb-level formaldehyde sensing," *Heliyon*, vol. 9, no. 2, p. e13486, 2023.

[8] A. H. Al-Naggar, N. M. Shinde, J.-S. Kim, and R. S. Mane, "Water splitting performance of metal and non-metal-doped transition metal oxide electrocatalysts," *Coordination Chemistry Reviews*, vol. 474, p. 214864, 2023.

[9] Lichchhavi, A. Kanwade, and P. M. Shirage, "A review on synergy of transition metal oxide nanostructured materials: Effective and coherent choice for supercapacitor electrodes," *Journal of Energy Storage*, vol. 55, p. 105692, 2022.

[10] A. A. Mohd Raub et al., "Synthesis and characterization of ZnO NRs with spray coated GO for enhanced photocatalytic activity," *Ceramics International*, 2022.

[11] P. G. Ramos Apestegui, L. Sanchez Rodas, and J. Rodríguez, "A review on improving the efficiency of photocatalytic water decontamination using ZnO nanorods," *Journal of Sol-Gel Science and Technology*, vol. 102, 2022.

[12] X. Peng, J. Xu, H. Zang, B. Wang, and Z. Wang, "Structural and PL properties of Cu-doped ZnO films," *Journal of Luminescence*, vol. 128, no. 3, pp. 297-300, 2008.

[13] A. A. Mohd Raub et al., "Characterization of ZnO/rGo Nanocomposite and its Application for Photocatalytic Degradation," *Journal of Nanoelectronics and Optoelectronics*, vol. 18, pp. 1147-1155, 2023.

[14] B. Tatykayev et al., "Synthesis of Core/Shell ZnO/rGO Nanoparticles by Calcination of ZIF-8/rGO Composites and Their Photocatalytic Activity," *ACS Omega*, vol. 2, no. 8, pp. 4946-4954, 2017.

Simulation-Based Approach to Detecting Pulmonary Embolism Using Capacitive Micromachined Ultrasonic Transducers

Hussnain Shahid
Institute of Microengineering and
Nanoelectronics (IMEN),
Universiti Kebangsaan Malaysia,
43600 Bangi, Selangor, Malaysia
p138478@siswa.ukm.edu.my

Dilla Duryha Berhanuddin*
Institute of Microengineering and
Nanoelectronics (IMEN),
Universiti Kebangsaan Malaysia,
43600 Bangi, Selangor, Malaysia
dduryha@ukm.edu.my

Rhonira Latif
Institute of Microengineering and
Nanoelectronics (IMEN),
Universiti Kebangsaan Malaysia,
43600 Bangi, Selangor, Malaysia
rhonira@ukm.edu.my

Poh Choon Ooi*
Institute of Microengineering and
Nanoelectronics (IMEN), Universiti
Kebangsaan Malaysia, 43600 Bangi,
Selangor, Malaysia
pcooi@ukm.edu.my

Tehseen Batool
Department of Physics,
Government College University
Faisalabad (GCUF),
Punjab, Pakistan
tehseenbatool640@gmail.com

Abstract—A severe cardiovascular condition known as pulmonary embolism is commonly underdiagnosed because of deficiencies in the diagnostic techniques available today. This study hypothesizes a capacitive micromachined ultrasonic transducer (CMUT) as a non-invasive diagnostic tool capable of identifying changes in ultrasonic wave penetration results by emboli. A CMUT is proposed for this application. Preliminary findings point to considerable promise for enhancing the accuracy of diagnoses as well as the care and results of patients. This study aims to provide a novel approach to pulmonary embolism detection. In compliance with the Malaysian MySTIE framework, this study advances Sustainable Development Goal 3: Well-being and Good Health.

Keywords— *CMUT, COMSOL 5.6, pulmonary embolism, diagnosis, ultrasonic, simulation, lung disease*

I. INTRODUCTION

Accurate diagnosis of pulmonary embolism (PE) is vital, yet it remains challenging due to its ambiguous clinical presentation, often indistinguishable from other prevalent medical emergencies [1]. Recent advancements have significantly enhanced lung imaging capabilities, yet the detection of PE via non-invasive and non-destructive means remains complex [2], [3]. This complexity arises from the necessity for expert analysis and interpretation, coupled with the absence of an immediate and comprehensive treatment system for PE [4]. An initial potential spark has been noted through simulation, helping carefully craft capacitive micromachined ultrasonic transducer (CMUT) size and geometry for specific frequencies. The COMSOL 5.6 is used for simulation, incorporating skin, fat, muscle, and the lung's material properties from its database.

The blood clot information is input from literature by generating a new material within COMSOL 5.6 [5].

This project was fully supported by the Fundamental Research Grant Scheme (FRGS) under the funding code FRGS/1/2022/TK07/UKM/02/7.

Optimizing results at 600kHz [6], a CMUT is advised with the obtained specifications. The study aims to demonstrate the effectiveness of capacitive micromachined ultrasonic transducers (CMUTs) in detecting changes in lung density presenting the emboli, which are critical for identifying abnormalities in lung function. This study aligns with Sustainable Development Goal 3 (Good Health and Well-being) by aiming to improve diagnostic methods and escalates healthcare access and quality, particularly in emergencies. Additionally, it supports the Malaysian Science, Technology, Innovation, and Economy framework by fostering health innovation [7]. The results indicate significant potential for improving diagnostic precision and patient treatment outcomes. This study advances semiconductor electronics and MEMS technology, specifically in medical diagnostics using CMUTs.

II. METHODOLOGY

COMSOL 5.6, with its robust simulation capabilities, allows for the precise modeling of complex biological structures and interactions. This study leverages COMSOL 5.6 to explore the behavior of ultrasonic waves in lung tissue, focusing on the impact of blood clot emboli to obtain specifications for CMUT.

Fig. 1 illustrates the flow of the methodology process. For simplicity, a 2D model was used, and then the frequency domain in pressure acoustics was used, with a geometry representing skin (0.02 cm), fat (1.30 cm), muscle (1.00 cm), and lung (3.00 cm). Materials were assigned respective sound velocities air (330 m/s), skin as soft tissue (1540 m/s), fat (1450 m/s), muscle (1590 m/s), and a new material for blood clots [8]. A monopole point source transmitted ultrasonic waves, ranging from 580 kHz to 600 kHz in 10 kHz steps. The results focused on the 600 kHz frequency, where a significant increase in ultrasonic wave penetration was observed in the lung tissue when blood clot embolus was present.

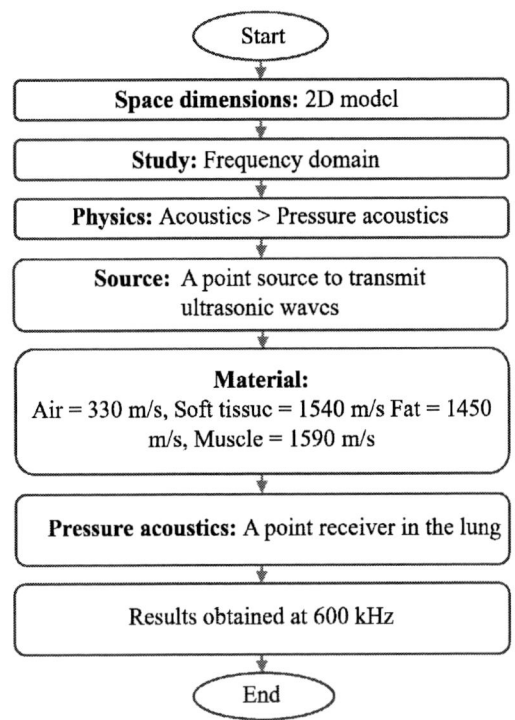

Fig. 1. Flowchart of the simulation study

III. RESULTS AND DISCUSSION

Fig. 2 and 3 illustrate the sound waves passing through pleural distances. Fig. 2, represents the reference scenario without emboli at 600 kHz, and shows a smooth gradient of sound pressure levels decreasing from high (red) to low (blue) as ultrasonic waves propagate through skin, fat, muscle, and the lung. In contrast, Fig. 3, which includes emboli at the same frequency, displays irregular pressure patterns with a noticeable increase in ultrasonic pressure, indicated by the intensified red color. The increased penetration of ultrasonic waves in the lung is due to the emboli.

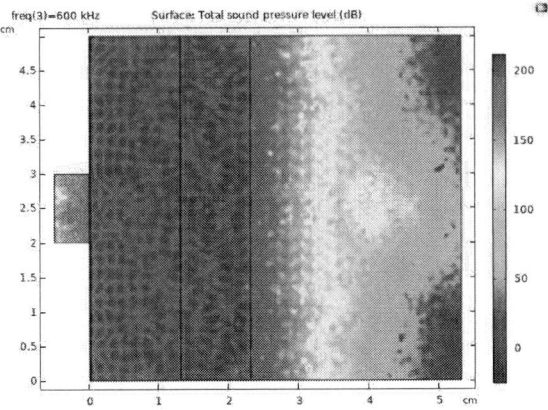

Fig. 2. Ultrasonic pressure (dB) response without emboli

The comparison of Fig. 2 and 3 highlights the effect of emboli on ultrasonic wave propagation, enabling the diagnosis of emboli through distinct differences in pressure distribution and wave behavior. To enhance diagnostic capability, it is recommended to use CMUT for transmitting and receiving ultrasonic waves, as shown in Fig. 3. CMUT offers improved sensitivity and resolution in detecting emboli.

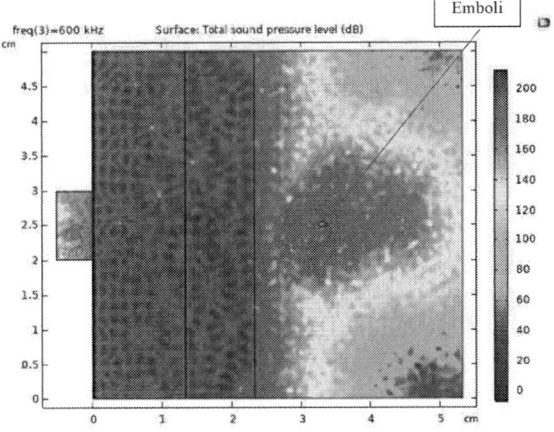

Fig. 3. Ultrasonic pressure (dB) response with emboli

To explain further, Fig. 4 demonstrates the sound pressure penetration comparison plot of Fig. 2 and 3. The red line represents the infected lung's response and the black line denotes the normal one. The maximum pressure at 600 kHz is 178.3 decibels (dB) and 116.8 dB respectively.

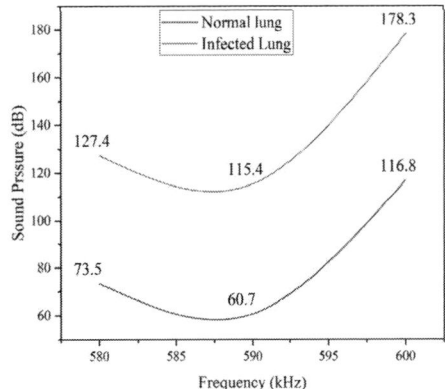

Fig. 4. Ultrasonic wave penetration plot in the lung

Further, the reflection of ultrasonic waves is compulsory for device fabrication. However, the change will be the same, as the increased penetration will decrease the reflection linearly, following the relation in equation (1).

$$P = 1\text{-}R \tag{1}$$

Where R defines the reflection and P represents penetration. This means that R is the complementary value of P, such that their sum always equals 1. As P increases, R decreases correspondingly, and vice versa, maintaining their total at 1.

The R relies on the impedance of the medium. If Z_1 is the acoustic impedance of the first medium, and Z_2 is the acoustic impedance of the second medium [9] then the reflection is,

$$R = (Z_2 - Z_1 / Z_2 + Z_1)^2 \qquad (2)$$

Equation (2) measures the reflection from muscle boundaries to reach the lung and vice versa. The plot for the net change in reflection/penetration measurement using the formula to convert decibel sound pressure level to pascals (Pa),

$$P = P_0 \cdot 10 \, (dB) / 20 \qquad (3)$$

P_0 is the reference sound pressure, typically 20×10^{-6} Pa [10]. For normal lungs, pressure penetration at 2.18cm deep in the lung is 116.8 dB and with an embolus, it is measured 178.3 dB respectively. Using the (3) for conversion, the net change in pressure is 16.4 kPa.

The acquired data can be used as the sensitivity requirement for CMUT. Additionally, data from a previous study [11] was used to estimate the size of the membrane. This involved examining the frequency relationship with the diameter of CMUT membrane testing in the 80 kHz to 2 MHz range. As a result, a membrane size of 87.6μm was determined for a frequency of 600 kHz.

IV. CONCLUSION

Using a CMUT with a frequency of 600 kHz, a membrane size of 87.6 micrometers, and a sensitivity of up to 16.4 kPa shows promise for improving the diagnosis of PE. To validate this diagnostic approach, further study is needed to refine the model data according to human anatomy. The suggested CMUT can potentially be used for PE diagnosis in humans with its reduced distance between the top electrode and bottom and low collapse voltage requirement. Other parameters, such as optimized geometry, material, thickness, array size, and a specific design, will be introduced after obtaining complete features from the model.

ACKNOWLEDGMENT

We would like to thank Dr. Sarwisan Krishna for his invaluable insights and engaging discussion on pulmonary embolism.

REFERENCES

[1] C. Falster et al., "Comparison of international guideline recommendations for the diagnosis of pulmonary embolism," The Lancet Haematology, vol. 10, no. 11. Elsevier Ltd, pp. e922–e935, Nov. 01, 2023. doi: 10.1016/S2352-3026(23)00181-3.

[2] L. Zhao, T. C. Fong, and M. A. L. Bell, "Detection of COVID-19 features in lung ultrasound images using deep neural networks," Communications Medicine, vol. 4. no. 1, Mar. 2024, doi: 10.1038/s43856-024-00463-5.

[3] F. Melazzini, M. Reduzzi, S. Quaglini, F. Fumoso, M. V. Lenti, and A. Di Sabatino, "Diagnostic Delay of Pulmonary Embolism in COVID-19 Patients," Front Med (Lausanne), vol. 8, Apr. 2021, doi: 10.3389/fmed.2021.637375.

[4] H. Fayiad, H. Moussa, Y. Nosair, and A. I. Mostafa, "Predictive accuracy of years score in diagnosis of pulmonary embolism," The Egyptian Journal of Bronchology, vol. 18, no. 1, Mar. 2024, doi: 10.1186/s43168-024-00269-y.

[5] V. M. Nahirnyak, S. W. Yoon, and C. K. Holland, "Acousto-mechanical and thermal properties of clotted blood," J Acoust Soc Am, vol. 119, no. 6, pp. 3766–3772, Jun. 2006, doi: 10.1121/1.2201251.

[6] P. C. Pedersen and H. S. Ozcan, "Ultrasound properties of lung tissue and their measurements," Ultrasound Med. Biol., vol. 12, no. 6, pp. 483–499, 1986, doi: 10.1016/0301-5629(86)90220-6.

[7] "Trailblazing the Way for Prosperity, Societal Well-Being & Global Competitiveness". www.akademisains.gov.my Available: https://www.akademisains.gov.my/10-10-mystie/

[8] A. Fenster et al., "Ultrasonic imaging of the human body," Reports on progress in physics, vol. 62, no. 5, pp. 671–722, 1999, doi: 10.1088/0034-4885/62/5/201.

[9] C. W. Chung, J. S. Popovics, and L. J. Struble, "Using ultrasonic wave reflection to measure solution properties," Ultrason Sonochem, vol. 17, no. 1, pp. 266–272, 2010, doi: 10.1016/j.ultsonch.2009.07.004.

[10] Poser, B. "Amplitude, Intensity and Loudness", www.ling.upenn.edu/. Available at: https://babel.ling.upenn.edu/phonetics/old_website_2015/archive/docs/Amplitude.pdf, 2002.

[11] P. Butaud et al., "Towards a better understanding of the CMUTs potential for SHM applications," Sensors Actuators A Phys., vol. 313, p. 112212, Jul. 2020.

Effect of Seed Layer Cycles on ZnO Nanowires Characteristics for Piezoelectric Nanogenerator Applications

Fong Jun Xian
Faculty of Electrical Engineering
Universiti Teknologi Malaysia
UTM Skudai, 81310
Johor Bahru, Malaysia
fongxian@graduate.utm.my

Suhana Mohamed Sultan
Faculty of Electrical Engineering
Universiti Teknologi Malaysia
UTM Skudai, 81310
Johor Bahru, Malaysia
suhanasultan@utm.my

Izzaty Mohd Zambri
Faculty of Electrical Engineering
Universiti Teknologi Malaysia
UTM Skudai, 81310
Johor Bahru, Malaysia
zatyzambri@gmail.com line

Rafidah Petra
Faculty of Engineering
Universiti Teknologi Brunei
Jalan Tungku Link Gadong
BE1410
Brunei Darussalam
rafidah.petra@utb.edu.bn

Khoo Wei How
Faculty of Electrical Engineering
Universiti Teknologi Malaysia
UTM Skudai, 81310
Johor Bahru, Malaysia
iamalvin1210@gmail.com

Abstract— This study investigates a low-temperature and cost-effective method for fabricating ZnO-based piezoelectric nanogenerators (PENGs). The hydrothermal method is employed to grow Zinc Oxide (ZnO) nanowires on flexible Indium-Tin-Oxide coated polyethylene terephthalate (ITO-PET) substrates, predeposited with seed layer for 5, 8 and 10 seed layer cycles. The aim is to assess the effect of different seed layer cycles on the alignment and dimensions of the ZnO nanowire grown. Field-emission Scanning Electron Microscopy (FESEM) is utilized to examine the alignment and dimensions of the nanowire arrays, while energy-dispersive X-ray Spectroscopy (EDX) analyses are employed to assess the elementary compositions. The alignment of ZnO nanowires improved with an increase in the number of seed layer cycles. In addition, it is found that an increased number of seed layer cycles resulted in an increase in the average diameter of ZnO nanowires from 89.5 nm to 122.31 nm, while the length was decreased from 831.67 nm to 548.67 nm. This may be attributed to variations in grain size and crystallinity of nanoparticles on the seed layers grown. These results contribute to the understanding of the growth process of ZnO nanowires for PENG applications and hold potential for advancements in green energy harvesting technologies.

Keywords— ZnO nanowires, PENG, low-temperature, hydrothermal method, flexible substrate, seed layer

I. INTRODUCTION

Green technology emerged as a transformative solution in response to the rapid depletion of non-renewable energy resources. Leveraging abundant green renewable energy sources, this technology has revolutionized power generation. Piezoelectric nanogenerators (PENGs) are one of the technologies which quickly gained traction across the globe. PENG is an electric power generator in nanoscale integrating piezoelectric material which harvests the ambient mechanical strains and converts them into electrical energy. The key feature of PENG is its capability to be manufactured in small dimensions and compact structure. This characteristic enables PENG to be sensitive to minute strains which can be triggered by small mechanical energy while providing desirable output power to activate electronic devices. Examples of mechanical energy that can be captured are vivo energy, vibration caused by transportation, body movement, airflow, etc. This development has been found to be a promising technology that helps in advancing technology in various industries, including the Internet of Things (IoT), wireless electronics, and wearable and implantable biomedical devices. Its unique characteristic has been discovered as a novel technology to replace bulk batteries and become the main green power supplier, which aligns with the trend of miniaturization of current technology [1].

Piezoelectric material is the materials experiencing piezoelectric effect functioning in conversion of mechanical strain to electrical power. These materials encompass various categories, including organic, inorganic, composite, and bioinspired natural materials. Among them, inorganic piezoelectric materials or piezoceramics are extensively utilized in PENGs due to their high piezoelectric coefficients. A great amount of piezoelectric charges can be generated when piezoceramic experiences the same amount of strains as compared to other categories [1]. However, brittleness is the significant drawback of piezoceramics, which limits their strength and resilience. Zinc Oxide (ZnO) is one of the piezoceramics that is widely used in PENGs due to its outstanding performance. Wurtzite structure of ZnO, which is a non-centrosymmetric crystal structure inherently equipped with piezoelectric properties naturally. When subjected to mechanical deformation, the net dipole moment shifted within the ZnO unit cell and no longer maintained zero. Displacement of positive and negative ions inside ZnO conducts polarization of ZnO immediately and generates charges on its crystal surface. Internal electric fields generated by opposite charges at each surface of ZnO exhibit piezo potential, enabling electrical charges to be transmitted through an external circuit [2].

Piezoelectric nanogenerators (PENGs) are commonly structured in a sandwich format with the piezoelectric material nestled between top and bottom metal electrodes and a substrate attached to the bottom electrode. This configuration offers several advantages, especially shielding the delicate piezoelectric material from external influences that could compromise its functionality[1]. To optimize the overall performance of PENG, flexible substrate not only functions as protection against piezoelectric materials but also increases overall flexibility. The augmented flexibility enables PENGs to endure greater mechanical strain without breaking. By withstanding greater strain, more piezoelectric charges can be induced, resulting in greater output power generation. Furthermore, top and bottom electrodes, which are bonded firmly to the piezoelectric material, may efficiently capture the induced piezoelectric charges and transmit them through an external circuit [3].

However, flexible substrates typically composed of polymers with low melting points have become a significant challenge in fabricating high-performance PENG. Conventional growth process for piezoelectric materials, such as Chemical Vapour Deposition (CVD) and wet oxidization, often requires complex, high operational costs, and high temperature poses a risk of substrate damage [2][4]. As a result, there is a critical need for the development of a low-temperature fabrication process that allows for the growth of piezoelectric material onto flexible substrate while preserving their structural integrity. Addressing this challenge is essential for the widespread adoption of PENGs in various applications.

The objective of this research is to grow ZnO nanowires on Indium-Tin-Oxide coated polyethylene terephthalate (ITO-PET) substrate using a low-temperature hydrothermal process known for its cost-effectiveness and compatibility with flexible substrate[5]. The hydrothermal process encompasses 2 key stages, which are the seed layer deposition process and the process of growing ZnO on the seed layer [2]. The pre-deposited seed layers aim to enhance the alignment and adhesion of the ZnO nanowires, thereby facilitating their growth at lower temperatures [2]. In this paper, ZnO nanowires were grown on ITO-PET by using the hydrothermal method. By systematically depositing different numbers of cycles of seed layers prior to ZnO growth, it is aimed to identify the optimal conditions for achieving the desired nanowire alignment. Field-Emission Scanning Electron Microscopy (FESEM) and Energy Dispersive X-ray Spectroscopy (EDX) analyses were employed to observe the morphology of the grown ZnO nanowires and analyze the elemental composition of the ITO-PET substrates.

II. Experimental

A. ZnO Seed Layer Deposition

The preparation of the seed solution started with the dissolving of 25 mmol of zinc acetate dihydrate into 50 mL of ethanol. This precursor underwent stirring in an ultrasonic bath at room temperature for 30 minutes to ensure the formation of an agglomerate-free solution. Subsequently, the solution was heated to 75°C and stirred for an additional 10 minutes on a hot plate using a magnetic stirrer. A 2 x 2 cm² of ITO-PET substrate was cleaned through an ultrasonication

Fig. 1. Samples with different seed cycles submerged in the growth solution.

bath using isopropyl alcohol and deionised water for 5 mins each. The substrate was completely dried at 90°C in a hot air oven. Then, the seed layer was deposited on a cleaned ITO-PET by using the spin coating technique. The seed solution was spin-coated by covering the substrate with fifteen drops of the seed solution. Spin coating was applied at a speed of 1000 rpm for 60 s [6]. Then, the substrate was post-annealed at 100°C for 2 mins on a hot plate to evaporate any remaining solvent and organic materials. Seed layers were created by repeating the spin coating and post-annealing processes for 5, 8, and 10 cycles.

B. Growing ZnO Nanowires

A 100 mL of precursor was prepared by dissolving 25 mmol of Zinc Nitrate Hexahydrate and 25 mmol of Hexamethylenetetramine in 100 mL of deionized water. The precursor was heated to 75°C on a hot plate and stirred for 30 mins to fully dissolve the materials. Then, the substrates with seed layer were submerged in the growth solution and heated for 9 hours in a hot air oven at 90°C. Finally, the sample with grown ZnO nanowires was rinsed with deionized water and thoroughly dried in a hot air oven at 100°C. Fig.1 shows the substrate being fixed and immersed in growth solution before heating in a hot air oven.

C. Characterization

Field-emission Scanning Electron Microscopy (FESEM) and Energy Dispersive X-ray Spectroscopy (EDX) analysis was conducted to confirm the formation of seed layers and ZnO nanowires. FESEM images of the grown ZnO nanowires were captured at various magnifications to examine their morphologies. Additionally, EDX analysis was employed to analyse the elemental composition of the ITO-PET substrates.

III. RESULTS AND DISCUSSION

A. Elementary Compositions

Fig. 2 depicts the EDX spectrum of substrates with seed layer only and ZnO nanowires grown on various numbers of seed layer cycles. In Fig. 2(a), the presence of Carbon (52.7%), Oxygen (32.6%) and Zinc (6.0%) confirmed the formation of seed layer on the ITO substrate. The significant amount of carbon and oxygen suggested the presence of ethanol used in producing the seed layer solution, while the presence of zinc nanoparticles is indicated by the zinc peak on the seed layer. From Fig. 2(b) to Fig. 2(d), the presence of Zn and O as the main elements confirmed that ZnO has successfully grown on the substrate. Additionally, minor peaks of Indium (In) and Carbon (C) were discovered due to the ITO substrate and ethanol-based seed layer. Trace amounts of Ag, Ti, and Si may be attributed to contaminants

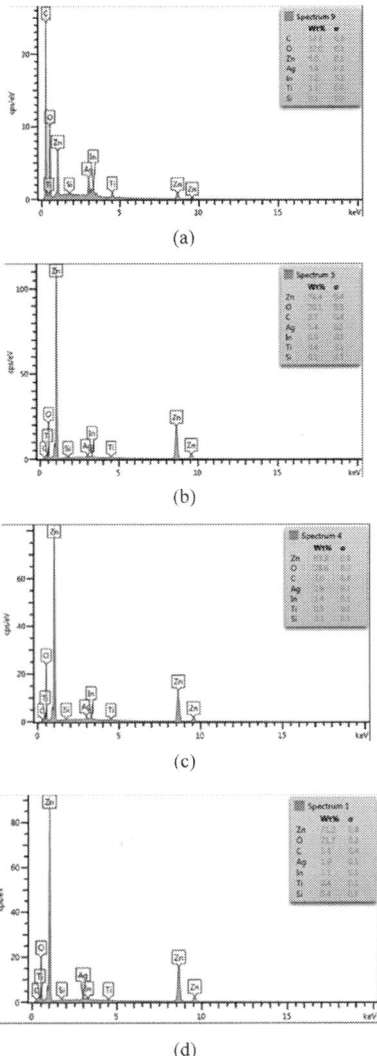

(a)

(b)

(c)

(d)

Fig. 2. EDX spectra for element composition for substrates with (a) 5 cycles of seed layer only, with ZnO grown on (b) 5 cycles, (c) 8 cycles, (d) 10 cycles of seed layer.

introduced during the experimental process. The EDX analysis of the grown ZnO revealed a weightage of zinc greater than 60%, higher than that of oxygen, indicating the formation of zinc-rich ZnO nanowires. This characteristic suggests n-type ZnO with excess free electrons was formed, contributing to enhanced conductivity performance [7][8].

Figure 3 (a)-(c) displayed the top-view images of prepared ZnO nanowires on 5, 8, and 10 seed layer cycles. The images were captured by using FESEM measurement at varying magnifications to assess the effect of seed layer cycles on ZnO nanowires' characteristics.

B. FESEM Analysis

As observed from the top view of SEM images shown in Fig 3(a)-(c), the alignment of ZnO nanowires improved with an increase in the number of seed layer cycles. Morphological analysis of ZnO nanowires was conducted at a higher magnification of FESEM. The results of the average diameter

(a)

(b)

(c)

Fig. 3. FESEM images of substrate with ZnO grown on seed layer of (a) 5, (b) 8 and (c) 10 cycles

and length of the ZnO nanowires corresponding to various numbers of seed layer cycles were summarized in Table I.

The findings revealed that the number of seed layer cycles significantly affects the length and diameter of ZnO nanowires. Increased seed layer cycles were associated with an increase in the average diameter of ZnO nanowires from 89.5 nm to 122.31 nm, while the length was decreased from 831.67 nm to 548.67 nm. The aspect ratio is determined as the ratio of length to diameter.

As the number of seed layer cycles increases, the aspect ratio decreases. The observation is in line with results reported by H.Ghayour et al that suggested the thicker seed layer features larger grain size and particle size, contributing to the increased diameter of ZnO nanowires [9]. As a template for growing ZnO, a seed layer with a larger particle size will facilitate the growth of ZnO nanowires in greater average diameter. Furthermore, it was found that thicker seed layers occupied a greater concentration of Zn nanoparticles, and larger grain size enhanced the crystallinity, thereby improving the alignment of ZnO nanowires[9]. By maintaining the constant growth time and temperature of 90°C for 9 hours during the hydrothermal growing process, the length of ZnO nanowires, which was strongly dependent on these 2 factors, has a significant drop [10].

IV. CONCLUSIONS

In summary, using a low-temperature hydrothermal method, ZnO nanowires were grown on ITO-PET substrates with different numbers of seed layer cycles. The effect of seed layer cycles on the alignment of ZnO nanowires was investigated. The changes in diameter, length and aspect ratio

TABLE I. Average diameter and length of ZnO Nanowires grown on ITO-PET substrate with different numbers of seed layer cycles.

Number of seed layer cycles	Average diameter of ZnO nanowires, D (nm)	Average length of ZnO Nanowires, L (nm)	Aspect ratio of ZnO Nanowires, L/D
5	89.25	831.67	9
8	94.05	642.67	6
10	122.31	548.67	4

of ZnO nanowires array on different number of seed layer cycles were examined and analyzed by using EDX and FESEM characterization method. Based on the results, an increased number of seed layer cycles led to increased diameter and decreased length of ZnO nanowires. From the EDX analysis, the successful formation of seed layers and the growth of ZnO nanowires were confirmed with a majority weightage of Zn and O. Higher weightage of Zn indicated that Zinc-rich nanowires (n-ZnO) with an excess of free electrons. While the FESEM result showed a higher number of seed layer cycles will provide better alignment of the nanowires array and increase the average diameter of ZnO nanowires. These findings contribute to our understanding of the growth process and provide valuable insights for optimizing the ZnO nanowire morphologies towards piezoelectric nanogenerator applications.

ACKNOWLEDGMENT

This work was supported by the UTM Fundamental Research Grant with UTM cost centre no. Q.J130000.3823.23H25.

REFERENCES

[1] Sezer, N., & Koç, M. (2021). A comprehensive review on the state-of-the-art of piezoelectric energy harvesting. Nano energy, 80, 105567.

[2] Le, A. T., Ahmadipour, M., & Pung, S. Y. (2020). A review on ZnO-based piezoelectric nanogenerators: Synthesis, characterization techniques, performance enhancement and applications. Journal of Alloys and Compounds, 844, 156172.

[3] Hu, D., Yao, M., Fan, Y., Ma, C., Fan, M., & Liu, M. (2019). Strategies to achieve high performance piezoelectric nanogenerators. Nano energy, 55, 288-304.

[4] ElZein, B., Salah, N., Barham, A. S., Elrashidi, A., Al Khatab, M., & Jabbour, G. (2023). Influence of Temperature on the Growth of Vertically Aligned ZnO Nanowires in Wet Oxygen Environment. Crystals, 13(6), 876.

[5] Zhang, J., Zhao, G., Li, Y., Ai, T., Wu, C., Jia, J., ... & Wang, Y. (2019). Study on the electrical properties of nano ZnO/PET-ITO heterojunction prepared by hydrothermal method. Journal of Electron Spectroscopy and Related Phenomena, 235, 68-72.

[6] Kim, H. G., Kim, E. H., & Kim, S. S. (2021). Growth of ZnO nanorods on ITO film for piezoelectric nanogenerators. Materials, 14(6), 1461.

[7] Galdámez-Martinez, A., Santana, G., Güell, F., Martínez-Alanis, P. R., & Dutt, A. (2020). Photoluminescence of ZnO nanowires: a review. Nanomaterials, 10(5), 857.

[8] Paiman, S., Ling, T. H., Husham, M., & Sagadevan, S. (2020). Significant effect on annealing temperature and enhancement on structural, optical and electrical properties of zinc oxide nanowires. Results in Physics, 17, 103185.

[9] Ghayour, H., Rezaie, H. R., Mirdamadi, S., & Nourbakhsh, A. A. (2011). The effect of seed layer thickness on alignment and morphology of ZnO nanorods. Vacuum, 86(1), 101-105.

[10] Song, J., & Lim, S. (2007). Effect of seed layer on the growth of ZnO nanorods. The Journal of Physical Chemistry C, 111(2), 596-600.

A Low Power Scan Cell Design using FreePDK3

Ahmad Awalluddin Mohd Ghazali
Malaysia-Japan International Institute of Technology
Universiti Teknologi Malaysia
Kuala Lumpur, Malaysia
ahmadawalluddin04@gmail.com

Chia Yee Ooi
Malaysia-Japan International Institute of Technology
Universiti Teknologi Malaysia
Kuala Lumpur, Malaysia
ooichiayee@utm.my

Hau Sim Choo
Intelligent Circuit Engineering
Selangor, Malaysia
choo.hau.sim@icesb.com

Nordinah Ismail
Malaysia-Japan International Institute of Technology
Universiti Teknologi Malaysia
Kuala Lumpur, Malaysia
nordinah.kl@utm.my

Siti Rahmah Aid
Malaysia-Japan International Institute of Technology
Universiti Teknologi Malaysia
Kuala Lumpur, Malaysia
sitirahmah.aid@utm.my

Abstract— Test power is always a significant challenge for test engineers as the power that is dissipated has a serious negative impact on the computer chip reliability. This paper introduces a low power scan cell with efficient state preserving and gating logic to mitigate the negative impact from the test power consumption and unnecessary switching in scan-based designs, particularly during shifting mode. Our experimental results have demonstrated that our proposed scan achieves significant reduction of ~48% in total average power compared to the conventional scan cells, as well as reduction of ~38% in most of the shift cycles when compared to one of the prevalent existing gating architectures, FLS. In summary, this innovative approach presents a promising solution for reducing test power consumption and unnecessary switching in scan-based designs, particularly during shift mode, while maintaining efficiency and minimizing impact on peak power during capture cycles.

Keywords—low power scan, design for testability, test power, gating scan cell.

I. Introduction

Design for testability (DFT) is an integrated circuit development method that gives testability features to a hardware design. The added feature makes it easy to detect the defect inside the product. However, excessive test power consumption is inevitable and may cause premature circuits degradation and aging. In the current state of literature, clock gating is one of the most famous methods implemented to reduce the transition in the scan chain [1][2]. However, the need for good clocking control over the segments of the scan chains become a big issue for this to work. Poor clocking control results in clock skew and multiple flop propagation. To lower the average test power, this study suggests designing a low power scan cell with state maintaining and gating logic. This is how the remainder of the paper is structured. We go over previous research on low power scan cell designs in Section II. In Section III, we propose a low power scan cell design. Section IV presents experimental results that illustrate how well the recommended approach lowers test power. We conclude in Section V.

II. Previous Works

Nilima and Ravi proposed a By-Passable Scan Data Retention Flip-Flop (BPS-DRFF) that consists of two secondary latches in addition to a master latch [3]. One of the slave latch act as the bridge between the scan cell and the combinational circuit, while another slave latch functioned to propagate from once scan cell to other scan cell during the shifting mode. Therefore, the propagated test pattern are gated from propagating into combinational circuits during shift mode as the first slave latch retain the data in sleep mode, leading to a substantial decrease in test power usage. The drawback of the work is the additional latch resulting in area overhead. Ahmed and Nourani suggested gating technique for the first level supply (FLS), which connects a pull-down network transistor to each scan chain output to disconnect the combinational block of a CUT from the scan chain [14]. In addition, a pull-up network transistor is inserted to force the first level gates in the combinational block from floating. But instead of preserving the previous state, the output of the first-level gates is fixed to a set to the VDD value, FLS does not stop the propagated signal from reaching the combinational logic.

Kim et al and Solanki et al suggested a logic BIST that generates test patterns with a low power linear feedback shift register (LFSR) that dissipate less power than a typical LFSR, so the switching activity is minimized [4] [5]. Nagesh and Chandra proposed a new low power test pattern generator (LP-TPG) that implements a LFSR to minimize the median and maximal power of a circuit during testing [6]. The main goal of low power BIST is to lower test pattern switching activity at primary inputs.

Similarly, Lee, Choi and Kang proposed an input control method to apply a control pattern (CP) at the primary input to minimize or even eliminate the propagation via the combinational circuit, thereby reducing the transition count of the combinational component of a full-scan circuit during shift mode [7].

Cao and Jiao proposed a new technique for creating power-aware test patterns that improves the target circuit given to the ATPG tool by using dummy hardware circuit functions to define the test mode power restrictions and create low power test patterns [8]. Furthermore, neither the pattern creation tool nor the process needs to be altered to use the suggested approach.

Xiang et al proposed the implementation of low power BIST that generates low power test patterns and applies test compression techniques [9]. Kolanchinathan and Kumar introduced a low power scan-based BIST that utilized an algorithm called reseeding and weighted pseudo random test data test generation [10]. This scheme disables scan cells during dummy random testing. During deterministic BIST, the LFSR is configured to a shorter length therefore in a single cycle, a limited number of scan chains are enabled.

Shivakumar and Senthilpari proposed a Test Controller with Toggle/Hold logic, based on the Model Predictive Control (MPC) concept [11]. This method reduces the number of tests by eliminating unnecessary test patterns to be injected into the CUT. This further reduces power consumption during testing.

Jambagi et al and Ravi et al examined X-filling approaches for reducing the power of LOS testing. In essence, the suggested approach found "X-bits," or bits that don't care, in test vectors and then used a variety of X-filling strategies to make the maximum power throughout the cycle of launch to capture equivalent to the power used in functional mode [12][13].

Tehranipour and Ahmed present a comparison of industrial performance test-bound strategies using ISCAS 89 Benchmark circuits with heuristic approaches for decreasing scan power using X-filling [14]. The algorithm produces the test cube generation and performs circuit analysis to identify whether it is and/or dominant or neither of it. Then X-filling technique is performed and the fault coverage is checked. From different types of X-filling technique, Weighted Switching Activity (WSA) and fault coverage are assessed where the best fault coverage and less WSA test vector to the fault list.

Gonda et al proposed a design of Test Pattern Generator for BIST that utilizes a mathematical calculation of a portion of initial pseudo-primary seed bits by utilizing Galois operations [2]. A weighted pattern of maximum length with minimal transition activity is chosen by the author. The Weighted Multiplexer (Mux) is integrated into the proposed method which acts as a phase shifter. As a result, it is instrumented to limit switching transition while the test-per-scan method is in process.

III. Proposed Design of Low Power Scan Cell

We presented a novel scan cell design that dissipate low power during scan shift. When the design of scan cell is implemented for the design for testability (DFT), the scan cell performs with minimal power dissipation by: i) disconnecting the duplicate transitions that spread from the scan cells to combinational circuit, and ii) lowering the scan cell's internal switching activity during shift mode. Fig. 1 shows the circuit level of the proposed scan cell structure. The slave latch has the state preserving and gating ability. In shift mode, the gating logic blocks the data transition from reaching the combinational circuit. Instead, the state preserving logic holds the previous state such that the signal at output Q stays unchanged. This controls the peak power and average power dissipation.

The suggested gating logic consists of a transmission gate and two inverters that are connected serially. The output of the second inverter is feedback to the transmission gate which is constructed as a mini latch. This mini latch acted as the state preserving to avoid any transition in the combinational circuit during data shifting. SE (Serial Enable) signal was connected to the mini latch which dictates when the scan chains were gated from the combinational circuit during data shifting and also being transparent during respond retrieval operation. While shifting mode operation, the path between master flip-flop and Q was disconnected. This is due to the level high signal on SE which makes the transmission gate detach from the circuit under test. We use the flip-flop's inverted output to move data from one scan cell to the others and it is reportedly to perform better in the term of speed. Fig. 2 shows the layout of the scan cell using FreePDK3. To evaluate our proposed scan cell, we select conventional scan cell and FLS scan cell to be compared with the proposed scan cell. Fig. 3 and Fig. 5 show the schematic of the conventional scan cell and FLS scan cell respectively. Fig. 4 and Fig. 6 show the layout of the conventional scan cell and FLS scan cell respectively.

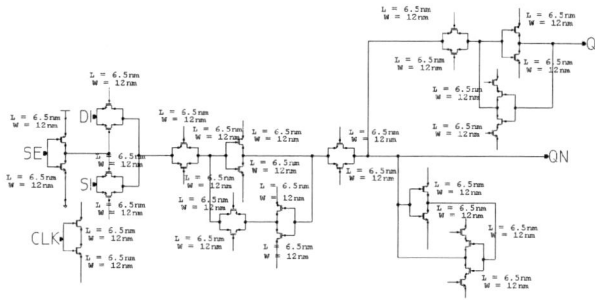

Fig. 1. Schematic of the proposed scan cell (L=6.5nm and W=12nm).

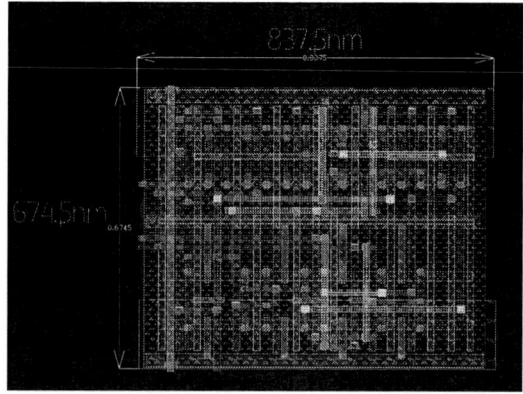

Fig. 2. Layout of the proposed scan cell (width=837.5nm and height=674.5nm).

979-8-3503-7832-0/24 $31.00 © 2024 IEEE

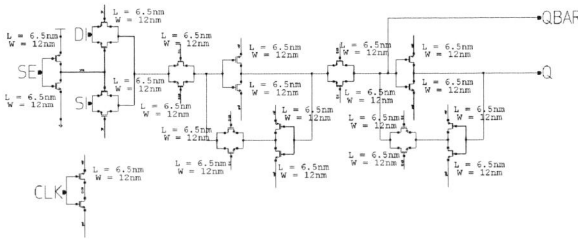

Fig. 3. Schematic of conventional scan cell (L=6.5nm and W=12nm).

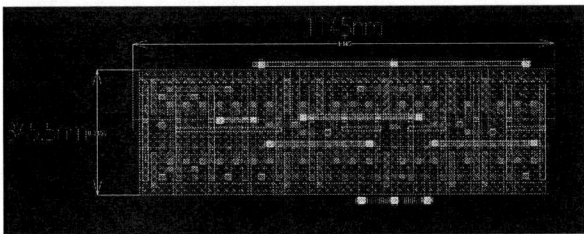

Fig. 4. Layout of conventional scan cell (width=1145nm and height=345.5nm).

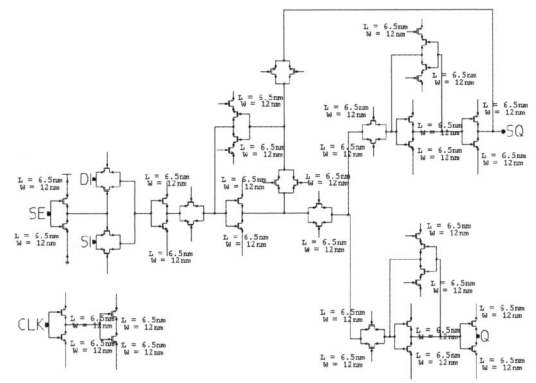

Fig. 5. Schematic of FLS scan cell (L=6.5nm and W=12nm).

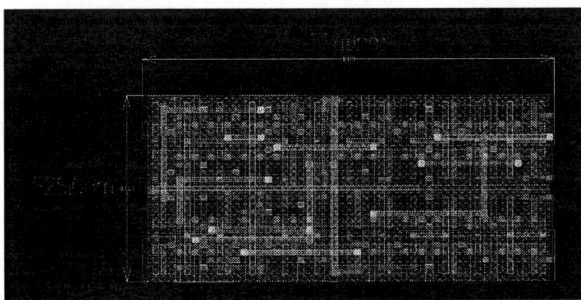

Fig. 6. Layout of FLS scan cell (width=1149nm and height=525.5nm).

IV. RESULTS AND DISCUSSION

To test the effectiveness of the proposed low power design, we conducted an experiment to test the power dissipation and compared its performance with the conventional scan cell design and FLS scan cell. All the designs were using NCSU Free PDK 3nm technology and the simulation was performed in HSPICE.

A. Power analysis on the scan cell

Power analysis was first done on the single scan cell alone to evaluate its shift power, which is an average power dissipation in shift mode, and its peak power. The experiment includes both pre-layout and post-layout simulation of the scan cells for multiple clock cycles to record the average power dissipation and capture peak power dissipation. The simulation were done through the spice simulation using HSPICE simulator. We set 0.8 V as the test voltage and voltage source. The temperature that the simulation performed were set at 25° degree Celsius. Table I tabulates the average power dissipation data of the proposed design scan cell over the data of the other scan cells. Conventional scan cell reportedly dissipates power almost twice of the proposed design scan cell as the data shows that our proposed scan cell achieved 46.7% reduction of the power. This is due to the spurious transitions in the master latch and also the slave latch. FLS scan cell could only manage to reduce the power for 15.4-16.3% compared to the conventional scan cell. When comparison is made between the proposed scan cell and FLS, the proposed scan cell has about 38% reduction of the power over FLS. In short, the proposed scan cell performs better in average power dissipation domain compared to the other scan cells. Fig. 7 shows the same trend of average power consumption in the form of chart.

TABLE I.   AVERAGE POWER CONSUMPTION COMPARITION WITH EXISTING SCAN CELLS

Shift cycle	#1		#2		#3		#4	
Average power / Method	Power (nW)	Red %	Power (nW)	Red %	Power (nW)	Red %	Power (nW)	Red %
Conventional	3.4209	-	3.4172	-	3.4261	-	3.4282	-
FLS	2.8857	15.6	2.8920	15.4	3.5013	2.2	2.8710	16.3
Proposed	1.7915	47.6	1.7880	47.7	1.7953	47.6	1.7947	47.6

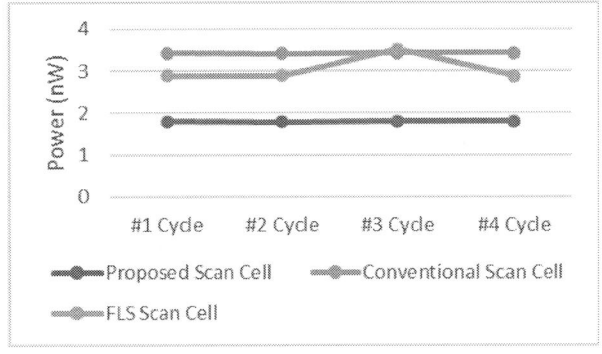

Fig. 7. Comparison of the average power consumption of the scan cells in the chart form.

Table II tabulates the capture peak power values of the proposed design, along with other scan cell. We observed a slight reduction in peak power captured consumption incurred in our proposed scan cell. This is because in capture or normal mode where we retrieve the response from the circuit under test, those three scan cells function in the similar manner. Fig. 8 depicts the same data but in the form of chart.

TABLE II. COMPARISON OF PEAK POWER CONSUMPTION WITH EXISTING SCAN CELLS

Clock cycle	#1		#2		#3		#4	
Peak power / Method	Power (nW)	Red %	Power (nW)	Red %	Power (nW)	Red %	Power (nW)	Red %
Conventional	55.041	-	54.132	-	54.746	-	54.201	-
FLS	53.953	2.0	53.969	0.3	54.454	0.5	54.589	-0.7
Proposed	53.993	1.9	53.967	0.3	53.942	1.5	54.085	0.2

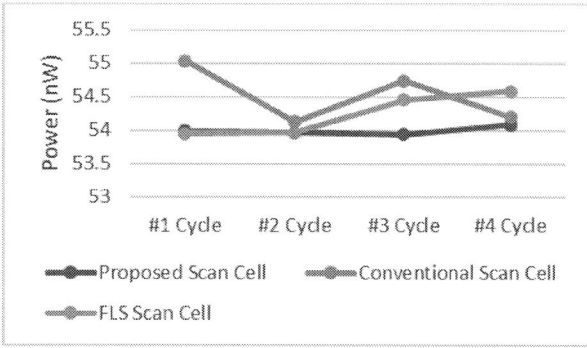

Fig. 8. Comparison of peak power consumption of the scan cells in chart form.

V. CONCLUSION

The proposed scan cell has been implemented using Free PDK 3nm. Experimental results demonstrated that the state preserving logic and gating logic are effective to reduce the average test power consumption in shift mode. This scan cell will contribute towards improving chip reliability and product yield.

ACKNOWLEDGMENT

This work was supported by UTM Fundamental Research Grant numbered 22H19.

REFERENCES

[1] Sravani Kapavarpu, Srinivas Cheedarla, "An efficient and low power sram testing using clock gating", International Research Journal of Engineering and Technology(IRJET).

[2] A. Gonda and B. S. H, "Design and Implementation of BIST Architecture for low power VLSI Applications using Verilog," 2023 International Conference on Smart Systems for applications in Electrical Sciences (ICSSES), Tumakuru, India, 2023, pp. 1-4, doi: 10.1109/ICSSES58299.2023.10199398.

[3] Nilima Warade, T. Ravi, "Implementation and Utilization of LBIST for 16 bit ALU", International Journal of Innovative Technology and Exploring Engineering (IJITEE) , Volume-8 Issue-10, August 2019.

[4] Youbean Kim, Myung-Hoon Yang, Yong Lee and Sungho Kang, "A New Low Power Test Pattern Generator using a Transition Monitoring Window based on BIST Architecture," 14th Asian Test Symposium (ATS'05), Calcutta, India, 2005, pp. 230-235, doi: 10.1109/ATS.2005.12.

[5] P. N. Solanki, N. Ranpura and Y. D. Parmar, "DFT Methodologies for Reducing Shift Power of Compression Architecture for 28NM ASIC," 2018 2nd International Conference on Trends in Electronics and Informatics (ICOEI), Tirunelveli, India, 2018, pp. 1-4, doi: 10.1109/ICOEI.2018.8553799.

[6] B. Nagesh and B. S. N. Chandra, "Design of Efficient Scan Flip-Flop," 2021 International Conference on Recent Trends on Electronics, Information, Communication & Technology (RTEICT), Bangalore, India, 2021, pp.146-150, doi: 10.1109/RTEICT52294.2021.9573924.

[7] S. Lee, K. Cho, S. Choi and S. Kang, "A New Logic Topology-Based Scan Chain Stitching for Test-Power Reduction," in IEEE Transactions on Circuits and Systems II: Express Briefs, vol. 67, no. 12, pp. 3432-3436, Dec. 2020, doi: 10.1109/TCSII.2020.3004371.

[8] X. Cao, H. Jiao and E. J. Marinissen, "A Bypassable Scan Flip-Flop for Low Power Testing With Data Retention Capability," in IEEE Transactions on Circuits and Systems II: Express Briefs, vol. 69, no. 2, pp. 554-558, Feb. 2022, doi: 10.1109/TCSII.2021.3096885.

[9] Dong Xiang, et al, "Low-Power Scan-Based Built-In Self-Test Based on Weighted Pseudorandom Test Pattern Generation and Reseeding", IEEE Transaction on Very Large Scale Integration (VLSI) System, Vol. 25, No. 3, March 2017.

[10] Kolanchinathan V P, Dinesh Kumar T R, Jaishree P, Niranjana M, Sowmiya M, Thresha V, & Pooja sri K. (2023). Low-Power and Space-Efficient Built In Self-Test Architecture With MSIC Test Pattern Generator. Journal of Population Therapeutics and Clinical Pharmacology, 30(9), 308–314.

[11] V. Shivakumar, C. Senthilpari and Z. Yusoff, "A Low-Power and Area-Efficient Design of a Weighted Pseudorandom Test-Pattern Generator for a Test-Per-Scan Built-in Self-Test Architecture," in IEEE Access, vol. 9, pp. 29366-29379, 2021, doi: 10.1109/ACCESS.2021.3059171.

[12] S. B. Jambagi and S. S. Yellampalli, "Exploration of Various Test Pattern Generators for Power Reduction in LBIST," 2017 International Conference on Current Trends in Computer, Electrical, Electronics and Communication (CTCEEC), Mysore, India, 2017, pp. 710-713, doi: 10.1109/CTCEEC.2017.8455062.

[13] Srivaths Ravi, V. R. Devanathan and R. Parekhji, "Methodology for low power test pattern generation using activity threshold control logic," 2007 IEEE/ACM International Conference on Computer-Aided Design, San Jose, CA, USA, 2007, pp. 526-529, doi: 10.1109/ICCAD.2007.4397318.

[14] N. Ahmed, M. H. Tehranipour and M. Nourani, "Low power pattern generation for BIST architecture," 2004 IEEE International Symposium on Circuits and Systems (ISCAS), Vancouver, BC, Canada, 2004, pp. II-689, doi: 10.1109/ISCAS.2004.1329365.

Exploring the Characterization of Electrodeposited MoS2 as a Hole Transport Layer in Methylammonium Perovskite Solar Cells

Ahmad Muhajer Abdul Aziz
Faculty of Electronic and Computer Technology and Engineering
Universiti Teknikal Malaysia Melaka (UTeM), Malaysia.
m022210016@student.utem.edu.my

Muhammad Idzdihar Idris
Micro and Nano Electronic Research Group (MiNE)
Faculty of Electronic and Computer Technology and Engineering
Universiti Teknikal Malaysia Melaka (UTeM), Malaysia.
idzdihar@utem.edu.my

Zul Atfyi Fauzan Mohammed Napiah
Micro and Nano Electronic Research Group (MiNE)
Faculty of Electronic and Computer Technology and Engineering
Universiti Teknikal Malaysia Melaka (UTeM), Malaysia.
zulatfyi@utem.edu.my

Radi Husin Ramlee
Faculty of Electronic and Computer Technology and Engineering
Universiti Teknikal Malaysia Melaka (UTeM), Malaysia.
radihusin@utem.edu.my

Muhammad Noorazlan Shah Zainudin
Faculty of Electronic and Computer Technology and Engineering
Universiti Teknikal Malaysia Melaka (UTeM), Malaysia.
noorazlan@utem.edu.my

Mohd Marzaini Mohd Rashid
School of Physics
Universiti Sains Malaysia (USM) Malaysia.
marzaini@usm.my

Mohd Iskandar Dzulkarnain M Rummaja
Faculty of Electronic and Computer Technology and Engineering
Universiti Teknikal Malaysia Melaka (UTeM), Malaysia.
iskandardzulkarnain964@gmail.com

Abstract— The study explores the characterization of electrodeposited Molybdenum Disulfide (MoS2) as a hole transport layer (HTL) in Methylammonium Perovskite Solar Cells. MoS2, a two-dimensional material, is meticulously deposited to ensure uniformity and then optimized to enhance its hole transport capabilities for efficient charge extraction. Comprehensive characterization techniques, including photovoltaic measurements, impedance spectroscopy, and morphology studies, evaluate the impact of MoS2 on the performance of Methylammonium Perovskite Solar Cells. The findings suggest that electrodeposited MoS2 exhibits promising potential as a high-performance HTL, offering insights into improving the efficiency and stability of Methylammonium Perovskite Solar Cells. This study contributes to advancing the understanding and development of efficient and stable solar cell technologies through the integration of MoS2 as a hole transport layer in Methylammonium Perovskite Solar Cells.

Keywords— Molybdenum Disulfide, Methylammonium, Perovskite Solar Cells, electrodeposition, hole transport layer

I. INTRODUCTION

In recent years, Perovskite Solar Cells have emerged as a promising candidate for next-generation photovoltaic devices due to their high-power conversion efficiency and low fabrication cost [1]. However, one of the key challenges in perovskite solar cell technology is the development of efficient Hole Transport Layers (HTLs) that can effectively extract and transport positive charges from the perovskite layer to the electrode [2], [3]. In this regard, Molybdenum Disulfide (MoS2) has gained significant attention as a potential HTL material due to its excellent electrical conductivity and high charge carrier mobility [4]. MoS2's flexibility and mechanical stability make it an ideal HTL for perovskite solar cells, significantly enhancing their performance.[5]. MoS2-based HTLs have demonstrated long-term stability and improved device lifetimes, further highlighting the potential of this material in the development of efficient perovskite solar cell technology [6], [7]. With further research and optimization, MoS2-based HTLs could play a crucial role in advancing the commercial viability of perovskite solar cells.

The use of MoS2 as an HTL in perovskite solar cells offers several advantages over traditional materials [8], [9]. Firstly, MoS2 has high charge carrier mobility, which allows for the efficient transport of electrons and holes within the device. This results in improved charge extraction and reduced recombination losses, leading to higher power conversion efficiencies. Also, MoS2 has a wide bandgap, allowing it to absorb a broad range of sunlight and convert it into electricity [10]. This makes it suitable for use in various lighting conditions and geographical locations. Furthermore, MoS2 has excellent thermal and chemical stability, making it a durable material for long-term use in solar cells [11]. It can withstand high temperatures and harsh environmental conditions without degradation, ensuring the longevity and reliability of the solar cells. Moreover, MoS2 is abundant and cost-effective compared to other materials used in solar cells, making it a viable option for large-scale production and widespread adoption [12].

979-8-3503-7832-0/24 $31.00 © 2024 IEEE

II. METHODOLOGY

A. Fabrication of MoS2 using Electrodeposition method

A MoS2 thin film has been used as the HTL for PSCs. We have produced thin uniform MoS2 film directly on the glass/Indium Tin Oxide (ITO) substrate by using the electrodeposition method. As the substrate, ITO glass was pretreated by an ultrasonic bath with acetone, deionized water, and isopropyl alcohol for 15 min and dried by nitrogen gas.

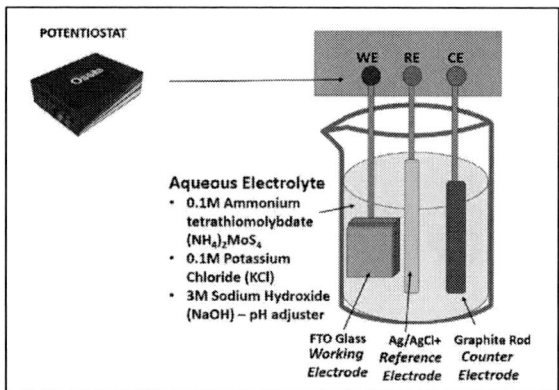

Fig. 1. Electrodeposition process of MoS2.

Figure 1 shows an example of the three-electrode system used in this project [13]. The platinum electrode, working electrode and calomel reference electrode were replaced with graphite rod, ITO-coated glass and Ag/AgCl. The system was connected to a potentiostat to supply a potential or voltage of -1 V into the electrochemical deposition system. 0.1M Ammonium tetrathiomolybdate $(NH_4)_2MoS_4$ added 0.1M Potassium Chloride (KCl) solution to 30 mL deionized (DI) water for the deposition process. The aqueous precursor solution was observed to be light brown. The MoS2 was electrodeposited in an aqueous solution at 24°C. After the deposited MoS2 onto ITO-coated glass substrates was acquired, the samples were annealed at 450°C for 30 minutes [14], [15], [16], [17].

B. Fabrication of MAPbI3 Perovskite Solar Cells precursor

This method as shown in Figure 2 dissolved a precursor solution of 80 mg of Methylammonium Iodide in 4ml of Iso Propyl Alcohol (IPA). In addition, 600 mg of Lead Iodide (PbI2) was dissolved in 4 ml of Dimethylformamide (DMF) and Dimethylsulfoxide (DMSO). The ratio of DMF:DMSO is 1:1 for a 4 ml solution. The PbI2 solution in DMF and DMSO will also become clear yellowish after stirring for 1 hour with a temperature of 80 °C [18], [19].

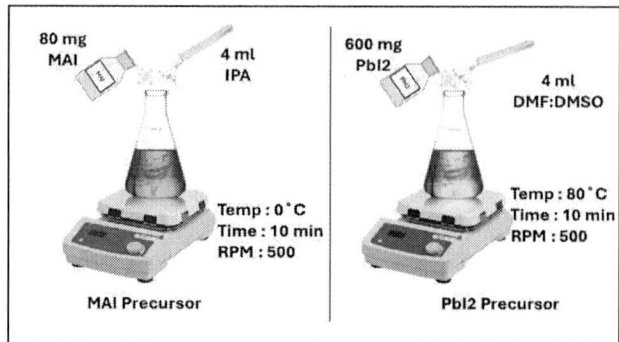

Fig. 2. Synthesis of precursor solution.

The PbI2 solution was heated at a temperature of 80 °C until it completely dissolved before starting the deposition process. On the other hand, the glass substrate was heated at 100 °C for 15 minutes. To prevent a temperature drop during the spin coat process, the warmed stage is crucial. The hot PbI2 solution was then immediately deposited onto the hot glass substrate and spin for 40 seconds at 4000 rpm. The spin-coated glass was then dried for 10 minutes at 80 °C on a hot plate. Subsequently, the MAI solution was immediately dropped and spun at 4000 rpm for 40 seconds on the PbI2 thin film to form the perovskite thin layer. Finally, the thin film was dried at 100°C for 10 minutes on a hot plate to form a perfect perovskite film [20], [21].

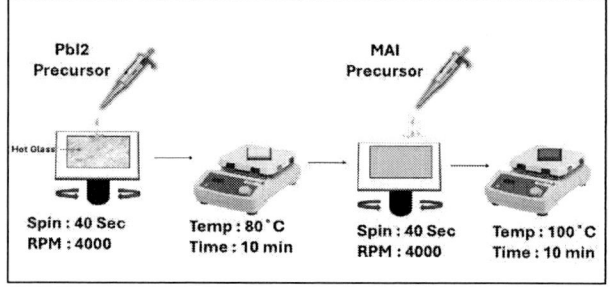

Fig. 3. The deposition process of MAPbI3 using the two-step spin coat method.

All the final preparation and deposition processes were entirely conducted in ambient air conditions. Figure 3 shows the summary of the deposited process of MAPbI3 using the two-step spin coat method.

III. RESULT OF CHARACTERIZATION

A. Ultraviolet-visible spectroscopy (UV-Vis) analysis

Ultraviolet-visible spectroscopy (UV-Vis) was employed to investigate the optical properties of the MoS2/MAPbI3 hybrid system, aiming to elucidate the interactions between MoS2 and Methylammonium Lead Iodide (MAPbI3) in the context of perovskite solar cell applications. Figure 4 displays the UV-Vis spectra of the MoS2/MAPbI3 composite, recorded in the wavelength range of 300 to 800 nm. The spectra were acquired with a scan rate of 200 nm/min and a resolution of 1 nm. Based on the general analysis, the results seem promising, especially for applications requiring broad and strong absorption of light.

The absorption spectrum provided indicates that electrodeposited MoS2 is a promising candidate for use as a

979-8-3503-7832-0/24 $31.00 © 2024 IEEE

HTL in MAPbI3 solar cells. The graph shows good absorption in the UV and visible range which is around 300 to 500, which is promising for photovoltaic applications and suggests that the MoS2/MAPbI3 composite can effectively absorb a significant portion of the solar spectrum, ensuring efficient utilization of sunlight without hindering the light absorption by the perovskite layer. Additionally, the broad absorption spectrum extending into the near-infrared region is advantageous for achieving high-efficiency solar cells.

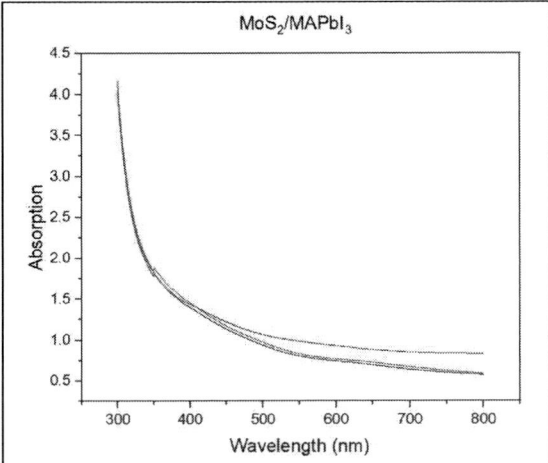

Fig. 4. UV-Vis results for MoS2/MAPbI3

The smooth curves in the absorption spectrum indicate consistent material quality, which is crucial for the performance of the HTL as it ensures effective charge transport and reduces recombination losses. MoS2 as an HTL can facilitate efficient hole transport and collection, improving the overall efficiency of the solar cell. Furthermore, the chemical stability and robustness of MoS_2 can enhance the durability and operational stability of perovskite solar cells. The cost-effectiveness and scalability of electrodeposition as a deposition method make it an attractive approach for producing large-area solar cells.

UV-Vis spectroscopic analysis of MoS2/MAPbI3 has revealed valuable insights into the optical properties of the hybrid system. The observed shifts and features in the spectra suggest electronic interactions between MoS2 and MAPbI3, emphasizing the potential of MoS2 as a contributor to the properties of MAPbI3 in perovskite solar cells. These findings underscore the significance of a comprehensive understanding of hybrid materials for advancing the development of efficient and stable photovoltaic devices. Given the absorption characteristics shown in the graph, electrodeposited MoS2 appears to be a good candidate for a hole transport layer in MAPbI3 perovskite solar cells. The broad absorption spectrum and consistent material quality of MoS_2 suggest effective support for capturing and converting sunlight in perovskite layers.

B. X-ray Diffraction (XRD) analysis

The X-ray diffraction (XRD) patterns depicted in Figures 5 and 6 offer a detailed examination of the crystalline phases present in the MoS2/MAPbI3 system, as detailed in the reference [22]. XRD analysis serves as a fundamental technique for elucidating the structural composition of materials, with diffraction peaks at specific 2θ angles providing critical insights into their crystallographic arrangements. Within this system, the observed diffraction

peaks closely correspond to the characteristic patterns of MoS2 and MAPbI3. Notably, the discernible peaks at 2θ angles of 12.7°, 22.9°, 28.3°, 39.5°, and 41.4° are precisely attributed to the (003), (101), (104), (110), and (018) planes of the MAPbI3 crystal lattice, respectively, offering valuable information regarding the spatial disposition of atoms within the MAPbI3 structure. Furthermore, supplementary peaks at 2θ angles of 14.1°, 32.9°, and 35.9° correspond to the (002), (100), and (102) planes of the MoS2 crystal lattice, providing additional insights into the atomic orientation within the MoS2 component. This meticulous analysis confirms the presence and characterizes the crystalline attributes of both MoS2 and MAPbI3 constituents within the system.

The peaks match well with standard reference patterns, confirming the phase purity of the materials and ensuring minimal defects or impurities that could hinder HTL performance. Furthermore, the strong peaks corresponding to specific planes ((104) for MAPbI3 and (002) for MoS2 suggest a preferred orientation in the crystal growth, enhancing the charge transport properties of the HTL. MoS2 high crystallinity and preferred orientation facilitate efficient hole transport from the perovskite layer to the electrode, reducing recombination losses and improving solar cell efficiency. The well-defined crystal structure of MAPbI3 suggests it can form a good interface with the MoS2 HTL, promoting effective charge separation and transport. MoS2's stability and scalability make it a practical choice for large-scale perovskite solar cell production.

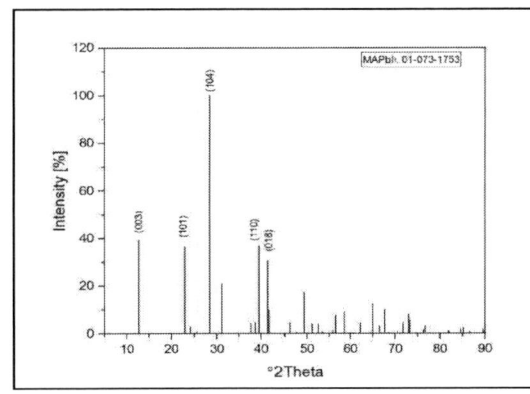

Fig. 5. XRD spectra of MAPbI3.

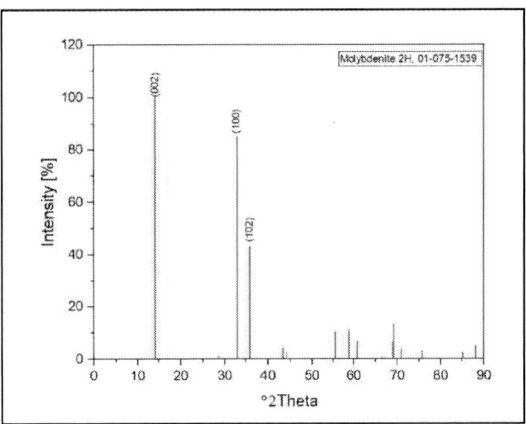

Fig. 6. XRD spectra of MoS2

IV. CONCLUSION

In conclusion, the fabrication and optimization of electrodeposited MoS2 as a hole transport layer in methylammonium perovskite solar cells represent a significant stride towards enhancing the efficiency and stability of next-generation solar energy conversion devices. Through meticulous experimentation and optimization processes, this study has demonstrated the potential of MoS2 as an effective and reliable material for facilitating hole transport within perovskite solar cells. The successful integration of electrodeposited MoS2 has not only improved the overall performance of the solar cells but has also laid the foundation for further advancements in the field.

The insights from this research contribute valuable knowledge to the ongoing efforts to develop sustainable and efficient photovoltaic technologies. The optimized MoS2-based hole transport layer offers a promising avenue for addressing key challenges associated with perovskite solar cells, such as long-term stability and scalability. As we strive towards a more sustainable energy future, the findings of this study underscore the importance of continued research and innovation in material science and solar cell technology. The outcomes presented here are anticipated to inspire further investigations, fostering collaborative efforts to bring us closer to the realization of efficient, cost-effective, and commercially viable perovskite solar cells.

ACKNOWLEDGMENT

This study is funded by the Ministry of Higher Education (MOHE) of Malaysia through the Fundamental Research Grant Scheme (FRGS), No. FRGS/1/2022/TK07/UTEM/02/17. The Authors also would like to thank Universiti Teknikal Malaysia Melaka (UTeM) for all the support. The author also acknowledges the provider of the free version of GPVDM software.

REFERENCES

[1] S. Muhammad, A. T. Nomaan, M. I. Idris, and M. Rashid, 'Structural, optical and electrical investigation of low-temperature processed zinc oxide quantum dots based thin films using precipitation-spin coating on flexible substrates', Physica B Condens Matter, vol. 635, p. 413806, Jun. 2022, doi: 10.1016/j.physb.2022.413806.

[2] S. Muniandy et al., 'The Effect of Different Precursor Solutions on the Structural, Morphological, and Optical Properties of Nickel Oxide as an Efficient Hole Transport Layer for Perovskite Solar Cells', Pertanika J Sci Technol, vol. 31, no. 4, Jun. 2023, doi: 10.47836/pjst.31.4.26.

[3] A. M. A. Aziz et al., 'Investigation of The Performance Impact of Active Layer Parameter Variations on Inverted Perovskite Solar Cells Using GPVDM', in 2023 IEEE Regional Symposium on Micro and Nanoelectronics (RSM), IEEE, Aug. 2023, pp. 130–133. doi: 10.1109/RSM59033.2023.10326977.

[4] P. N. A. Fahsyar et al., 'Correlation of simulation and experiment for perovskite solar cells with MoS2 hybrid-HTL structure', Applied Physics A, vol. 127, no. 5, p. 383, May 2021, doi: 10.1007/s00339-021-04531-8.

[5] D. Wang et al., 'MoS2 incorporated hybrid hole transport layer for high performance and stable perovskite solar cells', Synth Met, vol. 246, pp. 195–203, Dec. 2018, doi: 10.1016/j.synthmet.2018.10.012.

[6] P. N. A. Fahsyar, N. A. Ludin, N. F. Ramli, P. I. Zulaikha, S. Sepeai, and A. S. H. Md Yasir, 'Stabilizing high-humidity perovskite solar cells with MoS2 hybrid HTL', Sci Rep, vol. 13, no. 1, p. 11996, Jul. 2023, doi: 10.1038/s41598-023-39189-0.

[7] A. M. Abdul Aziz, M. I. Idris, Z. A. F. Mohammed Napiah, M. N. Shah Zainudin, M. Rashid, and L. Bradley, 'The development of low-cost spin coater with wireless IoT control for thin film deposition', Indonesian Journal of Electrical Engineering and Computer Science, vol. 34, no. 3, p. 1519, Jun. 2024, doi: 10.11591/ijeecs.v34.i3.pp1519-1529.

[8] N. A. Abd Malek et al., 'Ultra-thin MoS2 nanosheet for electron transport layer of perovskite solar cells', Opt Mater (Amst), vol. 104, p. 109933, Jun. 2020, doi: 10.1016/j.optmat.2020.109933.

[9] Q. Van Le, J.-Y. Choi, and S. Y. Kim, 'Recent advances in the application of two-dimensional materials as charge transport layers in organic and perovskite solar cells', FlatChem, vol. 2, pp. 54–66, Apr. 2017, doi: 10.1016/j.flatc.2017.04.002.

[10] R. Singh et al., 'Perovskite solar cells with an MoS 2 electron transport layer', J Mater Chem A Mater, vol. 7, no. 12, pp. 7151–7158, 2019, doi: 10.1039/C8TA12254G.

[11] V. P. Pham and G. Y. Yeom, 'Recent Advances in Doping of Molybdenum Disulfide: Industrial Applications and Future Prospects', Advanced Materials, vol. 28, no. 41, pp. 9024–9059, Nov. 2016, doi: 10.1002/adma.201506402.

[12] A. Capasso et al., 'Few - Layer MoS 2 Flakes as Active Buffer Layer for Stable Perovskite Solar Cells', Adv Energy Mater, vol. 6, no. 16, Aug. 2016, doi: 10.1002/aenm.201600920.

[13] Y. Nakayasu, H. Kobayashi, S. Katahira, T. Tomai, and I. Honma, 'Rapid, one-step fabrication of MoS2 electrocatalysts by hydrothermal electrodeposition', Electrochem commun, vol. 134, p. 107180, Jan. 2022, doi: 10.1016/j.elecom.2021.107180.

[14] I. Rummaja, 'Analysis of the electrodeposited ZnO as photo anode for solar cells', PRZEGLĄD ELEKTROTECHNICZNY, vol. 1, no. 2, pp. 125–128, Feb. 2023, doi: 10.15199/48.2023.02.20.

[15] M. A. Hossain, B. A. Merzougui, F. H. Alharbi, and N. Tabet, 'Electrochemical deposition of bulk MoS2 thin films for photovoltaic applications', Solar Energy Materials and Solar Cells, vol. 186, pp. 165–174, Nov. 2018, doi: 10.1016/j.solmat.2018.06.026.

[16] C. Y. Chang, K. S. Anuratha, Y. H. Lin, Y. Xiao, P. Hasin, and J. Y. Lin, 'Potential-reversal electrodeposited MoS2 thin film as an efficient electrocatalytic material for bifacial dye-sensitized solar cells', Solar Energy, vol. 206, pp. 163–170, Aug. 2020, doi: 10.1016/j.solener.2020.06.001.

[17] Md. A. Hossain, B. A. Merzougui, F. H. Alharbi, and N. Tabet, 'Electrochemical deposition of bulk MoS2 thin films for photovoltaic applications', Solar Energy Materials and Solar Cells, vol. 186, pp. 165–174, Nov. 2018, doi: 10.1016/j.solmat.2018.06.026.

[18] S. Muniandy, M. I. Bin Idris, Z. A. F. Bin Mohammed Napiah, and M. Rashid, 'The effect of pH level and annealing temperature on NiO thin films as Hole Transport Material in Inverted Perovskite Solar Cells', in 2022 IEEE International Conference on Semiconductor Electronics (ICSE), IEEE, Aug. 2022, pp. 13–16. doi: 10.1109/ICSE56004.2022.9863126.

[19] E. Smecca et al., 'Two-step MAPbI3 deposition by low-vacuum proximity-space-effusion for high-efficiency inverted semitransparent perovskite solar cells', J Mater Chem A Mater, vol. 9, no. 30, pp. 16456–16469, 2021, doi: 10.1039/d1ta02535j.

[20] G. Gordillo, O. G. Torres, M. C. Abella, J. C. Peña, and O. Virguez, 'Improving the stability of MAPbI3 films by using a new synthesis route', Journal of Materials Research and Technology, vol. 9, no. 6, pp. 13759–13769, Nov. 2020, doi: 10.1016/j.jmrt.2020.09.095.

[21] L.-C. Chen, C.-H. Tien, Z.-L. Tseng, and J.-H. Ruan, 'Enhanced Efficiency of MAPbI3 Perovskite Solar Cells with FAPbX3 Perovskite Quantum Dots', Nanomaterials, vol. 9, no. 1, p. 121, Jan. 2019, doi: 10.3390/nano9010121.

[22] P. Joensen, E. D. Crozier, N. Alberding, and R. F. Frindt, 'A study of single-layer and restacked MoS 2 by X-ray diffraction and X-ray absorption spectroscopy', Journal of Physics C: Solid State Physics, vol. 20, no. 26, pp. 4043–4053, Sep. 1987, doi: 10.1088/0022-3719/20/26/009.

Piezoelectric Nanogenerator based on Graphene and MXene Heterostructure

Lijie Kou
Institute of Microengineering and Nanoelectronics (IMEN)
Universiti Kebangsaan Malaysia (UKM), Computing and Information Science College, Fuzhou Institute of Technology, Fuzhou 350506, People's Republic of China Bangi
Selangor, Malaysia
p118051@siswa.ukm.edu.my

Rawhan Haque
Institute of Microengineering and Nanoelectronics (IMEN)
Universiti Kebangsaan Malaysia (UKM), Bangi
Selangor, Malaysia
p127643@siswa.ukm.edu.my

Aniq Shazni Mohammad Haniff
Institute of Microengineering and Nanoelectronics (IMEN)
Universiti Kebangsaan Malaysia (UKM), Bangi
Selangor, Malaysia
aniqshazni@ukm.edu.my

Chang Fu Dee
Institute of Microengineering and Nanoelectronics (IMEN)
Universiti Kebangsaan Malaysia (UKM), Bangi
Selangor, Malaysia
cfdee@ukm.edu.my

Poh Choon Ooi*
Institute of Microengineering and Nanoelectronics (IMEN)
Universiti Kebangsaan Malaysia (UKM), Bangi
Selangor, Malaysia
pcooi@ukm.edu.my

Abstract— **This study investigates the integration of graphene and MXene heterostructures as nanofillers in polyvinylidene fluoride (PVDF) matrices to enhance the performance of piezoelectric nanogenerators (PENGs). The combination of nitrogen, sulfur, and phosphorus tri-doped graphene (NSPG) and titanium carbonitride (Ti$_3$CNT$_x$) MXene significantly improves the electroactive β-phase content in PVDF, leading to enhanced piezoelectric output. The β-phase fraction in the nanocomposites increased with the ratio of NSPG to Ti$_3$CNT$_x$, reaching 88% for a 1:3 ratio. The resulting PENG devices exhibited higher open circuit voltage and short-circuit current values compared to pure PVDF and reference samples. The highest measured peak-to-peak output values were 14.6 V for V_{oc} and 1.48 μA for I_{sc}. The synergistic effect of the NSPG-Ti$_3$CNT$_x$ heterostructure resulted in an output power of 8.8 μW, five times higher than that of a pure PVDF device. These findings demonstrate the potential of this nanofiller combination to improve PENG efficiency and pave the way for the development of biomechanical energy harvesting devices.**

Keywords— *Heterostructure, Ti$_3$CNT$_x$ MXene, Tridoped-graphene, Piezoelectric, Nanogenerator*

I. INTRODUCTION

Electronic devices are getting smaller, more efficient, and using less power. Piezoelectric nanogenerators (PENGs) capture mechanical energy from the environment and convert it into electricity to power low-power gadgets [1], [2]. Piezoelectric polymers such as polyvinylidene fluoride (PVDF) and its copolymers polyvinylidene fluoride trifluoroethylene are flexible and lightweight, making them ideal for various applications. Nevertheless, PVDF is popular due to its strong mechanical, chemical, and thermal resilience and comparatively inexpensive cost. Techniques such as annealing, electric polarization, and adding nanofillers can efficiently modify the PVDF's piezoelectric coefficient [3], [4].

PVDF is a semi-crystalline polymer with different phases, including α, β, γ, δ, and ε. The β-phase is crucial for PENG's performance due to its strong polarization and piezoelectric sensitivity. Increasing the β-phase content enhances PVDF's piezoelectric characteristics [5]. Methods for achieving this include electrical poling, mechanical stretching, and adding nanofillers, which can increase the interaction and interfacial area between the polymer and the nanofiller, leading to improved polarization and dielectric response.

The two-dimensional (2D) materials titanium MXenes and graphene have lately been widely exploited as nanofillers to modify the structure and properties of composite nanomaterials due to their high specific surface areas and unique electrical properties [6], [7]. Compared to pure graphene, graphene is purposefully doped with nitrogen (N), sulfur (S), or phosphorus (P) atoms to improve performance and create unique features. The intrinsic chemical and electrical characteristics of carbon can be modified by functionalization of heteroatoms, and active sites can be introduced into graphene with slight modification to the 2D system through doping [8], [9]. The introduction of distinct electronic energy levels through elemental doping contributes to a rise in carrier concentration and an improvement in conductivity [8].

979-8-3503-7832-0/24 $31.00 © 2024 IEEE

Another feature is that transition metal carbides and nitrides with different surface terminations stacked in a multilayered structure make up 2D titanium carbonitrides (Ti_3CN) MXene, a member of the MXene family. Following the etching process, the material is called Ti_3CNT_x, where "T_x" stands for either -F, -OH, or -O surface termination group. MXene is a remarkable conductor with strong interfacial coupling and quick charge transfer kinetics that strengthens composites [10]. However, when Ti_3CNT_x is incorporated into devices, it is easily subjected to a layer structure stacking phenomenon because of the high activity of its surface functional groups. In order to solve problems with layer stacking and device stability, the hydrophilic groups on the graphene surface allow for efficient mixing with MXene [11] to produce composites with good dispersibility and metallic conductivity [12], [13], [14].

Given that Ti_3CNT_x is a metallic material with effective charge transfer kinetics, it could potentially enhance piezoelectric performance when combined with tri-doped graphene. This combination is expected to improve the PVDF's electroactive β-phase and overall piezoelectric output. Additionally, the use of graphene and MXene in PENG devices may increase strain distribution, mechanical characteristics, and piezoelectric response, ultimately boosting energy conversion efficiency for human motion-driven PENGs.

II. EXPERIMENTAL

A. Nanofillers and nanocomposite solutions preparation

PVDF powder (average Mw~534 000), magnesium chloride powder ($MgCl_2$), and N, N-dimethylformamide (DMF) solvent were purchased from Sigma Aldrich. No additional purification was required for any of the purchased components. Graphite was electrochemically exfoliated to create N, S, and P tri-doped electrochemically exfoliated graphene (NSPG) [15], [16], [17]. The preparation of Ti_3CNT_x powders involved ball milling and annealing Ti_3AlCN MAX phase powder under a constant argon flow [18], [19]. To prepare the PVDF solution, 1 g of PVDF was dissolved in 10 ml of DMF solvent. Next, 5 mg of $MgCl_2$ was added to the solution to improve its chemical polarity and stirred for two hours at 60 °C. After that, a 0.5 mg/ml NSPG suspension was prepared by sonicating 5 mg of NSPG with 10 ml of DMF for 4 hours at room temperature (RT). Then, to create a dispersion with a concentration of 1 mg/ml, 5 mg of Ti_3CNT_x was added to 5 ml of DMF and sonicated for 4 hours at RT. Combining the obtained nanofiller dispersions in different ratios, 2:1, 1:1, 2:2, and 1:3, the blend of NSPG and Ti_3CNT_x was created. To create a heterostructure nanofiller dispersion, the nanofiller mixture was once more sonicated for four hours at RT. Lastly, the PVDF solution was mixed with the produced nanofiller dispersions at a 4:1 volume ratio and left at RT for four hours. In that order, the nanocomposites were designated as PMIX1, PMIX2, PMIX3, and PMIX4. In addition, reference samples were labelled as PP, PNSPG, and PTi_3CN. These samples included pure PVDF, PVDF with NSPG, and PVDF with Ti_3CNT_x.

B. Device fabrication

The polyethylene terephthalate (PET) substrate coated with indium tin oxide (ITO) measuring 2 x 2 cm² was cleaned with ethanol and treated with oxygen plasma at 50 W for 5 min to increase its surface hydrophilicity prior to deposition. After that, PMIX1 was spin-coated for 50 s at 2000 rpm on two cleaned substrate pieces to create a corresponding ~ 4 μm nanocomposite layer. This was followed by 90 s of drying on a hot plate heated to 80 °C. The formed nanocomposite layers were instantly layered and subjected to an additional heat treatment for 30 minutes at 80 °C to form an 8 μm thick layer D-I device. External connections were formed using aluminum foil. The same fabrication methods were performed to fabricate D-II, D-III, and D-IV utilizing PMIX2, PMIX3, and PMIX4, respectively. Note that reference devices comprised of PVDF with NSPG, Ti_3CN, and pure PVDF were also fabricated and labeled as R-NSPG, R-Ti, and R-P, respectively. The solenoid tapper configuration and and the device structure are displayed in Fig. 1.

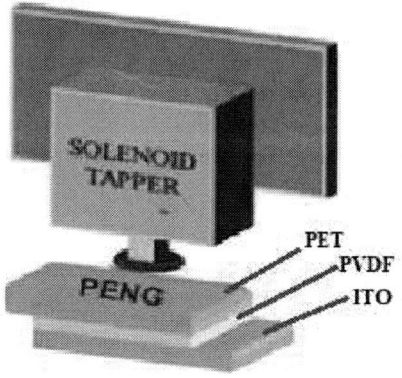

Fig. 1. Solenoid tapper setup with a PENG device.

III. RESULTS AND DISCUSSION

The material's chemical properties were examined using Raman spectroscopy and X-ray photoelectron spectroscopy (XPS). From the XPS scan, the presence of essential elements, namely titanium (Ti), carbon (C), and nitrogen (N), in MXene was confirmed. Moreover, the detection of fluorine and oxygen from the analysis ascertains the functionalization in the MXene structure. The morphology of MXene and NSPG multilayers was examined using high-resolution transmission electron microscopy (HRTEM).

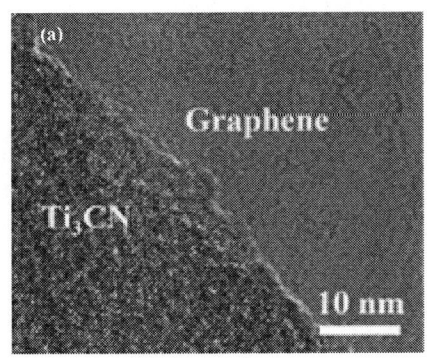

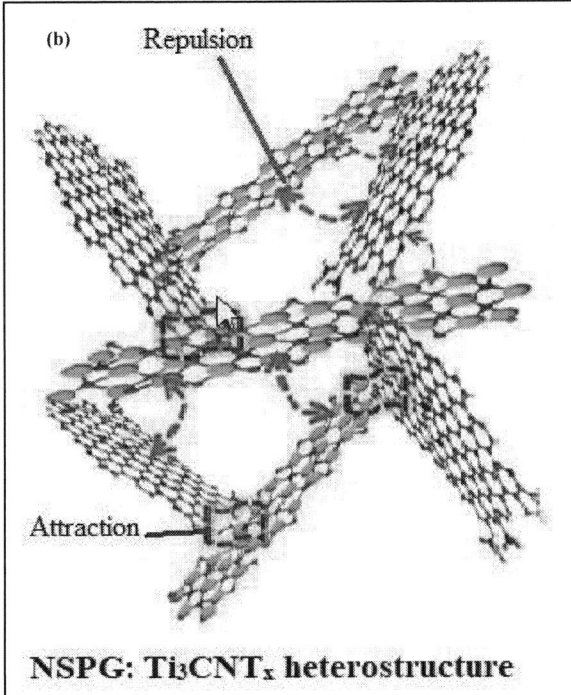

Fig. 2. (a) HRTEM image of graphene and Ti₃CN integration. (b) Schematic diagram of NSPG: Ti₃CNTₓ heterostructure formation.

Utilizing a Perkin Elmer 2000 apparatus, Fourier transform infrared spectroscopy (FTIR) was employed to assess the crystallinity and structure of piezoelectric layers, as shown in Fig. 3(a). The β-phase percentage of the pure PVDF films was discovered to be 45%. The β-phase content was increased to 57% for the R-NSPG sample and 60% for the R-Ti sample. However, compared to the reference samples, the β-phase content for PMIX1, PMIX2, PMIX3, and PMIX4 was assessed to be 64%, 75%, 88%, and 79%, respectively. Fig. 3(b) shows the generated open circuit voltage (V_{oc}) was then measured with a picoammeter (Keithley 6487), and the generated short-circuit current (I_{sc}) was observed with a digital oscilloscope (Keysight DSO9404A). The measured peak-to-peak V_{oc} for R-NSPG, R-Ti, R-P, D-I, D-II, D-III, and D-IV were 5.1, 6.0, 2.5, 9.5, 13.0, 14.6, and 13.1 V, respectively. Conversely, the corresponding maximum I_{sc} attained by R-NSPG, R-Ti, R-P, D-I, D-II, D-III, and D-IV were 0.70, 0.92, 0.21, 1.01, 1.15,

1.48, and 1.32 μA, respectively. Compared to pure PVDF, including NSPG and Ti₃CNTₓ significantly increases the output voltage and current. The optimum device attained a power output of 8.8 μW.

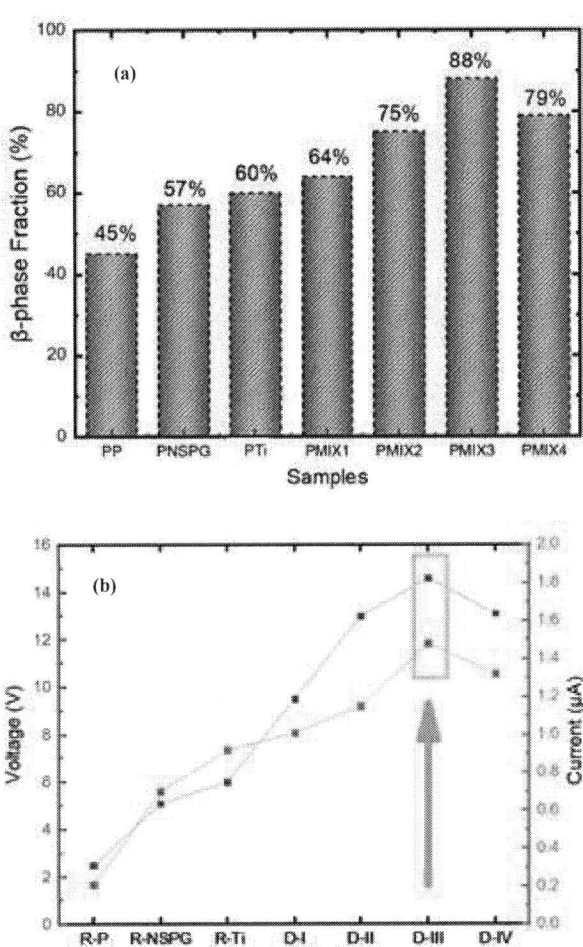

Fig. 3. (a) β-phase fraction of samples, (b) V_{oc} and I_{sc} of the fabricated devices.

IV. CONCLUSION

The complement of the NSPG-Ti₃CNTₓ heterostructure improved PENG performance. Utilized as nanofillers in the PVDF matrix, the NSPG, and Ti₃CNTₓ heterostructure formed by interfacial van der Waals force interactions. The V_{oc} reached synergistically to 14.6 V, five times higher than a pure PVDF device when the nanofiller ratio of NSPG to Ti₃CNTₓ was 1:3. The formation of an electroactive β-phase in PVDF molecular chains has been effectively promoted due to the strong cross-interfacial coupling of the interface functional groups of the two nanomaterials. These encouraging findings imply that, when combined in the right proportion, NSPG and Ti₃CNTₓ composition can improve nanogenerator efficiency and pave the way for the development of biomechanical energy harvesting.

979-8-3503-7832-0/24 $31.00 © 2024 IEEE

ACKNOWLEDGMENT

This project received support from the Fundamental Research Grant Scheme (FRGS) under funding code FRGS/1/2022/TK08/UKM/02/13.

REFERENCES

[1] A. Ali *et al.*, 'Advancements in piezoelectric wind energy harvesting: A review', *Results Eng.*, vol. 21, p. 101777, Mar. 2024, doi: 10.1016/j.rineng.2024.101777.

[2] B. Upendra, B. Panigrahi, K. Singh, and G. Sabareesh, 'Recent advancements in piezoelectric energy harvesting for implantable medical devices', *J. Intell. Mater. Syst. Struct.*, vol. 35, no. 2, pp. 129–155, Jan. 2024, doi: 10.1177/1045389X231200144.

[3] L. Yao, Z. Zhang, Q. Zhang, Z. Zhou, H. Yang, and L. Chen, 'Modified organic polystyrene microspheres embedded into P(VDF-TrFE) with lotus-leaf microstructure enables high performance triboelectric nanogenerator', *Nano Energy*, vol. 86, p. 106128, Aug. 2021, doi: 10.1016/j.nanoen.2021.106128.

[4] C. Zhang, Y. Li, W. Kang, X. Liu, and Q. Wang, 'Current advances and future perspectives of additive manufacturing for functional polymeric materials and devices', *SusMat*, vol. 1, no. 1, pp. 127–147, Mar. 2021, doi: 10.1002/sus2.11.

[5] S. Mirjalali *et al.*, 'A review on wearable electrospun polymeric piezoelectric sensors and energy harvesters', *Macromol. Mater. Eng.*, vol. 308, no. 1, pp. 1–27, Jan. 2023, doi: 10.1002/mame.202200442.

[6] S. Nashruddin, J. Abdullah, M. Mohammad Haniff, M. Mat Zaid, O. Choon, and M. Mohd Razip Wee, 'Label Free Glucose Electrochemical Biosensor Based on Poly(3,4-ethylenedioxy thiophene):Polystyrene Sulfonate/Titanium Carbide/Graphene Quantum Dots', *Biosensors*, vol. 11, no. 8, p. 267, Aug. 2021, doi: 10.3390/bios11080267.

[7] K. Nasrin, V. Sudharshan, M. Arunkumar, and M. Sathish, '2D/2D Nanoarchitectured Nb_2C/Ti_3C_2 MXene Heterointerface for High-Energy Supercapacitors with Sustainable Life Cycle', *ACS Appl. Mater. Interfaces*, vol. 14, no. 18, pp. 21038–21049, May 2022, doi: 10.1021/acsami.2c02871.

[8] J. Liu, Y. Zhu, X. Chen, and W. Yi, 'Nitrogen, sulfur and phosphorus tri-doped holey graphene oxide as a novel electrode material for application in supercapacitor', *J. Alloys Compd.*, vol. 815, p. 152328, Jan. 2020, doi: 10.1016/j.jallcom.2019.152328.

[9] U. Yaqoob, A. S. M. I. Uddin, and G.-S. Chung, 'A novel tri-layer flexible piezoelectric nanogenerator based on surface- modified graphene and $PVDF-BaTiO_3$ nanocomposites', *Appl. Surf. Sci.*, vol. 405, pp. 420–426, May 2017, doi: 10.1016/j.apsusc.2017.01.314.

[10] X. Wu, Z. Wang, M. Yu, L. Xiu, and J. Qiu, 'Stabilizing the MXenes by Carbon Nanoplating for Developing Hierarchical Nanohybrids with Efficient Lithium Storage and Hydrogen Evolution Capability', *Adv. Mater.*, vol. 29, no. 24, p. 1607017, Jun. 2017, doi: 10.1002/adma.201607017.

[11] X. Shi, S. Chen, H. Zhang, J. Jiang, Z. Ma, and S. Gong, 'Portable Self-Charging Power System via Integration of a Flexible Paper-Based Triboelectric Nanogenerator and Supercapacitor', *ACS Sustain. Chem. Eng.*, vol. 7, no. 22, pp. 18657–18666, Nov. 2019, doi: 10.1021/acssuschemeng.9b05129.

[12] Z. Ma, X. Zhou, W. Deng, D. Lei, and Z. Liu, '3D Porous MXene (Ti_3C_2)/Reduced Graphene Oxide Hybrid Films for Advanced Lithium Storage', *ACS Appl. Mater. Interfaces*, vol. 10, no. 4, pp. 3634–3643, Jan. 2018, doi: 10.1021/acsami.7b17386.

[13] J. Yan *et al.*, 'Flexible MXene/Graphene Films for Ultrafast Supercapacitors with Outstanding Volumetric Capacitance', *Adv. Funct. Mater.*, vol. 27, no. 30, p. 1701264, Aug. 2017, doi: 10.1002/adfm.201701264.

[14] M. Liu, Z. Wang, P. Song, Z. Yang, and Q. Wang, 'Flexible MXene/rGO/CuO hybrid aerogels for high performance acetone sensing at room temperature', *Sens. Actuators B Chem.*, vol. 340, p. 129946, Aug. 2021, doi: 10.1016/j.snb.2021.129946.

[15] P. Yu, S. E. Lowe, G. P. Simon, and Y. L. Zhong, 'Electrochemical exfoliation of graphite and production of functional graphene', *Curr. Opin. Colloid Interface Sci.*, vol. 20, no. 5–6, pp. 329–338, Oct. 2015, doi: 10.1016/j.cocis.2015.10.007.

[16] S. Yang *et al.*, 'Organic Radical-Assisted Electrochemical Exfoliation for the Scalable Production of High-Quality Graphene', *J. Am. Chem. Soc.*, vol. 137, no. 43, pp. 13927–13932, Nov. 2015, doi: 10.1021/jacs.5b09000.

[17] K. S. Rao, J. Sentilnathan, H. Cho, J. Wu, and M. Yoshimura, 'Soft Processing of Graphene Nanosheets by Glycine-Bisulfate Ionic-Complex-Assisted Electrochemical Exfoliation of Graphite for Reduction Catalysis', *Adv. Funct. Mater.*, vol. 25, no. 2, pp. 298–305, Jan. 2015, doi: 10.1002/adfm.201402621.

[18] M. Naguib *et al.*, 'Two-Dimensional Transition Metal Carbides', *ACS Nano*, vol. 6, no. 2, pp. 1322–1331, Feb. 2012, doi: 10.1021/nn204153h.

[19] K. Hantanasirisakul *et al.*, 'Effects of Synthesis and Processing on Optoelectronic Properties of Titanium Carbonitride MXene', *Chem. Mater.*, vol. 31, no. 8, pp. 2941–2951, Apr. 2019, doi: 10.1021/acs.chemmater.9b00401.

[20] R. Z. Auliya *et al.*, 'Exploration of 2D Ti_3C_2 MXene for all solution processed piezoelectric nanogenerator applications', *Sci. Rep.*, vol. 11, no. 1, p. 17432, Aug. 2021, doi: 10.1038/s41598-021-96909-0.

Thermal-Aware Test Scheduling with Floor planning for Three-Dimensional Stacked Integrated Circuit

Ganesan Patmanathan
Department of Electronic System Engineering, Malaysia-Japan International Institute of Technology,
University Technology of Malaysia,
Kuala Lumpur, Malaysia
ganesan.sp1999@graduate.utm.my

Chia Yee Ooi
Department of Electronic System Engineering, Malaysia-Japan International Institute of Technology,
University Technology of Malaysia,
Kuala Lumpur, Malaysia
ooichiayee@utm.my

Nordinah Ismail
Department of Electronic System Engineering, Malaysia-Japan International Institute of Technology,
University Technology of Malaysia,
Kuala Lumpur, Malaysia
nordinahismail@gmail.com

Siti Rahmah Aid
Department of Electronic System Engineering, Malaysia-Japan International Institute of Technology,
University Technology of Malaysia,
Kuala Lumpur, Malaysia
sitirahmah.aid@utm.my

Abstract—Testing a three-dimensional stacked integrated circuit (3D-SIC) remains a challenging problem, as generating an optimized test schedule to minimize test time is complicated due to the numerous variables involved. Accessing upper dies is only feasible through the bottom die, necessitating the extension of Test Access Mechanisms (TAMs) via Through-Silicon Vias (TSVs). Limited primary I/O pins, TSVs, and TAM width require efficient resource allocation. Thermal management is crucial due to high core power consumption and uneven distribution, which pose the risk of overheating. Advanced concurrent test scheduling is essential to effectively allocate resources and maintain power and temperature limits. This research proposes thermal-aware test scheduling optimization combined with floor planning for 3D-SICs, aiming to minimize test schedule time while addressing resource and power constraints. Experimental results using several ITC'02 benchmark circuits demonstrate an average estimated improvement of 0.2% in test schedule time when utilizing test scheduling with floor planning compared to test scheduling without floor planning.

Keywords—3D-SIC, test scheduling, test time, floor planning, thermal-aware

I. INTRODUCTION

Integrating entire electronic systems onto a single chip is the prevailing trend. Despite advancements in integrated circuit (IC) technology enabling smaller feature sizes, conventional System-on-Chips (2D-ICs) are growing wider due to increased demand for semiconductor devices with enhanced features. However, this growth poses challenges to circuit performance due to lengthy horizontal interconnects between cores. To address this issue, three-dimensional stacked integrated circuit (3D-SIC) technology has emerged. It involves stacking circuit components into multiple layers with vertical interconnections (TSVs), thereby reducing horizontal interconnect lengths and offering improved functionality, bandwidth, and a smaller footprint [1-4].

Despite the advantages of 3D-SICs, testing is challenging due to vertically stacked dies. Test data transmission between the test pins and the cores on different dies, relies on test access mechanisms (TAMs) and TSVs [1]. However, due to the limited availability of test resources such as test pins,

TAMs, and TSVs, a test schedule with efficient allocation and utilization of these resources are essential, but finding an optimal solution is inherently complex [2-4]. Consequently, researchers use heuristics and approximation algorithms to create near-optimal test schedules, aiming to minimize test schedule time effectively [5-7].

Testing concurrently multiple cores in a 3D-SIC can reduce test schedule time but increase power consumption, risking system-level power limits and core damage [2-4]. Concurrent testing of adjacent cores with higher power density (hot modules) can create hotspots, compromising chip reliability [8]. Therefore, researchers frequently employ power and thermal-aware test scheduling strategies to manage test concurrency, ensuring that the overall power consumption and temperature of the 3D-SIC remain within system limits [8-10]. Furthermore, the placement of cores within the stack and their test duration impact heat dissipation capabilities. Cores further from the heat sink dissipate heat slower, regardless of their power consumption. Longer test times generate more heat, exacerbating potential hotspots. Therefore, test scheduling must consider core heat dissipation, ensuring adequate space around cores for heat dissipation to prevent hotspot formation and irreversible damage to the 3D-SIC [8-9].

Heat from hot modules transfers to cooler adjacent cores, so placing them next to non-hot modules can prevent hotspot formation and allow for more concurrent testing [11]. However, when positioned far from the heat sink, hot modules may dissipate heat less effectively, raising the 3D-SIC temperature. Ideally, hot modules should be closer to the heat sink, often at the top of the stack, but this increases costs due to additional TSVs. Therefore, relying solely on resource, power, and thermal-aware test scheduling is inadequate; thermal-aware floor planning that considers TAM placement is crucial for optimizing test schedule quality, an aspect often overlooked in previous research.

II. METHODOLOGY

A. Problem Formulation

Given a 3D-SIC to be designed with (i) the maximum allowable number of stacks, (ii) the maximum TAM width limit, Wmax, (iii) the maximum TSV limit, TSVmax, (iv) the number of cores, Ci, (v) the number of I/O terminals per Ci, (vi) the number of internal scan chains per Ci, (vii) the length of each internal scan chain, (viii) the number of test patterns associated with each Ci, (ix) maximum power limit, Pmax, (x) maximum thermal limit, Tmax. A co-optimization of thermal-aware test scheduling and floor planning techniques was performed to obtain a test schedule for 3D-SIC with minimal test time under Pmax, Tmax, and resource constraints.

The study's methodology, depicted in Fig. 1, illustrates the co-optimization process for thermal-aware test scheduling with floor planning, taking into account resource and power constraints. The main objective is to develop a test scheduling technique for 3D-SICs aimed at minimizing overall test time..

B. Pareto Test Cubes

To support modular testing and optimize the utilization of TAM widths and core I/O pins, several wrapper designs were created for each core using the Design Wrapper heuristic based on the Best Fit Decreasing (BFD) algorithm, as proposed in [12]. Test times for each wrapper design were calculated, resulting in a range of test times for each core across various TAM widths. Pareto-optimal points were identified based on these calculations. Each core under a specific Pareto optimal wrapper design was represented as a 3-dimensional cube, with dimensions corresponding to TAM width, power consumption, and test time.

C. Ant Colony Optimization

The Ant Colony Optimization (ACO) algorithm was used to select pareto cubes for 3D-SIC floor planning and test

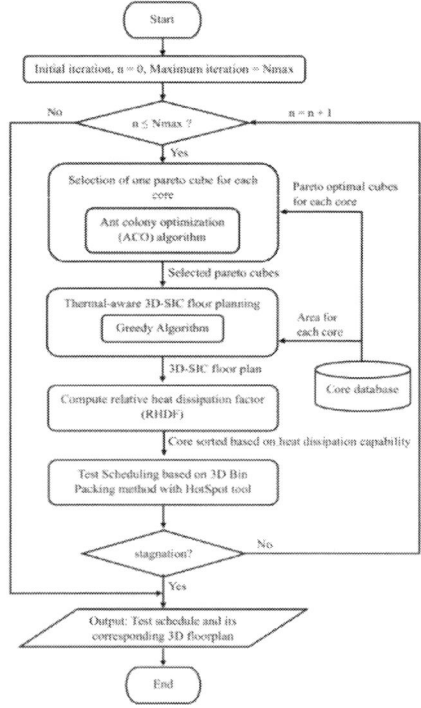

Fig. 1. Co-optimization procedure for thermal-aware test scheduling and floor planning in 3D-SICs

scheduling. In ACO, each core was denoted as Ci ($1 \leq i \leq N$), where N represented the total number of cores in the 3D-SIC. Each core Ci was associated with a collection of pareto-optimal cubes (Ri). For each cube (ri) in the set Ri, the dimensions are represented: the height as the allocated TAM width for the jth wrapper configuration, wij; the width as the corresponding test time, T(wij); and the third dimension as the corresponding peak power, P(wij).

ACO leveraged both pheromone trail values and heuristic favourability to find solutions. In this study, the heuristic favourability, denoted as Wprefer, represented the preferred TAM width for Ci. Initially, the pareto cube from the set Ri with the highest TAM width, which represented the lowest test time, was selected as Wprefer for each Ci. Subsequently, the Wprefer value for each Ci was updated to its corresponding pareto cube from the set of cubes of the global best ant (Rglobal), which resulted in the lowest test schedule time, to prevent the search from becoming too focused on local optima. The pheromone trail value, denoted as $\tau(i,j)$, represented the attractiveness of choosing a pareto cube with TAM width j for Ci. The probability of selecting TAM width j for Ci was calculated based on (1), where α was the relative importance of $\tau(i,j)$, and β was the relative importance of TAM width, wij [13].

$$prob(i,j) = \frac{[\tau(i,j)]^{\alpha} \cdot \beta}{\sum_{j=1}^{W_{max}}([\tau(i,j)]^{\alpha} \cdot \beta)} \qquad (1)$$

In the ACO algorithm, a set number of ants were randomly placed within test cubes during initialization. Each ant selected test cubes for cores iteratively. After N-1 iterations, the ants had chosen a test cube for each core. These cubes were used for thermal-aware 3D-SIC floor planning, generating multiple floor plans. Each floorplan underwent thermal-aware test scheduling with power and TAM constraints, producing test schedule time. The set of cubes resulting in the shortest test schedule time was denoted as Rglobal, while Rlocal signified the optimal solution discovered during the current iteration of the ACO, primarily used for updating the pheromone trail. This process continued until reaching the maximum cycle count or stagnation, where all ants selected the same cubes with identical test times.

D. Thermal-aware 3D Floor planning

The set of cubes chosen by each ant undergoes a thermal-aware floor planning process using a greedy algorithm. The purpose of generating a floor plan for each ant is to explore different configurations and find the most optimal layout considering both thermal factors and TAMs. The inputs include the number of 3D-SIC layers and the maximum layer area, as well as information for each core such as its area (ai), power density (pij), and TAM width (wij). Subsequently, the hot modules of the 3D-SIC are identified by calculating the average power density (p_avg) of all cores in the set. A core with a power density value greater than the average is designated as the hot module within the set. To mitigate hotspots and manage peak temperatures, special attention is given to the placement of the hot module, strategically arranging its position [11].

In this research, the optimal placement of the hot module involves several considerations. First, it is important to avoid placing hot modules in close proximity to each other to prevent localized temperature build-up and potential thermal issues within the chip. Secondly, positioning the hot module closer to the heat sink is beneficial for efficient heat dissipation and helps maintain temperatures within acceptable limits. Lastly, placing the hot module at the edge of the layer can enhance heat dissipation by increasing exposure to ambient air. Adhering to these guidelines can improve the overall thermal performance and reliability of the 3D-SIC

Moreover, this research considers a scenario where the heat sink is placed on the top layer of the 3D-SIC. Some hot modules may necessitate a larger TAM width. Positioning these hot modules closer to the heat sink would increase the number of required TSVs for test access. Therefore, it is crucial to carefully balance the trade-off between power density and TAM width to achieve advantages in two key areas: reducing the overall temperature of the system and optimizing TSV utilization. A trade-off function (TFi) that describes this trade-off between TAM width (wij) and power density (pij) is then calculated for each hot module using (2), where γ and δ represent the relative importance of wij and pij, respectively.

$$TFi = (\gamma * wij) + (\delta * pij) \qquad (2)$$

E. Relative Heat Dissipation Factor

For each ant, parameters such as the number of layers in the 3D-SIC, the maximum surface area of each layer, the thickness of the 3D-SIC, thermal interface material (TIM) thickness, the thickness of each die layer, the placement of each core within the 3D-SIC, the area of each core, and the power trace value of each core are derived from the preceding subsections. These parameters serve as inputs for computing the Relative Heat Dissipation Factor (RHDF) for each C_i in the 3D-SIC. The RHDF values for each core are computed for each ant (each floorplan) based on (3) [8].

$$RHDF = HHDF * VHDF * Power\ trace \qquad (3)$$

The Horizontal Heat Dissipation Function (HHDF) for each C_i in a layer is determined by measuring its distance from the four edges of the layer in the horizontal plane. The HHDF is computed as the sum of these distances, reflecting the core's heat dissipation capability in the horizontal direction. Conversely, the Vertical Heat Dissipation Function (VHDF) considers the layer's position within the stack relative to the heat sink. It quantifies heat dissipation by calculating the distance of each layer from the heat sink. The Relative Heat Dissipation Factor (RHDF) for each C_i is then derived from the product of HHDF and VHDF, providing an index that evaluates the core's heat dissipation capability based on its vertical and horizontal positioning within the stack. Additionally, the RHDF accounts for the core's power trace value and the duration of testing, represented by the number of clock cycles applied during testing. This comprehensive assessment ensures an accurate evaluation of each core's relative heat dissipation capability [8].

F. Thermal-aware Test Scheduling

After obtaining a set of cubes from the ACO algorithm and generating the 3D-SIC floorplan, the cores (cubes) for each ant are sorted in a non-increasing order based on their RHDF values from the previous subsection. This sorting step

prioritizes cores with lower heat dissipation capability, which may generate higher heat during testing mode, in the scheduling process. Subsequently, the 3D bin packing algorithm takes a set of cubes and the corresponding RHDF list for an ant, along Wmax, Pmax, and Tmax as inputs. The primary objective of the algorithm is to optimize the scheduling of these cubes while ensuring that, throughout the generated schedule, the total TAM width usage does not exceed Wmax, the total power consumption remains below Pmax, and the maximum temperature remains below Tmax.

The scheduling process begins by selecting the first cube from the RHDF list and placing it in a 3D bin. Subsequently, the next cube is chosen from the RHDF list, and the adjacency exclusion principle is applied. This principle ensures that the next cube can only be scheduled if it is not adjacent in the 3D-SIC floorplan to the previously scheduled cube within the same test session. This approach allocates sufficient space around the tested core to effectively dissipate generated heat, preventing heat accumulation and hotspot formation. If the next cube is adjacent to the previously scheduled cube within the same test session, it is placed in a temporary list to be scheduled in the subsequent test session.

Before scheduling the next selected cube based on the adjacency exclusion principle, the algorithm checks the availability of unallocated TAM width compared to the TAM requirement of the cube to determine if test concurrency is feasible. Next, it verifies the total power consumption to ensure that Pmax is adhered to throughout the test session if the cube is scheduled. Additionally, the algorithm checks the maximum temperature of the test session using the HotSpot 6.0 tool [14] to ensure that Tmax is adhered to throughout the test session if the cube is scheduled. If the cube satisfies Wmax, Pmax, and Tmax, it is scheduled within the 3D bin. However, if the cube fails to meet any of these limits, it is placed in a temporary list for scheduling in the subsequent test session.

III. Result And Discussion

To implement and evaluate the proposed test scheduling and floor planning techniques for 3D-SICs, widely recognized ITC'02 SoC benchmark circuits were utilized [15]. These benchmarks were established as standardized representative designs commonly employed in integrated circuit testing research. This study utilized core information

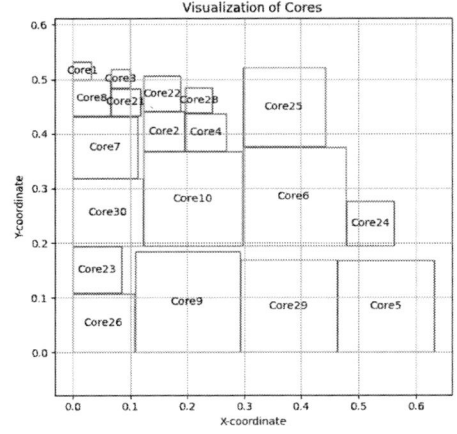

Fig. 2.   A stack of 3D-SIC with cores from the p93791 SoC.

from four benchmark SOCs: d695 (an academic benchmark), and the industrial SOCs p22810, p34392 and p93791. The 3D-SIC and thermal parameters used in this research were adopted from [8].

The thermal-aware floor planning based on a greedy algorithm able to place hot modules such that hot modules are not adjacent to each other but are next to non-hot modules, as shown in Fig. 2, where cores with red borders are hot modules, and cores with blue borders are non-hot modules. Fig. 2 shows a stack of 3D-SIC with cores from the p93791 SoC. However, the thermal-aware floor planning process, which uses a greedy algorithm, cannot guarantee that all hot modules are placed at the edge of the layer. This limitation may affect the optimal heat dissipation capabilities of the system. The reliance on a greedy algorithm means it may not always find the globally optimal solution, potentially resulting in less efficient thermal management. Despite these challenges, the optimal placement strategy employed in this research still facilitates efficient heat dissipation during testing. It helps maintain the temperature within acceptable limits, as heat was effectively dissipated to the adjacent cores and the heat sink. Additionally, the strategic arrangement of hot modules, considering factors like proximity to the heat sink and avoidance of proximity to each other, contributes to improved thermal performance.

Since heat dissipation of hot modules were improved by thermal-aware floor planning, more cores was able to be tested concurrently, as the temperature of 3D-SIC able to remain under the system-level temperature limit. Hence, the overall test schedule time of a 3D-SIC was reduced. Table 1 shows a comparison of test schedule times for the p93791 3D-SIC with and without floor planning. The schedule time was shorter when floor planning was applied.

IV. CONCLUSION

In conclusion, this paper proposed a co-optimization technique that integrates thermal-aware test scheduling and floor planning for 3D-SICs, addressing resource and power constraints to minimize total test schedule time. The method effectively reduces both test time and power consumption for the entire 3D-SIC while adhering to temperature and resource limitations. This integrated approach of test scheduling with floor planning surpasses the test scheduling approach without floor planning, proving that the proposed method results in enhanced and optimized testing for 3D-SICs. To address the limitations of the greedy algorithm in thermal-aware floor planning, future research could explore alternative algorithms with better global optimization capabilities, such as Genetic Algorithms, Simulated

Annealing, or Particle Swarm Optimization. Additionally, integrating more established floor planning techniques with the test scheduling method could enhance heat dissipation and overall system performance.

ACKNOWLEDGMENT

The authors would like to acknowledge that this research was supported by UTM Fundamental Research Grant numbered 22H19.

REFERENCES

[1] H. Vohra and A. Singh, "Test architecture optimization algorithms for coarse-grain partitioned 3D system-on-chip," Computers and Electrical Engineering, vol. 101, Article 108049, 2022, doi: 10.1016/j.compeleceng.2022.108049.

[2] S. Chatterjee, S. K. Roy, C. Giri, and H. Rahaman, "Frequency-scaled thermal-aware test scheduling for 3D ICs using machine learning based temperature estimation," Microelectronics Journal, vol. 128, Article 105535, 2022, doi: 10.1016/j.mejo.2022.105535.

[3] R. Karmakar, A. Agarwal, and S. Chattopadhyay, "Test Infrastructure Development and Test Scheduling of 3D-Stacked ICs under Resource and Power Constraints," in Proc. IEEE 24th Asian Test Symposium (ATS), 2015, pp. 73-78, doi: 10.1109/ATS.2015.20.

[4] S. Chatterjee, S. K. Roy, C. Giri, and H. Rahaman, "An Efficient Test Scheduling to Co-optimize Test Time and Peak Power for 3D ICs," in Proc. 2020 International Symposium on Devices, Circuits and Systems (ISDCS), 2020, pp. 1-6, doi: 10.1109/ISDCS49393.2020.9263015.

[5] X. Wu, Y. Chen, K. Chakrabarty, and Y. Xie, "Test-access mechanism optimization for core-based three-dimensional SOCs," Microelectronics Journal, 2010, vol. 41, pp. 601-615, doi:10.1016/j.mejo.2010.06.015.

[6] S. K. Roy, P. Ghosh, H. Rahaman, and C. Giri, "Session Based Core Test Scheduling for 3D SOCs," in Proc. 2014 IEEE Computer Society Annual Symposium on VLSI, 2014, pp. 196-201, doi: 10.1109/ISVLSI.2014.61.

[7] B. Noia, K. Chakrabarty, and Y. Xie, "Test-wrapper optimization for embedded cores in TSV-based three-dimensional SOCs," in Proc. 2009 IEEE International Conference on Computer Design, 2009, pp. 70-77, doi: 10.1109/ICCD.2009.5413172.

[8] I. Rawat and V. Singh, "Temperature efficient parallel test scheduling for higher order 3D stacked SoCs," in Proc. 2021 International Conference on Computing, Communication, and Intelligent Systems (ICCCIS), 2021, pp. 804-809, doi: 10.1109/ICCCIS51004.2021.9397211.

[9] I. Rawat, M. K. Gupta, and V. Singh, "Temperature and time efficient parallel test scheduling for 3D stacked SoCs," in Proc. 2015 IEEE International Conference on Research in Computational Intelligence and Communication Networks (ICRCICN), 2015, pp. 306-311, doi: 10.1109/ICRCICN.2015.7434255.

[10] C.-J. Shih et al., "Thermal-Aware Test Schedule and TAM Co-Optimization for Three-Dimensional IC," Active and Passive Electronic Components, 2012, doi: 10.1155/2012/763572.

[11] T. Ni et al., "Temperature-Aware Floor planning for Fixed-Outline 3D ICs," IEEE Access, 2019, vol. 7, pp. 139787-139794, doi: 10.1109/ACCESS.2019.2942839.

[12] V. Iyengar, K. Chakrabarty, and E. J. Marinissen, "Test wrapper and test access mechanism co-optimization for system-on-chip," in Proc. International Test Conf., 2001, pp. 1023-1032, doi: 10.1109/TEST.2001.966728.

[13] J. H. Ahn and S. Kang, "SoC Test Scheduling Algorithm Using ACO-Based Rectangle Packing," Computational Intelligence, 2006, pp. 655-660, doi: 10.1007/978-3-540-37275-2_82.

[14] R. Zhang, M. Stan, and K. Skadron, "HotSpot 6.0: Validation, Acceleration and Extension," University of Virginia Dept. of Computer Science Tech Report, 2015, doi: 10.18130/V3HV1Z.

[15] E. J. Marinissen, V. Iyengar, and K. Chakrabarty, "A set of benchmarks for modular testing of SOCs," in Proc. International Test Conf., 2002, pp. 519-528, doi: 10.1109/TEST.2002.1041802.

TABLE I.   COMPARISON OF TEST TIME FOR P93791 3D-SIC

TAM	Test schedule time without floor planning	Test schedule time with floor planning
16	1804033	1803812
24	1209389	1208901
32	1023984	1022134
40	778902	775981
48	619001	617410
56	520930	519781

Deep Learning (DL) based Computer Generated Hologram (CGH) for Beamsteering in Reconfigurable Holographic Switches

Clarence Augustine TH Tee
College of Physics & Electrical Information Engineering,
Zhejiang Normal University
321004 Zhejiang
People's Republic of China
catht@zjnu.edu.cn

Han Bin Sun
Photonics & Sensors Key Research Laboratory,
Zhejiang Normal University
321004 Zhejiang
People's Republic of China
hanbinsun@zjnu.edu.cn

W P Yeo
Photonics & Sensors Key Research Laboratory,
Zhejiang Normal University
321004 Zhejiang
People's Republic of China
svcat15@gmail.com

Burhanuddin Yeop Majlis
Institute of Microengineering and Nanoelectronics (IMEN)
Universiti Kebangsaan Malaysia
43600 UKM Bangi
Selangor Malaysia
burhan@ukm.edu.my

Muhamad Ramdzan Buyong
Institute of Microengineering and Nanoelectronics (IMEN)
Universiti Kebangsaan Malaysia
43600 UKM Bangi
Selangor Malaysia
muhdramdzan@ukm.edu.my

Ahmad Rifqi Md Zain
Institute of Microengineering and Nanoelectronics (IMEN)
Universiti Kebangsaan Malaysia
43600 UKM Bangi
Selangor Malaysia
rifqi@ukm.edu.my

Sheng Li
Zhejiang Institute of Optoelectronics,
321004 Zhejiang
People's Republic of China
shengli@zjnu.edu.cn

Le Song
State Key Laboratory of Precision Measurement Technology and Instruments,
Tianjin University
300072 Tianjin
People's Republic of China
songle@tju.edu.cn

Yelong Zheng
State Key Laboratory of Precision Measurement Technology and Instruments,
Tianjin University
300072 Tianjin
People's Republic of China
zhengyelongby@tju.edu.cn

Abstract— Beamsteering application via the dynamic holography with recorded dynamic holograms onto Spatial Light Modulators (SLM), has been used for optical interconnections especially in all-optical networks within the reconfigurable holographic switches. A novel AI based approach using Deep Learning (DL) methodology has been proposed for the generation of a new Computer Generated Hologram (CGH) with improved performance metrics i.e., peak signal-to-noise ratio (PSNR) and fast generation of prediction time, resulting in higher accuracy, controllable and precise predicted beamsteering application. The requirements of a high PSNR, fast generated CGH, low crosstalk, diffraction efficiency and polarization sensitivity for a Liquid Crystal (LC) 2D SLM for reconfigurable beamsteering among others necessitates the integration of the superior features of DL into CGH production i.e., proposed herewith the AI-based DL-CGH. The DL-CGH methodology proposed here has simplified input data channels, residual network and attention mechanism which performed better in comparison with Gerchberg-Saxton (G-S) and Holo-encoder CGH.

Keywords— *Beamsteering, AI-DL SE-HoloRes, PSNR, Prediction Time, Reconfigurable Holographic Switches, Squeeze-and-excitation Networks (SENet), Attention Mechanism, Residual (RES), Holo-encoder.*

I. INTRODUCTION

In recent years, artificial intelligence (AI)-deep learning (DL) has progressed seamlessly and unprecedented in all key technological advances, inclusive of the potential usage of AI-DL in Computer Generated Hologram (CGH) application with improved performance, new features and accelerate the calculation speed of holograms. Herewith, the proposed research on a novel AI based approach using Deep Learning

(DL) methodology for the generation of a new Computer Generated Hologram (CGH), aiming to improve performance metrics i.e., peak signal-to-noise ratio (PSNR) and fast generation of prediction time, resulting in higher accuracy, controllable and precise predicted beamsteering application. Beamsteering application is predominantly used in dynamic holography for optical interconnections constructed by recording dynamic holograms onto Spatial Light Modulators (SLM), forming crucial elements in all-optical networks especially within the reconfigurable holographic switches [1-5].

Common methods [6-9] that use iterative calculations to generate computational holograms, such as G-S [6] or non-convex optimization methods, require tedious iterative processes and are very time-consuming [10]. To enable the fast calculation of CGH, several non-iterative methods have been proposed, such as spatial multiplexing, error diffusion, and phase encoding. It can be expected that these methods are complex and increasing speed at the expense of image quality after reconstruction [11].

CGH method based on deep learning, divided into supervised and unsupervised learning. The former example is such as the multi-depth hologram generation network (MDHGN) [12] which network takes multiple images of different depths as inputs and calculates the complex hologram as an output, which reconstructs each input image at the corresponding depth. The later [13] is inspired by the iterative algorithm GS is using U-Net network coding with the given input. It is inspired by the iterative algorithm double phase amplitude encoding (DPAC) application with complementary two-dimensional binary grating sampling of two child hologram, and synthesizing them into optimized hologram [14].

Deep learning reduces the complexity and calculating time of the CGH in the reconfigurable holographic switches. Moreover, once the network training process is completed, deep learning can perform fast inference and even real-time reconstruction through a targeted trained network. In general, deep learning can significantly improve the outcomes of a variety of holographic tasks, reduce the complexity of imaging experiments, and facilitate the emergence of new functions. Deep learning has its potential problem also namely the lack of sufficient standard data. Due to the basic principle of computational holography, which usually requires a lot of datasets to obtain high efficient holograms, so a large number of accurately annotated datasets appear unlikely. Deep learning lacks interpretability, and the results from the same model will different interpretation for different fields. Regardless, the application of deep learning in CGH is promising, and most case studies conducted in the past five years have shown tremendous potential [15-17].

II. RELATED WORKS

A. Holo-encoder CGH

Automatic encoder (AE) [18] is one of the classic artificial neural network (ANN), which consists of the encoder and the decoder. The output data is reconstructed by the decoder and compared to the input data. The training network is based on the reconstructed error. The neural network uses a backpropagation algorithm, making the output value infinitely close to the input value. Another variant of AE is the holo-encoder CGH [7] where the recording and reconstruction of holograms can be viewed as image encoding and image decoding. The holographic display of the end-to-end model can be represented by an AE. Unlike the general AE, the decoding part of the holo-encoder CGH is of a determined optical propagation model with unlearn-able parameters.

B. Squeeze-and-Excitation Networks (SENet)

The SENet [19] models the interdependencies between channels through the adaptation of the global loss function of its network. In the forward convolution layer, although the output of each channel is equal, while weighting them, one can emphasize channels with high information values and suppress channels with relatively low information values since SENet basically consists of blocks of "SE-Block".

C. Residual Network (ResNet)

ResNet [20] is a module that identifies inefficient layers in the network. Assuming that, the existing network has multiple network layers, a few of which are optimal and many of which are redundant. The input and output of the redundant layers are almost identical, so there is a need to skip these redundant layers and train only the efficient network layer. The specific layer is the redundancy layer, in which the network will train at its own "intelligent" judgment.

III. IMPROVED SELF-ENCODING NETWORKS BASED ON ATTENTION MECHANISMS

The novel AI-DL SE-HoloRes based CGH for beamsteering in the reconfigurable holographic switches, is a AL-DL algorithm taking into account of the beneficial parts of SE-holo-encoder CGH and attention mechanism while hybridized with the algorithms of four SE-Block modules combining with ResNet for the four-skip connections as shown in Figure 1(a). The added modules enable the network to prioritize the order of information regions while reducing the network noise. Thus, improving the performance of the network in terms of the accuracy and anti-variability of the network. The improved encoder network structure has eight image resolution conversion modules, four of which are located in the downsampling stage and four in the upsampling stage, connected by skipping the link. At the downsampling process, the input images are convolved according to the set N 3 * 3 convolution kernel, and the N feature channels of target resolution are obtained through the rectified linear unit (ReLU) function and repeated again to obtain the target image of this layer. The target image of this layer is taken by the maximum pooling of 2 * 2 as input. Each downsampling layer reduces the picture in half and doubles the number of convolution cores.

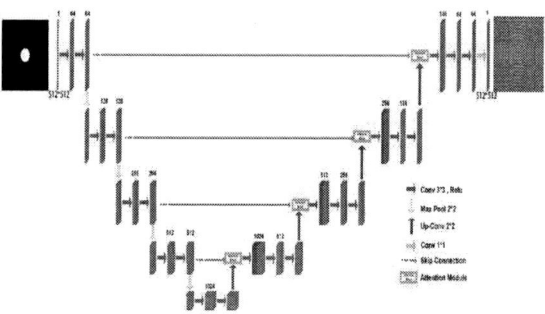

Fig. 1(a). The novel AL-DL SE-HoloRes CGH

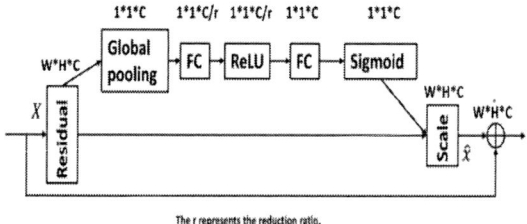

The r represents the reduction ratio.

Fig. 1(b). The AL-DL SE-HoloRes CGH Network Architecture

For the upsampled image resolution transformation module, features are extracted by deconvolution enlarged pictures and convolution. With the added attention mechanism, and for each up-sampled module, the SE-RES modules are added to connect them. The SE-RES module reduces the W of the global spatial information of W * H * C, H to a channel descriptor, viz., 1 * 1 * C. This 1 * 1 * C is the effect of the channel on the output. The larger the channel, in the subsequent training has a greater impact on the output, and further improves the quality of the network output, as shown in Figure 1 (b). The AL-DL SE-HoloRes implementation of the network structure is as follows:

```
class SE-Holo encoder(nn.Module):
  def __init__(self, n_in):
    super().__init__()
    self.down1 = ResDown(n_in=n_in, n_out=16)
```

```
self.down2 = ResDown(n_in=16, n_out=32)
self.down3 = ResDown(n_in=32, n_out=64)
self.down4 = ResDown(n_in=64, n_out=128)
self.SEBlock=SEBlock(128)
self.up4 = ResUp(n_in=128, n_out=64)
self.SEBlock = SEBlock(64)
self.up3 = ResUp(n_in=64, n_out=32)
self.SEBlock = SEBlock(32)
self.up2 = ResUp(n_in=32, n_out=16)
self.SEBlock = SEBlock(16)
self.up1 = ResUp(n_in=16, n_out=1)
self.norm = nn.BatchNorm2d(num_features=1)
```

Among them, the core idea of SE-Block module is to model the interdependence between channels, adapting the corresponding strength of the characteristics through the global loss function of the network. It facilitates the integration of global and local context and improves the efficiency and reliability of network learning. To address the gradient vanishing problem and stabilize the training, a residual block is added, by introducing the so-called skip connections, enabling the network to learn the residual between the input and the output, more effectively passing the gradient in the network. In the decoding part, the optical model simulating Fresnel diffraction is used [21], and the completed meter CGH of the network training is restored by calculating equation, Eq. 1. below:

$$I(x,y) = \left| F\left\{ \exp[i\phi(x_o, y_o)] \exp[i\frac{\pi}{\lambda z}(x_o^2 + y_o^2)] \right\} \right|^2 \quad (1)$$

Comparing with the angular spectrum method, the Fresnel transformation method involves a single Fourier transform and does not require zero-filling operation and saves a lot of computational resources. By exploiting automatic differentiation, the losses can be spread back to the encoder part, and the learnable parameters of U-Net can be updated during training.

$$L\left(\hat{I}, I\right) = L_{NPCC}\left(\hat{I}, I\right) + \tau L_{percep}\left(\hat{I}, I\right) + \gamma \|\nabla \Phi\|_p \quad (2)$$

Whereas:

$$L_{NPCC}(X,Y) = (-1) \times \frac{\sum_i^n (X_i - \overline{X})(Y_i - \overline{Y})}{\left\{ \sum_i^n (X_i - \overline{X})^2 \sum_i^n (Y_i - \overline{Y})^2 \right\}^{1/2}}$$

$$L_{percep}(X,Y) = \|P(X) - P(Y)\|_2^2$$

A combination of negative Pearson correlation coefficient (NPCC) [22] and perceived loss [23] as a network input and output loss function to train the network is used, as shown in Eq.2. An optional regularizer on the reconstructed phase pattern can impose constraints on the phase variation. NPCC loss ensures linear amplification and unbiased reconstruction, and improves the probability of convergence. Phase regulars can impose constraints on phase changes.

IV. EXPERIMENTAL RESULTS

A. Data set and Experimental Setup

Dataset: To reduce the effect of the dataset on the experimental results, adoption of the DIV 2K [24] dataset adopted by the holo-encoder CGH is used. DIV2K is a dataset for super-resolution tasks by down sampling the real-world images that are all realistic amplitude maps. It contains not only high images up to 4K resolution, but also a portion of low resolution images.

Experimental setting: In the experiment, Windows-based laboratory computer is used with an image processor model of NVIDIA GeForce RTX 3060. The experiment went through 20 cycles. The learning rate is set to 0.0001 using Adam optimizer.

B. Loss function

The interdependence between the characteristic channels of the attention module evolution network is added to improve the quality of the network generation so as the loss function in the training process can also improve. The residual network module has enabled the network compute the residual of the convolutional block in training, and skip the convolution block with small

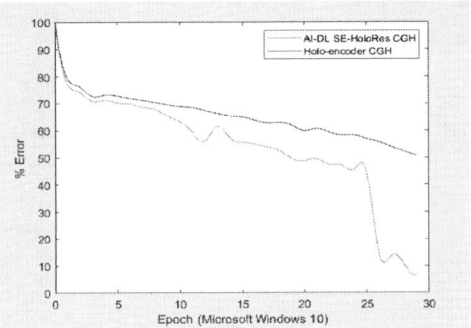

Fig. 2. Network training process: Loss function

output and input difference in subsequent training. The "Blue" curve represents the percentile (%) error of the Holo-encoder CGH [7] whereas the "Orange" curve represents AI-DL SE-HoloRes CGH. After the addition of the SE-RES module, AI-DL SE-HoloRes CGH outperforms Holo-encoder CGH in % error, dipping at approximately Epoch 25. At Epoch 25 approximately (number of the training rounds), the % error of AI-DL SE-HoloRes CGH drops from 50% to 10% error, better than Holo-encoder CGH. At approximately Epoch 27, the % error of AI-DL SE-HoloRes CGH peaks slightly to 15% and dips again to below 10% error after that. This could be due to the setting of the network learning rate which were too large, causing the "skipping of the optimal solution and shocks" effects when the network parameters were updated, leading to slight increment of % error at that particular Epoch 27.

C. Experimental results

The evaluation matrix: AI-DL SE-HoloRes CGH algorithms and architecture network was compared to GS [25] and Holo-encoder CGH [7] in terms of PSNR and generation prediction time (Table 1).

Table 1. PSNR & Prediction Time Performance Metrics

Model	PSNR	Prediction Time
GS (10 iterations)	22.7	3.1s
Holo-encoder CGH	26.7	0.5s
AI-DL SE-HoloRes CGH	28.6	0.5s

The hologram is simulated by computer to reproduce the image in the far field. The peak signal-to-noise ratio (PSNR) is measured. Based on the experimental results, the proposed algorithms and architecture network are superior than GS and Holo-encoder in terms of PSNR as shown in Table 1 i.e. higher PSNR means better quality of the reconstructed image for beamsteering in the reconfigurable holographic switches (Figure 3).

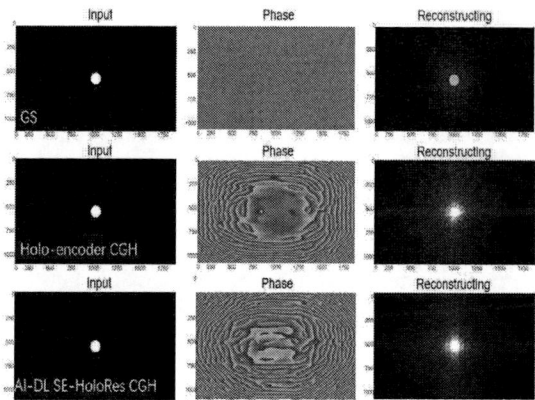

Fig. 3. Far-field beamsteering component: Experimental results of writing a CGH on a Liquid Crystals Spatial Light Modulator (LC-SLM) and the generated Fourier plane reconstruction of the hologram at the far-field.

V. CONCLUSION

The AI-DL SE-HoloRes CGH algorithms and network architecture has superior performance in terms of peak signal-to-noise ratio (PSNR) with nearly 26% and 7% better (on the basis of the same computing time) compared with the GS [25] and Holo-encoder CGH deep learning [7] networks respectively which translates to higher quality, accurate, efficient and effective reconstructed image for beamsteering in the reconfigurable holographic switches. Future works will be on crosstalk and diffraction efficiency.

ACKNOWLEDGMENT

The research work has been supported by the research team (and international collaborators) within the Photonics & Sensors Key Research Laboratory, part of the cross-disciplined Research Project "Multi-Dimensional Nano-Microelectronics Functional Devices & System Engineering (MINDS)", an interdisciplinary and foresight collaborative key research initiative. The work is funded and supported by the China National High Talent Foreign Expert Grant (G2021016010L), Zhejiang Normal University's Distinguished (Eminent) Professorship Grant and Zhejiang's "Pioneering and Leading Goose" R&D Programme Grant 2022C01030.

REFERENCES

[1] C. A. T. H. Tee, W. A. Crossland, T. D. Wilkinson, A. B. Davey, "Binary Phase Modulation Using Electrically Addressed Transmissive and Silicon Backplane Spatial Light Modulators", Opt. Eng. 39(9), pp. 2527-2534, Sept 2000, Optical Engineering Journal, USA.

[2] R. James, F. A. Fernández, S. E. Day, M. Komar~evic´, W. A. Crossland, "Modeling of the diffraction efficiency and polarization sensitivity for a liquid crystal 2D spatial light modulator for reconfigurable beam steering", J. Opt. Soc. Am. A, Vol. 24, No. 8 August 2007.

[3] J. A. Neff, R. A. Athale, S. H. Lee, "Two-Dimensional Spatial Light Modulators: A Tutorial", Invited Paper, PROCEEDINGS OF THE IEEE, VOL. 78, NO. 5, MAY 1990.

[4] B. Lofving, S. Hard, "Beam steering with two ferroelectric liquid-crystal spatial light modulators", OPTICS LETTERS, Vol. 23, No. 19, October 1, 1998.

[5] T. Sakano, K. Kimura, K. Noguchi, N. Naito, "256 3 256 Turnover-type free-space multichannel optical switch based on polarization control using liquid-crystal spatial light modulators", APPLIED OPTICS, Vol. 34, No. 14, 10 May 1995.

[6] Gerchberg, R. W.; Saxton, W. O. "A practical algorithm for the determination of the phase from image and diffraction plane pictures", Optik. 35: 237–246, 1972.

[7] J. Wu, K. Liu, X. Sui, and L. Cao, "High-speed computer-generated holography using an autoencoder-based deep neural network," Opt. Lett., OL, vol. 46, no. 12, pp. 2908–2911, Jun. 2021.

[8] T. Zeng, Y. Zhu, and E. Y. Lam, "Deep learning for digital holography: a review," Opt. Express, OE, vol. 29, no. 24, pp. 40572–40593, Nov. 2021.

[9] S.-C. Liu and D. Chu, "Deep learning for hologram generation," Opt. Express, OE, vol. 29, no. 17, pp. 27373–27395, Aug. 2021.

[10] T. Zhao and Y. Chi, 'Modified Gerchberg–Saxton (G-S) Algorithm and Its Application', Entropy, vol. 22, no. 12, Art. no. 12, Dec. 2020.

[11] G. Yang, B. Dong, B. Gu, J. Zhuang, and O. K. Ersoy, 'Gerchberg–Saxton and Yang–Gu algorithms for phase retrieval in a nonunitary transform system: a comparison', Appl. Opt., AO, vol. 33, no. 2, pp. 209–218, Jan. 1994.

[12] J. Y. Lee, J. S. Jeong, J. B. Cho, D. H. Yoo, B. G. Lee, B. G. Lee, "Deep neural network for multi-depth hologram generation and its training strategy", Optics Express 27137, Vol. 28, No. 18, 31 August 2020.

[13] K. Liu, J. Wu, Z. He, and L. Cao, "4K-DMDNet: diffraction model-driven network for 4K computer-generated holography," OEA, vol. 6, no. 5, pp. 220135–13, May 2023.

[14] Q. Liu, J. Chen, B. Qiu, Y. Wang, and J. Liu, "DCPNet: a dual-channel parallel deep neural network for high quality computer-generated holography," Opt. Express, OE, vol. 31, no. 22, pp. 35908–35921, Oct. 2023.

[15] R. Horisaki, R. Takagi, and J. Tanida, "Deep-learning-generated holography," Appl. Opt. 57(14), 3859–3863 (2018).

[16] B. Lee, J. Jeong, D. You, and J. Lee, "Learning-based synthesis of computer-generated hologram," in Digital Holography and Three-Dimensional Imaging, Imaging and Applied Optics Congress 2020, (Optical Society of America, 2020), HTu4B.1.

[17] J. Lee, J. Jeong, J. Cho, D. Yoo, and B. Lee, "Complex hologram generation of multi-depth images using deep neural network," in Digital Holography and Three-Dimensional Imaging, Imaging and Applied Optics Congress 2020, (Optical Society of America, 2020), JTh2A.12.

[18] D. E. Rumelhart, G. E. Hinton, R. J. Williams, "Learning internal representations by error propagation", pp. 318-362, MIT Press, 1988, DOI: https://doi.org/10.7551/mitpress/4943.0010001.

[19] J. Hu, L. Shen, S. Albanie, G. Sun and E. Wu, "Squeeze-and-Excitation Networks," in IEEE Transactions on Pattern Analysis and Machine Intelligence, vol. 42, no. 8, pp. 2011-2023, 1 Aug. 2020.

[20] K. He, X. Zhang, S. Ren, and J. Sun, 'Deep Residual Learning for Image Recognition', in 2016 IEEE Conference on Computer Vision and Pattern Recognition (CVPR), pp. 770-778, Jun. 2016.

[21] P. W. M. Tsang, Y.-T. Chow, and T.-C. Poon, 'Generation of phase-only Fresnel hologram based on down-sampling', Opt. Express, OE, vol. 22, no. 21, pp. 25208–25214, Oct. 2014.

[22] P. Sedgwick, 'Pearson's correlation coefficient', BMJ, vol. 345, p. e4483, Jul. 2012.

[23] Zhou, C.; Zhang, J.; Liu, J.; Zhang, C.; Fei, R.; Xu, S. PercepPan: Towards Unsupervised Pan-Sharpening Based on Perceptual Loss. Remote Sens. 2020, 12, 2318.

[24] E. Agustsson and R. Timofte, 'NTIRE 2017 Challenge on Single Image Super-Resolution: Dataset and Study', presented at the Proceedings of the IEEE Conference on Computer Vision and Pattern Recognition Workshops, pp. 126–135, 2017.

[25] P. Memmolo et al., 'Investigation on specific solutions of Gerchberg-Saxton algorithm', Optics and Lasers in Engineering, vol. 52, pp. 206-211, Jan. 2014.